T0189937

Fields Institute Monographs

VOLUME 36

Messoud Efendiev

Symmetrization and Stabilization of Solutions of Nonlinear Elliptic Equations

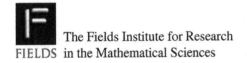
The Fields Institute for Research in the Mathematical Sciences

Springer

Messoud Efendiev
Institute of Computational Biology
Helmholtz Center Munich
Neuherberg, Bayern, Germany

ISSN 1069-5273   ISSN 2194-3079   (electronic)
Fields Institute Monographs
ISBN 978-3-030-07491-3   ISBN 978-3-319-98407-0   (eBook)
https://doi.org/10.1007/978-3-319-98407-0

Mathematics Subject Classification (2010): 35J61, 35Q92, 35J60, 35J70, 35J92, 37C45, 37L30

Cover illustration: Drawing of J.C. Fields by Keith Yeomans

This Springer imprint is published by the registered company Springer Nature Switzerland AG
The registered company address is: Gewerbestrasse 11, 6330 Cham, Switzerland

Preface

This book mainly deals with a systematic study of a dynamical system approach to investigate the symmetrization and stabilization properties (as $|x|$ tends to infinity) of nonnegative solutions of nonlinear elliptic (both degenerate and nondegenerate) problems in asymptotically symmetric unbounded domains. To this end, we use a trajectory dynamical systems approach and the concept of its attractor. Recall that a dynamical system (DS) is a system which evolves with respect to time. To be more precise, a DS $(S(t), \Phi)$ is determined by a phase space Φ which consists of all possible values of the parameters describing the state of the system and an evolution map $S(t) : \Phi \rightarrow \Phi$ which allows us to find the state of the system at time $t > 0$ if the initial state at $t = 0$ is known. Very often, in biology, ecology, mechanics, and physics, more generally in the modeling of life science problems, the evolution of the system is governed by systems of differential equations. If the system is described by ordinary differential equations (ODEs)

$$u'(t) = F(t, u(t)) \tag{1}$$

for some nonlinear function $F : \mathbb{R}^+ \times \mathbb{R}^N \rightarrow \mathbb{R}^N$, we have a so-called finite-dimensional DS. In that case, the phase space Φ is some (invariant) subset of $\mathbb{R}^N$ and the evolution operator $S(t)$ is defined by

$$S(t)y_0 := y(t), \tag{2}$$

where $y(t)$ solves (1). We also recall that, in the case where Eq. (1) is autonomous (i.e., does not depend explicitly on time), the evolution operators $S(t)$ generate a semigroup on the phase space Φ, i.e.

$$S(t_1 + t_2) = S(t_1)S(t_2), \qquad t_1, t_2 \in \mathbb{R}^+, \tag{3}$$

The qualitative study of DS in finite dimensions goes back to the beginning of the twentieth century with the pioneering works of Poincaré on the N-body problem and important contributions of Lyapunov on stability and of Birkhoff on minimal sets and the ergodic theorem. One of the most surprising and significant facts discovered at the very beginning of the theory is that even relatively simple equations can generate very complicated chaotic behaviors. Moreover, these types of systems are extremely sensitive to initial conditions (the trajectories with close but different initial data diverge exponentially). Thus, in spite of the deterministic nature of the system (we recall that it is generated by a system of ODEs, for which we usually have a unique solvability theorem), its temporal evolution is unpredictable on time scales larger than some critical time T_0 (which clearly depends on the error of approximation and on the rate of divergence of close trajectories) and can show typical stochastic behaviors. This fact was used by Lorenz to justify the so-called butterfly effect, a metaphor for the imprecision of weather forecasting. The theory of DS in finite dimensions has been extensively developed during the twentieth century, due to the efforts of many mathematicians (such as Anosov, Arnold, LaSalle, Sinai, Smale), and, nowadays, very much is known about the nontrivial dynamics of such systems, at least in low dimensions. In particular, it is known that very often, the trajectories of a chaotic system are localized, up to a transient process, in some subset of the phase space having a very complicated fractal geometric structure (e.g., locally homeomorphic to the Cartesian product of $\mathbb{R}^m$ and some Cantor set) which thus accumulates the nontrivial dynamics of the system (this is the so-called strange attractor). The chaotic dynamics on such sets are usually described by symbolic dynamics generated by Bernoulli shifts on the space of sequences. We also note that, nowadays, a mathematician has a large number of tools available for the extensive study of concrete chaotic DS in finite dimensions. In particular, we mention here different types of bifurcation theories (including the KAM theory and the homoclinic bifurcation theory with related Shilnikov chaos), the theory of hyperbolic sets, stochastic description of deterministic processes, Lyapunov exponents and entropy theory, dynamical analysis of time series, etc. In other words, in the mid-1970s of the twentieth century, we already had available highly developed finite-dimensional dynamical systems theory, and it has become vitally important to extend this theory to the evolution processes that are usually governed by partial differential equations (PDEs). In this case, the corresponding phase space Φ is some infinite-dimensional function space (e.g., $F := L^2(\Omega)$ or $F := L^8(\Omega)$ for some domain $\Omega \subset \mathbb{R}^N$, or Hölder space). Such DS are usually called infinite-dimensional. These classes of equations in the abstract setting can be written as

$$u'(t) = F(t, u(t)) \tag{4}$$

in an infinite-dimensional Banach space Φ. We emphasize that for an infinite-dimensional DS generated by PDEs, the situation becomes much more complicated. Indeed, a first important difficulty which arises here is related to the fact that the analytic structure of a PDE is essentially more complicated than that of an ODE and, in particular, in general we do not have a unique solvability theorem as in the case of ODEs, so that even finding the proper phase space and the rigorous construction of the associated DS can be a highly nontrivial problem. In order to indicate the level of difficulties arising here, it suffices to recall that, for the three-dimensional Navier-Stokes system (which is one of the most important equations of mathematical physics), the required associated DS has not yet been constructed. Nevertheless, there exists a large number of equations for which the problem of the global existence and uniqueness of a solution has been solved. Thus, the question of extending the highly developed finite-dimensional DS theory to infinite dimensions arises naturally. Note that at present there are numerous monographs on infinite-dimensional dynamical systems generated by nondegenerate parabolic and hyperbolic equations (both autonomous and nonautonomous) in bounded domains and its application in various areas of mathematical physics (see, e.g., [1, 35, 67, 92, 98] and the references therein). In those books, the long-time dynamics of solutions is described in terms of a global attractor (uniform attractor in the nonautonomous case). It is important to note that the PDE equations in the books mentioned above possess regular structure (nondegenerate diffusion and nondegenerate chemotaxis) and belonging to the so-called dissipative partial differential equations (sometimes partially dissipative, when the underlying domain is unbounded or with hyperbolic equations in a bounded domain). We emphasize once more that the phase spaces Φ in those books mentioned above are appropriate infinite-dimensional function spaces. Nevertheless, it was observed in experiments that, up to a transient process, the trajectories of the DS considered are localized inside "very thin" invariant subsets of the phase space having a complicated geometric structure which, thus, accumulates all the nontrivial dynamics of the system. It was conjectured that these invariant sets are, in some proper sense, finite-dimensional and that the dynamics restricted to these sets can be effectively described by a finite number of parameters. Thus (when this conjecture is true), in spite of the infinite-dimensional initial phase space, the effective dynamics (reduced to this invariant set) is finite-dimensional and can be studied by using the algorithms and concepts of the classical finite-dimensional DS theory. Indeed, the above finite-dimensional reduction principle of dissipative (or partially-dissipative) PDEs in bounded domains has been given a solid mathematical grounding (based on the concept of the so-called global and exponential attractors) over the last four decades, starting from the pioneering papers of Ladyzhenskaya (see [77]) and continued successfully in [1, 35, 67, 92, 98] (see also references therein), mainly for evolution PDEs with more or less regular structure (e.g., uniformly parabolic, nondegenerate parabolic and nondegenerate hyperbolic) leading to finite dimensionality of their attractors. Here, as mentioned above, boundedness of the underlying domains plays a decisive role in obtaining finite fractal dimensional attractors of the associated semigroups. In [48] (see also references therein), we systematically studied infinite-dimensional dynamical systems and their attractors generated by a quite large class of nondegenerate parabolic type PDEs (both for second order and for fourth order) in unbounded domains (both in weighted and in uniformly local phase spaces) and proved that the fractal dimension of their attractors is infinite, so that the reduction principle fails in general. In spite of this new feature, that is, the infinite dimensionality of attractors, we managed to find asymptotics of their Kolmogorov's ε-entropy which has logarithmic type asymptotics.

It is worth mentioning that, in contrast to nondegenerate evolution equations, very little was known about the long-time dynamics of degenerate parabolic equations such as porous-medium equations, parabolic p-Laplacian, doubly nonlinear equations, and degenerate diffusion with chemotaxis and ODE-PDE coupling. In the books [46, 47] that were published very recently, we discussed these classes of degenerate parabolic equations in bounded domains in connections with the modeling of life science problems and observed some very interesting new features from the dynamical systems viewpoint, related to the attractors of such equations which cannot be observed in nondegenerate cases, namely:

(a) Infinite dimensionality of the attractor
(b) Polynomial asymptotics of its epsilon-Kolmogorov entropy
(c) Differences in the asymptotics of the epsilon-Kolmogorov entropy depending on the choice of the underlying phase spaces

Furthermore, for the evolution equations that are described by parabolic or hyperbolic type equations, the well-posedness, as already mentioned above, has been proved for a quite large class of nonlinearities. However, the nonlinearities involving the abovementioned PDEs that provides nonuniqueness of solutions usually possess some "pathological" nature (see, for instance, [52]) and in turn lead to multivalued semigroups from the dynamical systems viewpoint. I would like especially to emphasize that, in contrast to such equations where non-unicity appears for "pathological" class of nonlinearities, elliptic boundary value problems in unbounded domains, interpreted as evolution equations, naturally lead to nonuniqueness of solutions even for simple polynomial nonlinearities which as a consequence leads to multivalued semigroups, where the role of time (i.e., $t \geq 0$) is played by one of the unbounded directions of the underlying domain, say $t := x_1$. Indeed, it was a temptation to apply such a strong dynamical systems tool in order to study asymptotic of solutions of elliptic boundary value problem in unbounded domains. In other words, to consider state of equilibrium of bio-physico-chemical-mechanical systems from the point of view of dynamical systems. In order to apply DS approach to elliptic equations of the form given in *Chaps. 3–7* of this book, one has to take the following difficulties into account: firstly, the Cauchy initial value problem for such equations (prescribing both the value of solutions $u(0)$ and its normal derivative $u'(0)$ on the initial surface $t = 0$) is not well-posed. More precisely, in general, its unique solution either does not exist or blows up in a finite time. Secondly, the boundary value problem for elliptic equations may have a solution, but in general it is not unique. Indeed, Hadamard was the first to notice that the initial value problem for elliptic equations is ill-posed (see [66, Bk.I, Ch. II, §18]). In [8] this difficulty was overcome, where the attractor of an elliptic equation was constructed by means of the theory of semigroup of multivalued mappings. Unfortunately, the theory of semigroups of multivalued mappings does not provide theorems on the finite dimensionality of the attractors. Indeed such theorems are an important part of the theory of semigroups of single-valued mappings in infinite-dimensional spaces, which we called above a reduction principle. Since we are mainly interested in the finite fractal dimension of the attractors of semigroups associated with elliptic problems (so-called finite-dimensional reduction discussed above for parabolic/hyperbolic type equations), such multivalued semigroup approach is not suitable for our purposes. Therefore, the usage of infinite-dimensional dynamical systems methods mentioned above for elliptic problems in unbounded domains as well as finite-dimensional reduction of their dynamics requires new ideas and tools. Fortunately, these difficulties have been overcome in several interesting particular cases. We first mention the pioneering work by Kirchgässner [72] on small solutions of elliptic equations in infinite cylinders. His idea was to construct invariant manifolds, where the elliptic initial value problem is well-posed and a flow, or at least a semiflow, is defined (see also [62]). This idea was extended to large solutions, later, in the "parabolic," convection dominated limiting case of large wave speeds $\gamma \in \mathbb{R}$ (see [31] and [87]). Without such a restriction, the case of elliptic equations in a strip was treated (see [9] and [82]). Note that in [31] a similar problem to [8] was considered in the scalar case, when the elliptic equation contains a small parameter in front of second derivative, while the coefficient in front of first derivative is different from zero. In [8], the elliptic BVP was studied with Cauchy initial conditions $u(0) = u_0$ and $u'(0) = u_1$. In this case, the semigroup corresponding to this elliptic equation consists of a single-valued mappings, but the set on which it is defined is not described in an explicit form. We strongly believe that in the dynamical setting the most natural boundary value problem is to prescribe the initial value $u(0)$ at the $t = 0$ and impose the condition of boundedness of $u(t)$ as t goes to infinity. We will consider throughout of this book this type of elliptic boundary value problem.

In this book, we overcome the abovementioned difficulties and some other ones (such as finite-dimensional reduction of dynamics) for an elliptic equation in unbounded domains by usage of the trajectory dynamical systems approach (the details are in *Chap. 3*) as well as symmetry and monotonicity properties of solutions of an elliptic equation (both nondegenerate and degenerate) in unbounded domains. In what follows, we use the following convention.

Convention 1 The attractor of a trajectory dynamical system associated to an elliptic equation is called an elliptic attractor.

We specify below some of the topics covered by this book that we are mainly interested in:

1. To construct single-valued dynamical systems for nonlinear elliptic equations in unbounded domains
2. To find appropriate functional phase space (possibly not weighted space) in unbounded domains and determine a topology in the phase space guaranteeing existence of a global attractor
3. To find assumptions on nonlinearities under which all solutions converge as $t :=$ x_1 goes to infinity to the same limiting one-dimensional profile, irrespectively of the "initial" value u_0

4. To provide assumptions on the data, that is, the geometry of underlying domains, assumptions on nonlinearity, etc., for equations under consideration that guarantee finite-dimensional attractors
5. To find new effects imposed by the dynamical systems approach to nonlinear elliptic equations and to their elliptic attractors which was not observed in the parabolic or hyperbolic PDEs previously considered from the dynamical systems viewpoint
6. To understand whether asymptotic symmetry of the domain implies symmetry of elliptic attractors
7. To understand how smoothness or nonsmoothness of underlying domains is inherited by elliptic attractors
8. To provide assumptions guaranteeing that omega-limit sets of a solution of an elliptic equation is a singleton

This book consists of seven chapters.

Chapter 1 consists of nine subsections. In *Sect. 1.1*, we introduce some functional spaces and study their properties that we will use in subsequent chapters. *Sections 1.2* and *1.3* are devoted to linear elliptic boundary value problems and the properties of Nemytskii operators in various functional spaces, respectively. *Sections 1.4* deals with the classical maximum principles as well as their versions for narrow and small domains, which are used in *Sects. 1.6, 1.7,* and *1.8*. In *Sect. 1.5*, using maximum principles, we obtain explicit and uniform bounds that ensure the boundedness and the asymptotic dissipation of the solutions of semilinear elliptic equations in bounded (or unbounded) domains. These estimates are crucial to construct the attractive basin of trajectories in the dynamical system approach to elliptic equations. We will use these estimates in the following chapters. *Sections 1.6* and *1.7* are devoted to the sweeping principle, the moving plane method as well as sliding methods both in bounded and in unbounded domains and their role in the study symmetry and monotonicity properties of nonnegative solutions of nonlinear elliptic equations in unbounded domains. We use this symmetry and monotonicity properties of nonnegative solutions of nonlinear elliptic equations in unbounded domains for the complete characterization of the elliptic attractor. *Sections 1.8* and *1.9* are devoted to variational solutions and elliptic regularity for the Neumann problem for the Laplace operator on an infinite edge. We will use the results of these subsections to study the effects of nonsmoothness of the domain to nonsmoothness of elliptic attractor in *Chap. 2*. I would like especially to emphasize that, in some cases (of course, for the convenience of the reader), I present some well-known results both in this chapter as well as in the following chapters on the one hand referring to original sources and on the other hand giving proofs with preservation of the original notations of the authors which hopefully will make the book readable and self-consistent.

Chapter 2 deals with the general theory of trajectory dynamical system and its attractor and its application for the study the asymptotics of solutions of semilinear elliptic boundary value problems in unbounded domains. *Chapter 2* consists of 14 subsections. In *Sects. 2.1* and *2.2*, we give definition of the fractal dimension and Kolmogorov's ε-entropy as well as their properties. Moreover, the definition of a global attractor for the semigroup as well as finite fractal dimensional reduction of its dynamics are also given in these sections. In *Sect. 2.3*, we give the complete characterization of asymptotic behavior of all bounded nonnegative solutions for semilinear elliptic boundary value problems in a two-dimensional rectangle. *Sections 2.4–2.12* are devoted to the trajectory dynamical systems approach to semilinear elliptic system in general unbounded domains, the existence of at least one solution, the regularity of solutions, basic definitions of trajectory attractors, and its regularity as well as to some examples of trajectory attractors and its generalizations. In *Sect. 2.13*, we study how nonsmoothness of underlying unbounded domain transfers to nonsmoothness of trajectory attractors. *Section 2.14* is devoted to the relation of nonautonomous parabolic equations in cylindrical domain with its elliptic counterpart via a traveling wave ansatz, and we study the latter from the dynamical systems viewpoint.

In *Chaps. 3* and *4*, we consider semilinear elliptic boundary value problems in the quarter-space and study asymptotic behavior of its nonnegative solutions when the underlying dimension of the quarter-space is less than or equal to three and four, respectively. Here the dimension of the underlying domain plays a very essential role, because the technique for each dimension providing asymptotic behavior of solutions in terms of the trajectory attractor depends heavily on the dimension of the quarter-space. Indeed we show that under very natural dissipativity assumptions, as well as sign condition on the nonlinearity (any polynomial nonlinearity satisfies this dissipativity condition), the trajectory attractor (that by definition captures all the asymptotic behavior of nonnegative solutions in the quarter-space) consists of all nonnegative solutions of the same semilinear elliptic equation, however, in the half-space. Using the symmetry and monotonicity results for nonnegative solutions in such domains (the technique in this chapter for the proving of symmetry and monotonicity results requires dimension restrictions and is based on the moving plane method considered in *Sect. 1.7*), we can describe the asymptotic behavior of solution and prove that the elliptic attractor is one-dimensional and any solution of the elliptic BVP in the quarter-space converges to a unique solution of a second-order ODE, indicating remarkable one-dimensional reduction of dynamics on the attractor. We emphasize that indeed such trajectory DS approach provides an elegant proof of a stabilization and symmetrization results that in general are really difficult to obtain using pure elliptic techniques, due to the unboundedness of the underlying domains in any directions and the possible appearance of oscillations during the limiting procedure. These chapters consists of four subsections, which deals with formulation of the problem (*Sects. 3.1* and *4.1*), a priori estimates and solvability results (*Sects. 3.2* and *4.2*), the existence of attractor for associated trajectory dynamical systems (*Sects. 3.3* and *4.3*), and symmetry and stabilization results (*Sects. 3.4* and *4.4*), respectively.

In *Chap. 5*, we consider the same elliptic boundary value problem as was studied in *Chaps. 3* and *4*. However, in this chapter, the dimension of the underlying domain is less than or equal to five. We aim at extending the results from dimensions three and four to the dimension five. Unfortunately we cannot apply the techniques from the previous chapters. We show how to avoid technical difficulties arising in this case of dimension five and obtain new type of asymptotics in contrast to previous chapters. Indeed, under the assumptions that all equilibria are stable, we prove that:

(a) ANY nonnegative bounded solution of a semilinear elliptic BVP in the quarter-space converges to a uniquely defined solution of a second-order ODE.
(b) ANY nonnegative bounded solution in the half-space converges to a uniquely defined constant solution.

In *Chap. 6*, we present completely new Liouville-type theorem for two classes of nonlinearities which give us the possibility to consider the same semilinear elliptic boundary value problem in any dimension. It is worth noting that these new Liouville-type results for any dimension are of independent interest for the solutions in half-spaces with homogeneous Dirichlet boundary conditions or in the whole space. In contrast to the results of the *Chap. 5*, one of the main points is that the results of *Chap. 6* hold in any dimensions without any assumption on a solution other than its boundedness. In *Chap. 5*, the main results were obtained in lower dimensions of the underlying domains and under the additional assumption that the solutions are stable, which among others requires C^1-differentiability of the nonlinearities. However, the assumptions on the nonlinearity $f(s)$ imposed in the *Chap. 5* are incompatible with the assumptions for the nonlinearity in *Chap. 6* which are crucial to prove new Liouville-type results. Using these new Liouville-type results, we completely characterize the limiting behavior of solutions of semilinear elliptic boundary value in the quarter-space as well as in the half-space based on the trajectory dynamical systems approach. It is worth noting that all of these results required completely new ideas, tools, and techniques.

Moreover in this *Chap. 6*, in contrast to the previous *Chaps. 3–5*, we compare two different approaches, namely, PDEs and dynamical systems approach discussing their individual advantages, as regards describing asymptotic profiles of solutions both in the quarter- and half-spaces in any dimension. To prove there exists a one-dimensional elliptic attractor, we use, among others, the sliding method which was discussed in *Sect. 1.7*. *Chapter 6* consists of four subsections. *Section 6.1* is devoted to the formulation of the problem as well as the introduction of two classes of nonlinearities: sign-changing and sign-preserving. In *Sect. 6.2*, we deal with the PDE approach to an elliptic boundary value problem both in the quarter- and half-spaces and study some properties of solutions based on, among others, sliding methods that are relevant for determining the asymptotic of solutions. *Section 6.3* deals with new type of Liouville results for bounded nonnegative solutions which on the one hand is of independent interest (it holds in any dimension without any assumption on a solution other than its boundedness) and on the other hand plays a decisive role in the study of uniqueness and one-dimensional symmetry of the limiting profiles of solutions as x_1 goes to infinity. In *Sect. 6.4*, we apply the trajectory DS approach to study symmetrization and stabilization properties (as x_1 goes to infinity) of nonnegative solutions for the equations in *Sect. 6.1* and compare this DS approach with the PDE approach developed in Sect. *6.2*.

Chapter 7 deals with the asymptotic behavior of solutions of quasilinear elliptic problems over the quarter-space, and with similar problems over the half-space, and consists of four subsections. In *Sect. 7.1*, we formulate the same type of elliptic boundary value problem for the general p-Laplacian as was done in *Chap. 6* for the standard Laplacian and discuss the difficulties that arise in this case. We introduce two classes of nonlinearities: sign-changing and sign-preserving nonlinearities satisfying new assumptions that take into account the p-Laplacian nature of the equations. Since there is no strong comparison principle in general for quasilinear elliptic problem, we have to adopt rather different approaches in many key steps previously done in *Chaps. 3–6*. In *Sect. 7.2*, we present some basic results which will be needed in our investigation of the half- and quarter-space problems for p-Laplacian. A crucial ingredient here is a simple weak sweeping principle which is a consequence of the weak comparison principle for the p-Laplacian. We show that in many situations, it is possible to use the weak sweeping principle to replace the moving plane or sliding method which are based on the strong comparison principle for the Laplacian case which were frequently used in *Chaps. 3–6*. Using the weak sweeping principles, we prove new Liouville-type results for p-Laplacian equation in the unbounded domains mentioned above. *Section 7.3* deals with the asymptotic behavior of solutions of p-Laplacian equations using these new Liouville-type results for p-Laplacian equations. *Section 7.4* is devoted to corresponding results in the case of p-Laplacian with those of the semilinear case, existence of solution, as well as exact multiplicity results.

I would like to thank many friends and colleagues who gave me suggestions, advice, and support. In particular, I wish to thank H. Berestycki, X. Cabre, N. Dancer, Y. Du, A. Farina, F. Hamel, H. Kielhöfer, K. Kirchgässner (deceased), H. Matano, A. Mielke, L. Nirenberg, M. Otani, J. Scheurle, C.A. Stuart, J.R.L. Webb, W.L. Wendland, J. Wu, E. Valdinochi, M.I. Vishik (deceased), S. Zelik, and A. Zhigun. Furthermore, I am greatly indebted to my colleagues at the Institute of Computational Biology in the Helmholtz Center Munich and Technical University of Munich, Alexander von Humboldt Foundation, as well as the Fields Monographs book series for their efficient handling of publication. This book was finished when I visited as a Dean's Distinguished Visiting Professor of the University of Toronto and in the Fields Institute. I would like to express my sincere gratitude for these institutions in Toronto for providing an excellent and unique scientific atmosphere. In particular, my thanks go to my colleagues, friends, and staffs in the Fields Institute, namely, Ian Hambleton, Huaxiong Huang, Esther Berzunza, Tanya Nebesna, Miriam Schoeman, Bryan Eelhart, and Tyler Wilson.

Last but not least, I wish to thank my family for permanently encouraging me during the writing of this book.

Neuherberg, Bayern, Germany   Messoud Efendiev

Contents

Chapter 1
Preliminaries

In this chapter, we present notation and generally known facts (mostly without proofs) that we use to state and derive the results of the subsequent chapters. For the sake of convenience, we introduce the following conventions:

- $\mathbb{N}$, $\mathbb{N}_0$, $\mathbb{Z}$, $\mathbb{R}$, $\mathbb{R}^+$ and $\mathbb{R}_0^+$ are sets of natural, non-negative integer, integer, real, positive real and non-negative real numbers respectively;
- $x^+ := \max\{x, 0\}$ returns the positive part of a number $x \in \mathbb{R}$;

$$
\text{sign}(x) := \begin{cases} 1 & \text{for } x > 0, \\ 0 & \text{for } x = 0, \\ -1 & \text{for } x < 0 \end{cases}
$$

  returns its sign;
- The integer and fractional parts of a number $x \in \mathbb{R}$ are the numbers $[x] := \max\{q \in \mathbb{Z} \mid x \geqslant q\}$ and $\{x\} := x - [x]$ respectively;
- By $|\cdot|$ we denote:
  - for a number x its absolute value $|x| = \max\{M, -M\}$;
  - for a vector $x = (x_1, \ldots, x_d) \in \mathbb{R}^d$ its Euclidean norm

$$
|x| := \left(\sum_{i=1}^{d} |x_i|^2 \right)^{\frac{1}{2}};
$$

  - for a multiindex $\alpha = (\alpha_1, \ldots, \alpha_d) \in \mathbb{N}_0^d$ for $d \in \mathbb{N}$ its absolute value $|\alpha| = \sum_{j=1}^{d} \alpha_j$;
  - for a Lebesgue measurable set its Lebesgue measure;
- By measure we always understand the Lebesgue measure;

© Springer Nature Switzerland AG 2018
M. Efendiev, *Symmetrization and Stabilization of Solutions of Nonlinear Elliptic Equations*, Fields Institute Monographs 36, https://doi.org/10.1007/978-3-319-98407-0_1

- In a topological space X, we denote by $cl_X(A)$ the closure of a set A in X. In $\mathbb{R}^d$ we use the notation $\overline{A}$ instead. ∂A denotes the topological boundary of A.
- In a linear space X, we define $x + A := \{x + a \mid a \in A\}$ for $x \in X$, $A \subset X$.

1.1   Functional Spaces and Their Properties

This section is devoted to the classical Lebesgue and Sobolev spaces and to some of their modifications. We refer to [2, 101] for a detailed analysis of the classical spaces.

Most of the presented spaces are normable (e.g., $L^p(\Omega)$ and $W^{s,p}(\Omega)$), some are metriziable, but not normable (e.g., $L^p_{loc}(\Omega)$ and $W^{s,p}_{loc}(\Omega)$), and some (e.g., $L^\infty_{w-*}(\Omega)$) are not even metriziable, though locally convex. Let us, therefore, before looking at concrete examples, briefly recall several facts originating from the general framework in locally convex and normed spaces. We refer to [85] (or some other standard textbook) for these as well as for the other facts from functional analysis that we use in this work.

Let X and Y be two locally convex spaces. The (continuous) dual space of X is denoted by X', its weak and weak-$*$ topologies by $\sigma(X, X')$ and $\sigma(X', X)$ respectively. $X \cong Y$ denotes the topological equivalence of X and Y, which means that they are homeomorphic.

Let F be an operator, not necessarily linear, between X and Y. F is said to be compact if it maps bounded subsets of X onto relatively compact subsets of Y. It is said to be closed if its graph $\Gamma(F) := \{(x, F(x)) \mid x \in X\}$ is closed in $X \times Y$.

We will often consider embeddings of a 'smaller' locally convex space into a 'larger' one:

Definition 1.1 (Embedding) Let X and Y be two locally convex spaces. An injective linear operator $\iota : X \to Y$ is called an embedding. If such an operator ι exists between X and Y, then X is said to be embedded in Y. Further, we say that

(1) X is continuously embedded in Y (and write $X \hookrightarrow Y$) if ι is a continuous operator;
(2) X is compactly embedded in Y (and write: $X \hookrightarrow\hookrightarrow Y$) if ι is a compact operator;
(3) X is densely and continuously embedded in Y (and write: $X \overset{d}{\hookrightarrow} Y$) if ι is a continuous operator and $\iota(X)$ is dense in Y;
(4) X is densely and compactly embedded in Y (and write: $X \overset{d}{\hookrightarrow\hookrightarrow} Y$) if ι is a compact operator and $\iota(X)$ is dense in Y.

An important property of the dense and continuous embeddings concerns duality:

Theorem 1.1 (Embedding of Dual Spaces) *Let X and Y be two locally convex spaces. Then*

$$X \xhookrightarrow{d} Y \Rightarrow Y' \xhookrightarrow{d} X'.$$ (1.1)

It is well known that every compact linear operator between Banach spaces is continuous. It is even weak-to-norm continuous, which means that it is continuous between $(X, \sigma(X, X'))$ and Y. A similar property holds for the weak-∗-to-norm continuity:

Theorem 1.2 (Weak-∗-to-Norm Continuity of a Compact Linear Operator)
Let X be a normed space, Y a Banach space and F a compact linear operator between X' and Y. Then F is a continuous operator between $(X', \sigma(X', X))$ and Y.

1.1.1 L^p Spaces

We assume in this subsection that Ω is a nonempty measurable subset of $\mathbb{R}^d$, $d \in \mathbb{N}$. Let us denote by $L^0(\Omega)$ the (linear) space of all equivalence classes of measurable functions on Ω. Each such class consists of functions that are equal almost everywhere in Ω. As usual, we identify the functions from one equivalence class and write u instead of $[u]$.

For $p \in [1, \infty]$ the function

$$|| \cdot ||_{L^p(\Omega)} : L^0(\Omega) \to [0, \infty],$$

$$||u||_{L^p(\Omega)} := \begin{cases} \left(\int_\Omega |u(x)| \, dx \right)^{\frac{1}{p}} & \text{for } p \in [1, \infty), \\ \operatorname*{ess\,sup}_{x \in \Omega} |u(x)| & \text{for } p = \infty \end{cases}$$ (1.2)

is called the L^p norm. It is a well defined norm on the space

$$L^p(\Omega) := \left\{ u \in L^0(\Omega) | \, ||u||_{L^p(\Omega)} < \infty \right\}.$$

Each L^p space equipped with the L^p norm is a Banach space. The space $L^2(\Omega)$ is a Hilbert space with the scalar product

$$(u, v)_{L^2(\Omega)} := \int_\Omega u(x)v(x) \, dx \text{ for } u, v \in L^2(\Omega).$$

We write $|| \cdot ||_p$ instead of $|| \cdot ||_{L^p(\Omega)}$ and even $|| \cdot ||$ instead of $|| \cdot ||_2$ and $(\cdot, \cdot)$ instead of $(\cdot, \cdot)_{L^2(\Omega)}$ to shorten the notation.

Some of the most important results about L^p spaces are:

Theorem 1.3 (Hölder Inequality) *Let $p, q \in [1, \infty]$ be such that $\frac{1}{p} + \frac{1}{q} = 1$. Then for all $u \in L^p(\Omega)$, $v \in L^q(\Omega)$, we have $uv \in L^1(\Omega)$, and the following inequality holds*

$$||uv||_1 \leqslant ||u||_p||v||_q.$$

Theorem 1.4 (Interpolation Inequality for $L^p(\Omega)$) *Let $1 \leqslant p_1 < p < p_2 \leqslant \infty$. Then for all $u \in L^{p_1}(\Omega) \cap L^{p_2}(\Omega)$ we have $u \in L^p(\Omega)$, and the following inequality holds*

$$||u||_p \leqslant ||u||_{p_1}^{1-\theta}||u||_{p_2}^{\theta},$$

where

$$\theta := \frac{\frac{1}{p_1} - \frac{1}{p}}{\frac{1}{p_1} - \frac{1}{p_2}}.$$

Theorem 1.5 (Dual Representation for $L^p(\Omega)$) *Let $p \in [1, \infty)$. Put*

$$p' := \begin{cases} \frac{p}{p-1} & \text{for } p \in (1, \infty), \\ \infty & \text{for } p = 1. \end{cases}$$

Then it holds that

$$(L^p(\Omega))' \cong L^{p'}(\Omega).$$

In particular, consider the linear operator

$$\iota_p : L^{p'}(\Omega) \to (L^p(\Omega))', \quad \iota_p(u)(v) := \int_\Omega u(x)v(x)\, dx \text{ for all } v \in L^p(\Omega).$$

ι_p *is an isometric isomorphism between $L^{p'}(\Omega)$ and $(L^p(\Omega))'$.*

Up to this point, we have considered the $L^\infty(\Omega)$ space, as well as other L^p spaces, equipped with the topology produced by the corresponding norm. For some of our applications (see below), the original topology appears too restrictive. We are forced to pass to weaker topologies where it is easier to prove compactness. We start with the following

Definition 1.2 ($L_{w-*}^\infty(\Omega)$ Space) We define $L_{w-*}^\infty(\Omega)$ to be the set of all $L^\infty(\Omega)$-functions equipped with the topology

$$\left\{ \iota_1^{-1}(O) \mid O \in \sigma((L^1(\Omega))', L^1(\Omega)) \right\},$$

where ι_1 is the isometric isomorphism defined in *Theorem 1.5*.

Some properties of $L_{w-*}^\infty(\Omega)$ are collected in *Theorem 1.6* and *Remark 1.1* below.

Theorem 1.6 (Properties of $L^\infty_{w-*}(\Omega)$)

(1) The space $L^\infty_{w-*}(\Omega)$ is a locally convex space;
(2) A subset of $L^\infty(\Omega)$ is bounded in $L^\infty_{w-*}(\Omega)$ if and only if it is norm bounded;
(3) The topology of $L^\infty_{w-*}(\Omega)$, if restricted to an $L^\infty(\Omega)$ ball, is completely metrizable;
(4) $L^\infty(\Omega)$ balls are compact in $L^\infty_{w-*}(\Omega)$.

Sketch of the proof Observe that ι_1 is not only an isometric isomorphism between $(L^1(\Omega))'$ and $L^\infty(\Omega)$, it is, due to *Definition 1.2*, also a linear homeomorphism between $((L^1(\Omega))', \sigma((L^1(\Omega))', L^1(\Omega)))$ and $L^\infty_{w-*}(\Omega)$. But every homeomorphism preserves metrizability and compactness properties, and every linear homeomorphism preserves locally convex structure and boundedness of subsets. Therefore, the properties (i)–(iv) are consequences of the corresponding properties of the space $((L^1(\Omega))', \sigma((L^1(\Omega))', L^1(\Omega)))$, which is the dual space of the infinitely dimensional separable Banach space $L^1(\Omega)$, equipped with the weak-$*$ topology, and the fact that compact metric spaces are complete. □

Remark 1.1 (Further Properties of $L^\infty_{w-*}(\Omega)$)

(1) The weak-$*$ topology is the topology of pointwise convergence, so that the topology of $L^\infty_{w-*}(\Omega)$ can be also obtained by means of the following convergence notion: A sequence $\{v_n\}_{n\in\mathbb{N}}$ converges in $L^\infty_{w-*}(\Omega)$ to a v if and only if

$$\int_\Omega u(x)v_n(x)dx \underset{n\to\infty}{\to} \int_\Omega u(x)v(x)\,dx \text{ for all } u \in L^1(\Omega);$$

(2) A metric for the restriction of the $L^\infty_{w-*}(\Omega)$ topology to a ball of radius R centered at 0 can be defined in the following way. Let $\{u_n\}_{n\in\mathbb{N}}$ be a dense subset of $L^1(\Omega)$ and let $\{B_n\}_{n\in\mathbb{N}}$ be a sequence of positive real numbers, such that

$$\sum_{n\in\mathbb{N}} B_n\|u_n\|_{L^1(\Omega)} < \infty.$$

Then the function defined by

$$d_*^{(\infty)}(v_1, v_2) := \sum_{n\in\mathbb{N}} B_n \left| \int_\Omega (v_1(x) - v_2(x))u_n(x)\,dx \right| \tag{1.3}$$

for all $v_1, v_2 \in L^\infty(\Omega)$, $\|v_1\|_\infty, \|v_2\|_\infty \leqslant R$, is an example of a metric which produces the relative topology;

(3) The property 4. from *Theorem 1.6* is equivalent to $L^\infty(\Omega) \hookrightarrow\hookrightarrow L^\infty_{w-*}(\Omega)$. This is due to the definition of compact embedding.

(4) For more information on compactness and metrizability in the weak-$*$ topology see [85].

Thus, the $L^\infty(\Omega)$ balls are metrizable compact subsets of $L^\infty_{w-*}(\Omega)$. The following intersections of the $L^\infty_{w-*}(\Omega)$ topology with the L^p norm topology for $p \in [1, \infty)$ are of independent interest. The definition is as follows:

Definition 1.3 (The Space $\mathcal{H}^p(\Omega)$) Let $p \in [1, \infty]$. We define $\mathcal{H}^p(\Omega)$ to be the set of all $L^\infty(\Omega)$-functions equipped, for $p = \infty$, with the topology of $L^\infty_{w-*}(\Omega)$ and, for $p \in [1, \infty)$, with the intersection of the topologies of the spaces $L^\infty_{w-*}(\Omega)$ and $L^p(\Omega)$.

Next theorem contains several properties of the spaces $\mathcal{H}^p(\Omega)$.

Theorem 1.7 (Properties of $\mathcal{H}^p(\Omega)$) Let $p \in [1, \infty)$. Then:

(1) The space $\mathcal{H}^p(\Omega)$ is a locally convex space;
(2) A subset of $L^\infty(\Omega)$ is bounded in $\mathcal{H}^p(\Omega)$ if and only if it is norm bounded;
(3) The topology of $\mathcal{H}^p(\Omega)$, if restricted to an $L^\infty(\Omega)$ ball, is completely metrizable;
(4) The topologies of $\mathcal{H}^p(\Omega)$ and $L^2(\Omega)$, if restricted to an $L^\infty(\Omega)$ ball, coincide.

Sketch of the proof Let $p \in [1, \infty)$.

(1) We observe that the set $\{l_u \mid u \in L^1(\Omega)\}$, where $l_u(u') := |u'(u)|$ for $u' \in (L^1(\Omega))'$, is an example of a system of seminorms on $(L^1(\Omega))'$ that generates $\sigma((L^1(\Omega))', L^1(\Omega))$ topology on $(L^1(\Omega))'$. Hence the locally convex structure of the space $L^\infty_{w-*}(\Omega)$ is given by the family $\{\omega_u \mid u \in L^1(\Omega)\}$, where $\omega_u(v) := \left|\int_\Omega u(x)v(x)\,dx\right|$ for all $u \in L^1(\Omega)$ and $v \in L^\infty(\Omega)$. Consequently, the family $\{\omega_u + \|\cdot\|_p \mid u \in L^1(\Omega)\}$ is an example of a system of norms that generates the locally convex structure of $\mathcal{H}^p(\Omega)$;
(2) A set is bounded in $\mathcal{H}^p(\Omega)$ if and only if it is bounded in each of the seminorms that defines it locally convex structure. In the particular case of $\mathcal{H}^p(\Omega)$ it follows with the proof of the property (1) that a set is bounded in $\mathcal{H}^p(\Omega)$ if and only if it is bounded in each of the norms ω_u and in the L^p norm, which is equivalent to the boundedness in both $L^\infty_{w-*}(\Omega)$ and $L^p(\Omega)$. The statement now follows with the property (2) from *Theorem 1.6* and the fact that, for Ω bounded, the L^∞ norm is stronger then any other L^p norm;
(3) It is a consequence of (4);
(4) Observe first that, due to the Hölder inequality and *Theorems 1.4*, we have

$$\frac{1}{\|1\|_{\frac{p}{p-1}}}\|v\|_1 \leqslant \|v\|_p \leqslant \|v\|_\infty^{1-\frac{1}{p}}\|v\|_1^{\frac{1}{p}} \quad \text{for all } v \in L^\infty(\Omega).$$

This shows that the topologies of $L^p(\Omega)$ and $L^1(\Omega)$, if restricted to an $L^\infty(\Omega)$ ball, coincide. To show the property (4) it then suffices to check that the restriction of the $L^\infty_{w-*}(\Omega)$ topology is weaker than, for example, the $L^2(\Omega)$ topology.

Since the space $L^2(\Omega)$ is dense in $L^1(\Omega)$, we may assume that $\{u_n\}_{n\in\mathbb{N}} \subset L^2(\Omega)$ in the definition of the metric $d_*^{(\infty)}$ from (1.3) and choose the sequence $\{B_n\}_{n\in\mathbb{N}}$ to be such that

$$\sum_{n\in\mathbb{N}} B_n \|u_n\|_{L^2(\Omega)} < \infty$$

holds. Consequently, we obtain with the Hölder inequality that

$$d_*^{(\infty)}(v_1, v_2) = \sum_{n\in\mathbb{N}} B_n \left| \int_\Omega (v_1(x) - v_2(x))u_n(x)\, dx \right|$$

$$\leqslant \sum_{n\in\mathbb{N}} B_n \|u_n\|_{L^2(\Omega)} \|v_1 - v_2\|_2.$$

This shows that, if restricted to an $L^\infty(\Omega)$ ball, the $L^\infty_{w-*}(\Omega)$ topology is weaker then the topology of $L^2(\Omega)$.

$\square$

Thus, the $L^\infty(\Omega)$ balls are metrizable subsets of $\mathcal{H}^p(\Omega)$, and, for $p \in [1, \infty)$, a subset of an $L^\infty(\Omega)$ ball is compact if an only if it is compact in $L^p(\Omega)$.

Sometimes, especially in case when Ω is unbounded, it is useful (see [35, 48]) to consider the local version of an L^p space, the space

$$L^p_{loc}(\Omega) := \left\{ u \in L^0(\Omega) \mid u \in L^p(K) \text{ for all compact sets } K \subset \Omega \right\}.$$

This space is not normable, though metrizable. Define the function

$$\|\cdot\|_{L^p_b(\Omega)} : L^p_{loc}(\Omega) \to [0, \infty],$$

$$\|u\|_{L^p_b(\Omega)} := \sup_{x_0 \in \mathbb{R}^d} \|u\|_{L^p(\Omega \cap B_{x_0}(1))},$$

where $B_{x_0}(1)$ is a unit ball in $\mathbb{R}^d$ centered at x_0. $\|\cdot\|_{L^p_b(\Omega)}$ is a norm on a subspace of $L^p_{loc}(\Omega)$, namely on the space

$$L^p_b(\Omega) := \left\{ u \in L^p_{loc}(\Omega) \mid \|u\|_{L^p_b(\Omega)} < \infty \right\}.$$

Note that $L^1_{loc}(\Omega)$ is the largest of the presented spaces of the L^p type.

1.1.2   Sobolev Spaces

From now on we assume Ω to be a nonempty domain (i.e. a nonempty open connected set) in $\mathbb{R}^d$. We denote by $D(\Omega)$ the (locally convex) space of all test functions over Ω. As a set, $D(\Omega)$ coincides with $C_0^\infty(\Omega)$, the set of all infinitely differentiable functions with compact support in Ω. The dual space of $D(\Omega)$, the space $D'(\Omega)$, is the space of distributions over Ω. Distributions of the form $v \to \int_\Omega u(x)v(x)\,dx$ for $u \in L^1_{loc}(\Omega)$ are called regular. In case of a regular distribution, we identify the distribution with the L^1_{loc} function that produces it.

For $u \in D'(\Omega)$ and $v \in D(\Omega)$ we denote by (u, v) the value of u on v. In case when $u \in L^2(\Omega)$ we recover the scalar product in $L^2(\Omega)$.

For a multiindex $\alpha = (\alpha_1, \ldots, \alpha_d)$ we define the differential operator of the order $|\alpha|$: $D^{(\alpha)} := \left(\partial_{x_1}^{\alpha_1}, \ldots, \partial_{x_d}^{\alpha_d}\right)$, where ∂_{x_k} is the partial distributional derivative along the variable x_k and $\partial_{x_k}^{\alpha_k} = (\partial_{x_k})^{\alpha_k}$. Recall that any distribution is infinitely differentiable in the distributional sense.

For $s \in \mathbb{N}_0$ and $p \in [1, \infty]$ the function

$$\|u\|_{W^{k,p}(\Omega)} := \begin{cases} \left(\sum_{|\alpha| \leqslant k} \|D^{(\alpha)}u\|_{L^p(\Omega)}^p\right)^{\frac{1}{p}} & \text{for } p \in [1, \infty), \\ \max_{|\alpha| \leqslant k} \|D^{(\alpha)}u\|_\infty & \text{for } p = \infty \end{cases}$$

is a well defined norm on the space

$$W^{k,p}(\Omega) := \left\{ u \in L^p(\Omega) \mid D^{(\alpha)}u \in L^p(\Omega) \text{ for } |\alpha| \leqslant k \right\},$$

Equipped with the $\| \cdot \|_{W^{k,p}(\Omega)}$ norm, the space $W^{k,p}(\Omega)$ is the classical Sobolev space of order k. All $W^{k,p}(\Omega)$ spaces are Banach spaces. The space $H^k(\Omega) := W^{k,2}(\Omega)$ is a Hilbert space with the scalar product

$$(u, v)_{H^k(\Omega)} := \sum_{|\alpha| \leqslant k} \left(D^{(\alpha)}u, D^{(\alpha)}v \right) \text{ for } u, v \in H^k(\Omega).$$

With $k = 0$ we recover the definitions of the corresponding L^p spaces. One of those subspaces of $W^{k,p}(\Omega)$ that play an important role in partial differential equations is the space $W_0^{k,p}(\Omega)$ for Ω bounded. It consists of functions that 'vanish on the boundary' in the sense of trace (see [2]). One of the equivalent ways to define these spaces is:

$$W_0^{k,p}(\Omega) := cl_{W^{k,p}(\Omega)} (D(\Omega))$$

for $k \in \mathbb{N}$ and $p \in [1, \infty]$. For Ω bounded the seminorm

$$|| \cdot ||_{W_0^{k,p}(\Omega)} : W^{k,p}(\Omega) \to [0, \infty),$$

$$||u||_{W_0^{k,p}(\Omega)} := \left(\sum_{|\alpha|=k} \left\| D^{(\alpha)} u \right\|_{L^p(\Omega)}^p \right)^{\frac{1}{p}}$$

is an equivalent norm on $W_0^{k,p}(\Omega)$. This is a consequence of the Poincaré inequality:

Theorem 1.8 (Poincaré Inequality) *Let $p \in [1, \infty]$ and let Ω be a smooth bounded domain in $\mathbb{R}^d$. Then there exists a positive constant $P(\Omega, p)$ that depends only on Ω and p and such that*

$$||u||_p \leqslant P(\Omega, p) ||Du||_p$$

holds for all $u \in W_0^{1,p}(\Omega)$.

The norm $|| \cdot ||_{W_0^{k,p}(\Omega)}$ is called the energy norm. On the space $H_0^k(\Omega) := W_0^{k,2}(\Omega)$ the bilinear form defined via

$$(u, v)_{H_0^k(\Omega)} := \sum_{|\alpha|=k} \left(D^{(\alpha)} u, D^{(\alpha)} v \right) \text{ for } u, v \in H_0^k(\Omega)$$

is a scalar product. The space $W_0^{k,p}(\Omega)$ is a closed subspace of $W^{k,p}(\Omega)$, thus it is a Banach space, while the space $H_0^k(\Omega)$ is a Hilbert space.

It is often useful to consider a class of 'in-between' spaces, that is, to extend the notion of classical Sobolev spaces of non-negative integer order k to the case $s \in \mathbb{R}_0^+ \backslash \mathbb{N}_0$. One of the possible contractions uses the Slobodeckij seminorm

$$[u]_{\theta,p} := \begin{cases} \left(\int_\Omega \int_\Omega \frac{|u(x)-u(y)|^p}{|x-y|^{p\theta+d}} \, dx \, dy \right)^{\frac{1}{p}} & \text{for } p \in [1, \infty), \\ \operatorname*{ess\,sup}_{x,y \in \Omega, x \neq y} \frac{|u(x)-u(y)|}{|x-y|^\theta} & \text{for } p = \infty \end{cases}$$

defined for $\theta \in (0, 1)$ and $p \in [1, \infty]$. For $s \in \mathbb{R}_0^+ \backslash \mathbb{N}_0$ and $p \in [1, \infty]$ the function

$$|| \cdot ||_{W^{s,p}(\Omega)} : W^{[s],p}(\Omega) \to [0, \infty],$$

$$||u||_{W^{s,p}(\Omega)} := \begin{cases} \left(||u||_{W^{[s],p}(\Omega)}^p + \sum_{|\alpha|=s} [D^{(\alpha)} u]_{\{s\},p}^p \right)^{\frac{1}{p}} & \text{for } p \in [1, \infty), \\ ||u||_{W^{[s],\infty}(\Omega)} + \max_{|\alpha|=s} [D^{(\alpha)} u]_{\{s\},\infty} & \text{for } p = \infty \end{cases}$$

is a well defined norm on the space

$$W^{s,p}(\Omega) := \left\{ u \in W^{[s],p}(\Omega) \middle| \left[D^{(\alpha)} u \right]_{\{s\},p} < \infty \text{ for } |\alpha| = [s] \right\}.$$

The spaces $W^{s,p}(\Omega)$ for $s \in \mathbb{R}_0^+ \backslash \mathbb{N}_0$ are called Sobolev-Slobodeckij spaces. These spaces are Banach spaces. The space $H^s(\Omega) := W^{s,2}(\Omega)$ is a Hilbert space with the scalar product

$$(u, v)_{H^s(\Omega)}$$

$$:= (u, v)_{H^{[s]}(\Omega)}$$

$$+ \sum_{|\alpha|=s} \int_\Omega \int_\Omega \frac{\left(D^{(\alpha)} u(x) - D^{(\alpha)} u(y) \right) \left(D^{(\alpha)} v(x) - D^{(\alpha)} v(y) \right)}{|x - y|^{2\theta + d}} \, dx dy.$$

If the domain Ω is suitably regular then, indeed, $W^{s_2,p}$ is a subset of $W^{s_1,p}$ for all $0 \leqslant s_1 < s_2 < \infty$.

Just as in case of integer order Sobolev spaces, we can define for $s \in \mathbb{R}_0^+ \backslash \mathbb{N}_0$ and $p \in [1, \infty]$ the space

$$W_0^{s,p}(\Omega) := cl_{W^{s,p}(\Omega)} \left(D(\Omega) \right).$$

Observe that for all $s \in \mathbb{R}^+$ and $p \in [1, \infty]$ the space $D(\Omega)$ is densely and continuously embedded in the space $W_0^{s,p}(\Omega)$ by means of the identity operator. This is because the convergence in $D(\Omega)$ is stronger than the convergence in $W_0^{s,p}(\Omega)$ and because $D(\Omega)$ is dense in $W_0^{s,p}(\Omega)$ (by definition). With (1.1) it follows that $\left(W_0^{s,p}(\Omega) \right)' \xrightarrow{d} D'(\Omega)$. Now, for $s \in \mathbb{R}^+$ and $p \in (1, \infty]$ set

$$W^{-s,p}(\Omega) := \left(W_0^{s,p'}(\Omega) \right)', \quad p' := \begin{cases} \frac{p}{p-1} & \text{for } p \in (1, \infty), \\ 1 & \text{for } p = \infty, \end{cases}$$

$$H^{-s}(\Omega) := W^{-s,2}(\Omega).$$

This is the way to define the Sobolev spaces of negative order. For $p \in (1, \infty)$ it also holds

$$\left(W^{-s,p}(\Omega) \right)' \cong W_0^{s,p'}(\Omega).$$

This is a consequence of

Theorem 1.9 (Reflexivity of $W^{s,p}(\Omega)$) *Let $s \in \mathbb{R}$ and $p \in (1, \infty)$. The space $W^{s,p}(\Omega)$ is reflexive.*

For all $s \in \mathbb{R}$ and $p \in [1, \infty]$ the number $\gamma = s - \frac{d}{p}$ is called the Sobolev number (corresponding to the pair s, p). The numbers s, p and γ can be used to compare a Sobolev space with another Sobolev space or with a Hölder space. This is the subject of

Theorem 1.10 (Sobolev Embedding Theorem) *Let Ω be smooth and bounded. Let $-\infty < s_1 < s_2 < \infty$, $1 \leqslant p_2 \leqslant p_1 \leqslant \infty$ and let γ_1 and γ_2 be the Sobolev numbers corresponding to the pairs s_1, p_1 and s_2, p_2 respectively. Then:*
(Part I)

$$\gamma_2 > \gamma_1 \Rightarrow W^{s_2, p_2}(\Omega) \hookrightarrow\hookrightarrow W^{s_1, p_1}(\Omega),$$

$$\gamma_2 = \gamma_1 \text{ and } p_1 < \infty \Rightarrow W^{s_2, p_2}(\Omega) \hookrightarrow W^{s_1, p_1}(\Omega),$$

the embedding being the identity operator. In both cases the Sobolev inequality

$$\|u\|_{W^{s_1, p_1}(\Omega)} \leqslant C_0(s_1, s_2, p_1, p_2)\|u\|_{W^{s_2, p_2}(\Omega)} \text{ for all } u \in W^{s_2, p_2}(\Omega) \qquad (1.4)$$

holds. The embedding constant $C_0(s_1, s_2, p_1, p_2)$ depends only on s_1, s_2, p_1, p_2 and the domain Ω.
(Part II)

$$\gamma_2 > \gamma_1 \text{ and } p_1 = \infty, \ s_1 > 0 \Rightarrow W^{s_2, p_2}(\Omega) \hookrightarrow\hookrightarrow C^{[s_1], \{s_1\}}(\overline{\Omega}),$$

the embedding being the identity operator and the Sobolev inequality

$$\|u\|_{C^{[s_1], \{s_1\}}(\overline{\Omega})} \leqslant C_1(s_1, s_2, p_2)\|u\|_{W^{s_2, p_2}(\Omega)}$$

holds. The embedding constant $C_1(s_1, s_2, p_2)$ depends only on s_1, s_2, p_2 and Ω.

Remark 1.2 In part II of the Sobolev embedding theorem, the Sobolev spaces are compared with the spaces of continuously differentiable functions $C^k(\overline{\Omega})$ and the Hölder spaces $C^{k,\theta}(\overline{\Omega})$. They are continuous versions of the Sobolev spaces $W^{k,\infty}(\Omega)$ and the Sobolev-Slobodeckij spaces $W^{k+\theta,\infty}(\Omega)$, respectively:

$$C^0(\overline{\Omega}) := C(\overline{\Omega}) := \left\{ u : \overline{\Omega} \to \mathbb{R} \mid u \text{ continuous on } \overline{\Omega} \right\},$$

$$C^k(\overline{\Omega}) := \left\{ u \in C(\overline{\Omega}) \mid D^{(\alpha)} u \in C(\overline{\Omega}) \text{ for } |\alpha| \leqslant k \right\}, \ k \in \mathbb{N},$$

$$C^{k,\theta}(\overline{\Omega}) := \left\{ u \in C^k(\overline{\Omega}) \mid \sup_{x,y \in \overline{\Omega}, x \neq y} \frac{\left| D^{(\alpha)} u(x) - D^{(\alpha)} u(y) \right|}{|x - y|^\theta} < \infty \right\},$$

$$\| \cdot \|_{C^{k,\theta}(\overline{\Omega})} := \| \cdot \|_{W^{k+\theta,\infty}(\Omega)}, k \in \mathbb{N}_0, \ \theta \in [0, 1).$$

As in case of L^p spaces, we have an interpolation inequality for a space 'in-between':

Theorem 1.11 (Interpolation Inequality for $W^{s,p}(\Omega)$**)** *Let* Ω *be smooth and bounded. Let* $s_1, s, s_2 \in (0, \infty)$ *and* $p_1, p, p_2 \in [1, \infty]$ *be such that*

$$s_2 > s \geqslant s_1,$$

$$\gamma_2 > \gamma > \gamma_1,$$

$$\theta := \frac{\gamma - \gamma_1}{\gamma_2 - \gamma_1} \in \left(\frac{s - s_1}{s_2 - s_1}, 1 \right),$$

where γ_1, γ *and* γ_2 *are the Sobolev numbers corresponding to the pairs* s_1, p_1, s, p *and* s_2, p_2 *respectively. Then the following interpolation inequality holds for all* $u \in W^{s_2, p_2}(\Omega)$:

$$||u||_{W^{s,p}(\Omega)} \leqslant I(s_1, s, s_2, p_1, p, p_2)||u||^{1-\theta}_{W^{s_1,p_1}(\Omega)}||u||^{\theta}_{W^{s_2,p_2}(\Omega)}. \tag{1.5}$$

The constant $I(s_1, s, s_2, p_1, p, p_2)$ *depends only on* s_1, s, s_2, p_1, p, p_2 *and the domain* Ω.

The following useful nonlinear version of the Sobolev inequality (1.4) is a consequence of Lemma 1.2 from [46].

Lemma 1.1 *Let* $s \in (0, 1)$, $p \in [1, \infty)$ *and* $q \in (1, \infty)$. *Then there exists a constant* $N(q)$ *that depends only on* q *and such that it holds*

$$||u||_{W^{\frac{s}{q}, sq}(\Omega)} \leqslant N(q) \left\| |u|^{q-1} u \right\|^{\frac{1}{q}}_{W^{s,p}(\Omega)} \quad \textit{for all } u \in W^{s,p}(\Omega).$$

We conclude this subsection with the definition of a local Sobolev space:

$$W^{k,p}_{loc}(\Omega) := \left\{ u \in L^p_{loc}(\Omega) \mid D^{(\alpha)} u \in L^p_{loc}(\Omega) \text{ for } |\alpha| \leqslant k \right\}.$$

Remark 1.3 Below, we give definitions of several Fréchet spaces which will be useful in the study of nonlinear elliptic systems in unbounded domains that we study in *Chap. 2*.

Let $\mathbb{V} \subset \mathbb{R}^N$ be some open set in $\mathbb{R}^N$. We denote by

$$W^{l,p}(\mathbb{V}) = \left\{ w \in D'(\mathbb{V}) \mid D^\alpha w \in L^p(\mathbb{V}) \right\},$$

$$H^{l,p}(\mathbb{V}) = W^{l,p}(\mathbb{R}^N)|_{\mathbb{V}}.$$

It is well-known that if $\mathbb{V}$ is a bounded domain with a 'smooth' boundary, then the spaces $H^{l,p}(\mathbb{V})$ and $W^{l,p}(\mathbb{V})$ coincide (see [26]).

Remark 1.4 Let $\mathbb{V}$ be a bounded subset in $\mathbb{R}^N$. By $W^{l,p}_0(\mathbb{V})$ we denote completion of $C^\infty_0(\mathbb{V})$ in the metric of $W^{l,p}(\mathbb{V})$. $\left[W^{1,2}_0(\mathbb{V}) \right]^* \subset D'(\mathbb{V})$ we denote by $W^{-1,2}(\mathbb{V})$.

We define the spaces $W_{loc}^{1,2}(\mathbb{V})$ for an arbitrary unbounded domain $\mathbb{V}$ as a Frechét subspace of $D'(\mathbb{V})$ endowed by the seminorms

$$||v, \mathbb{B}_{x_0}^R \cap \mathbb{V}|| := ||v||_{W^{1,2}(\mathbb{B}_{x_0}^R \cap \mathbb{V})}, \quad R \subset \mathbb{R}_+, \ x_0 \in \mathbb{R}.$$

Analogously, we define $L_{loc}^p(\mathbb{V})$, $W_{loc}^{-1,2}(\mathbb{V})$ for $1 \leqslant p \leqslant \infty$. We set

$$\theta(\mathbb{V}) := \left[W_{loc}^{1,2}(\mathbb{V}) \cap L_{loc}^r(\mathbb{V}) \right],$$

where $\mathbb{V}$ is an open set in $\mathbb{R}^N$, r is the same as in the assumption on f (see *Sect. 2.4*). Seminorms in $\theta(\mathbb{V})$ we define in the following way:

$$||v, \mathbb{B}_{x_0}^R \cap \mathbb{V}||_+ := \max \left\{ ||v||_{W^{1,2}(\mathbb{B}_{x_0}^R \cap \mathbb{V})}, \ ||v||_{L^r(\mathbb{B}_{x_0}^R \cap \mathbb{V})} \right\},$$

where $R \subset \mathbb{R}_+$, $x_0 \in \mathbb{R}^N$.

By $\theta_0(\mathbb{V})$ we denote completion of $C_0^\infty(\mathbb{V})$ in topology of $\theta(\mathbb{V})$.

Let $\mathbb{V}$ be an arbitrary open set in $\mathbb{R}^n$. For another open set $\mathbb{W}$ in $\mathbb{R}^n$, we define by $\theta_0(\mathbb{V}, \mathbb{W})$ the completion of $C_0^\infty(\mathbb{V})|_{\mathbb{V} \cap \mathbb{W}}$ in the space $\theta(\mathbb{V} \cap \mathbb{W})$. It is obvious that

$$\theta_0(\mathbb{V} \cap \mathbb{W}) \subset \theta_0(\mathbb{V}, \mathbb{W}) \subset \theta(\mathbb{V} \cap \mathbb{W}).$$

We denote by $\theta(\mathbb{V}, \mathbb{W})$ a completion of the set $\theta(\mathbb{V})|_{\mathbb{V} \cap \mathbb{W}}$ in the space $\theta(\mathbb{V} \cap \mathbb{W})$. Usually, in application to elliptic systems in an unbounded domain $\Omega \subset \mathbb{R}^n$, we will take $\mathbb{V} := \Omega$ and $\mathbb{W} := B(x_0, R)$.

Definition 1.4 Let $\mathbb{V}$ be an open set in $\mathbb{R}^n$. Then

$$\Xi(\mathbb{V}) := [W_{loc}^{-1,2}(\mathbb{V}) + L_{loc}^q(\mathbb{V})], \quad \frac{1}{q} + \frac{1}{r} = 1,$$

where r is the same as in the definition of $\theta(\mathbb{V})$.

The space $\Xi(\mathbb{V})$ consists of all $g \in D'(\Omega)$ that have a representation

$$g = g_1 + g_2, \ g_1 \in [W_{loc}^{-1,2}(\mathbb{V})]^k, \ g_2 \in [L_{loc}^q(\mathbb{V})]^k. \tag{1.6}$$

The system of seminorms in $\Xi(\mathbb{V})$ are given by

$$\left| g, \mathbb{V} \cap \mathbb{B}_{x_0}^R \right|_+ := \inf \left\{ \left\| g_1, \mathbb{V} \cap \mathbb{B}_{x_0}^R \right\|_{-1,2} + \left\| g_2, \mathbb{V} \cap \mathbb{B}_{x_0}^R \right\|_{0,q} \right\},$$

$$g = g_1 + g_2, \ g_1 \in [W_{loc}^{-1,2}(\mathbb{V})]^k, \ g_2 \in [L_{loc}^q(\mathbb{V})]^k, \ R \in \mathbb{R}_+, \ x_0 \in \mathbb{R}^n.$$

Analogously, we define $\Xi(\mathbb{V}, \mathbb{W})$ as the completion of $\Xi(\mathbb{V})|_{\mathbb{V} \cap \mathbb{W}}$ in the space $\Xi(\mathbb{V} \cap \mathbb{W})$.

$$v_0(\partial\mathbb{V}) := \theta(\mathbb{V})/\theta_0(\mathbb{V})$$

and the system of seminorms in $v_0(\partial\mathbb{V})$ are defined in the following way:

$$\|u_0\|_{v_0(\partial\mathbb{V} \cap B_{x_0}^R)} := \inf\{\|w\|_+, \ w \in \theta(\mathbb{V}), \ w|_{\partial\mathbb{V}} = u_0\},$$

where $R \in \mathbb{R}_+$, $x_0 \in \mathbb{R}^N$.

The following Propositions will be used in *Chap. 2*:

Proposition 1.1 *Let $\mathbb{V}$ be a bounded domain in $\mathbb{R}^n$. Then for any $0 \leqslant \delta \leqslant r - 1$ it holds that*

$$\theta_0(\mathbb{V}) \subset\subset L^{r-\delta}(\mathbb{V}). \tag{1.7}$$

Proof Indeed, any function $u \in \theta_0(\mathbb{V})$ can be extended (by zero) to the function $\tilde{u} \in \theta(\mathbb{V})$ with preserving norm, so that it suffices to prove (1.7) for a domain $\tilde{\mathbb{V}}$ with sufficiently smooth boundary ($V \subset\subset \tilde{\mathbb{V}}$). Due to the embedding $\theta(\tilde{\mathbb{V}}) \subset W^{\varepsilon, r-\delta}(\tilde{\mathbb{V}})$ for sufficiently small $\varepsilon > 0$ ($l_1 = 1, l_2 = 0, p_1 = 2, p_2 = r$)

$$u \in W^{l_1, p_1} \cap W^{l_2, p_2} \Rightarrow l = \theta l_1 + (1 - \theta)l_2, \quad \frac{1}{p} = \frac{\theta}{p_1} + \frac{1-\theta}{p_2}.$$

Then, $u \in W^{l,p}$ and (Gagliardo-Nirenberg)

$$\|u\|_{W^{l,p}} \leqslant C\|u\|_{W^{l_1, p_1}}^{\theta} \|u\|_{W^{l_2, p_2}}^{1-\theta}, \ \theta \in [0, 1].$$

Using this fact, we can state that

$$W^{\varepsilon, r-\delta} \subset\subset L^{r-\delta} \text{ for } \varepsilon << 1, \ \delta << 1.$$

$\square$

Proposition 1.2 *Let $\mathbb{V}$ be a bounded domain in $\mathbb{R}^n$. Then, the Laplace operator maps $W^{1,2}(\mathbb{V})$ to $W^{-1,2}(\mathbb{V})$. Moreover, it holds:*

$$\|\Delta u, \mathbb{V}\|_{-1,2} \leqslant \|u, \mathbb{V}\|_{1,2} \text{ for all } u \in W^{1,2}(\mathbb{V}). \tag{1.8}$$

Proof From the definition of $W^{1,2}(\mathbb{V})$ it follows that

$$|(\Delta u, \varphi)| = |(u, \Delta\varphi)| = |(\nabla u, \nabla\varphi)| \leqslant \|u, \mathbb{V}\|_{1,2}\|\varphi, \mathbb{V}\|_{1,2}$$

for all $\varphi \in D(\mathbb{V})$. Hence,

$$||\Delta u, \mathbb{V}||_{-1,2} = \sup \left\{ \frac{|(\Delta u, \varphi)|}{||\varphi, \mathbb{V}||_{1,2}} | \varphi \in D(\mathbb{V}) \right\} \leqslant ||u, \mathbb{V}||_{1,2}.$$

This proves *Proposition 1.2*.   □

Proposition 1.3 *Let* $\mathbb{V} \subset \mathbb{R}^n$ *be a bounded domain. Then, for any* $u \in \theta_0(\mathbb{V})$ *and* $g \in \Xi(\mathbb{V})$ *holds:*

$$|(u, g)| \leqslant ||u, \mathbb{V}||_+ |g, \mathbb{V}|_+,$$

where by $(\cdot, \cdot)$ *we denote the scalar product in* $[L^2(\mathbb{V})]^k$ *(naturally continuously extended to* $D'(\mathbb{V})$*).*

Proof Let $g = g_1 + g_2$, where $g_1 \in [W^{-1,2}(\mathbb{V})]^k$ and $g_2 \in [L^q(\mathbb{V})]^k$. Then

$$|(u, g)| \leqslant |(u, g_1)| + |(u, g_2)| \leqslant ||u, \mathbb{V}||_{1,2} ||g_1, \mathbb{V}||_{-1,2} + ||u, \mathbb{V}||_{0,r} ||g_2, \mathbb{V}||_{0,q}$$

$$\leqslant ||u, \mathbb{V}||_+ \left(||g_1, \mathbb{V}||_{-1,2} + ||g_2, \mathbb{V}||_{0,q} \right).$$

Taking in the last inequality infimum with respect to $g = g_1 + g_2$, we obtain the assertion of *Proposition 1.3*.   □

Remark 1.5 The following holds (see [7]):

$$[\mathbb{D}_1 \cap \mathbb{D}_2]^* = \mathbb{D}_1^* + \mathbb{D}_1^*,$$

where $\mathbb{D}_i \subset D'(\Omega)$, $i = 1, 2$ are some Banach spaces such that $\mathbb{D}_1 \cap \mathbb{D}_2$ is dense both in $\mathbb{D}_1$ and $\mathbb{D}_2$. Thus, in the case of a bounded domain $\Omega \subset \mathbb{R}^n$, the spaces $\theta_0(\Omega)$ and $\Xi(\Omega)$ are naturally conjugate.

Below we give several definitions, which we will be useful in the sequel.

Definition 1.5 Let X be a linear topological space [85, 105]. A set $\mathbb{B} \subset X$ is called bounded if there exists a neighbourhood of the origin (zero) E in X and $N = N(E)$ such that

$$\frac{1}{n} \mathbb{B} \subset E \text{ for } n \geqslant N.$$

Corollary 1.1 *Let* X *be a locally convex space [85, 105]. Then, a subset* $\mathbb{B} \subset X$ *is bounded if and only if, for every seminorm* $|| \cdot ||_p$ *from the definition of* X *topology, it holds:*

$$||\mathbb{B}||_p \leqslant C = C(|| \cdot ||_p).$$

Definition 1.6 Let X be a linear topological space and X^* be its dual space. A system of seminorms defining the (strong) topology in X^* is given [86] by

$$||T||_{\mathbb{B}} = \sup_{x \in \mathbb{B}} |Tx|,$$

where $\mathbb{B}$ is a bounded set in X, $T \in X^*$. The weak topology in X is given by a system of seminorms

$$||x||_T = |Tx|, \ T \in X^*, \ x \in X.$$

We denote by X^w the space X endowed with the weak topology. The weak-$*$ topology in X^* is given by the system of seminorms

$$||T||_x = |Tx|, \ x \in X, \ T \in X^*. \tag{1.9}$$

The following Lemma holds:

Lemma 1.2 ([86, 105]) *Let X be a Fréchet space (F-space). Then X is reflexive if and only if any bounded subset in X is precompact in X^w.*

Lemma 1.3 (Eberlein) *Let X be an F-space. Then, a set $\mathbb{B} \subset X$ is precompact in X^w if and only if $\mathbb{B}$ is sequentially compact in X^w, which means that any sequence $\{x_n\} \subset \mathbb{B}$ has a converging subsequence in X^w.*

Lemma 1.4 ([85]) *Let X be a separable locally convex space. Then, any bounded subset in X^* is metrisable in the weak-$*$ topology.*

Lemma 1.5 ([86]) *Let X be a locally convex space and let $\mathbb{B} \subset X$. It holds:*

1. *$\mathbb{B}$ is bounded in X^w if and only if $\mathbb{B}$ is bounded in X.*
2. *Let $\mathbb{B}$ be convex. Then, $\mathbb{B}$ is closed in X^w if and only if $\mathbb{B}$ is closed in X.*

Lemma 1.6 ([86]) *Let X be a locally convex space and let X_0 be a closed subspace of X. Then the weak topology in X_0 coincides with the topology induced by X^w.*

Let us now study the structure of the dual spaces introduced above.

Theorem 1.12 *Let X be one of the spaces $\theta(\Omega)$, $\theta_0(\Omega)$, $\Xi(\Omega)$. Then, for every $T \subset X^*$ there exists an $l \in \left(X(\Omega, \mathbb{B}^R_{x_0}) \right)^*$ such that*

$$Tx = \left\langle l, x_{\Omega \cap \mathbb{B}^R_{x_0}} \right\rangle, \ for \ all \ x \in X. \tag{1.10}$$

Each $l \in \left(X(\Omega, \mathbb{B}^R_{x_0}) \right)^*$ defines by (1.10) a continuous linear functional T on X. Here, $< \cdot, \cdot >$ denotes the canonical paring of $X(\Omega, \mathbb{B}^R_{x_0})$ and $\left(X(\Omega, \mathbb{B}^R_{x_0}) \right)^*$. The embedding given by (1.10) is continuous.

Proof Since T is continuous, the exists, due to [105] and the definition of continuous seminorms in X, a ball $\mathbb{B}_{x_0}^R \subset \mathbb{R}$ such that

$$|Tx| \leqslant C \left\|x, \Omega \cap \mathbb{B}_{x_0}^R\right\|_X \leqslant C \left\|x|_{\Omega \cap \mathbb{B}_{x_0}^R}\right\|_{X(\Omega, \mathbb{B}_{x_0}^R)}. \tag{1.11}$$

The functional T is well-defined and uniformly continuous in the subspace $X(\Omega)|_{\Omega \cap \mathbb{B}_{x_0}^R}$ of $X(\Omega, \mathbb{B}_{x_0}^R)$; $X(\Omega)|_{\Omega \cap \mathbb{B}_{x_0}^R}$ is dense in $X(\Omega, \mathbb{B}_{x_0}^R)$. Therefore, T can be uniquely extended to a continuous linear functional $l : X(\Omega, \mathbb{B}_{x_0}^R) \to \mathbb{R}$. A continuous embedding of $X(\Omega, \mathbb{B}_{x_0}^R) \subset X^*$ is obvious. This proves *Theorem 1.12*.

$\square$

Corollary 1.2 *Let X be the same as in Theorem 1.12. Then, a sequence $x_n \in X$ converges weakly to some $x \in X$ if and only if*

$$x_n|_{\Omega \cap \mathbb{B}_{x_0}^R} \rightharpoonup x|_{\Omega \cap \mathbb{B}_{x_0}^R} \tag{1.12}$$

in the space $X(\Omega, \mathbb{B}_{x_0}^R)$ for each $\mathbb{B}_{x_0}^R \subset \mathbb{R}^n$.

Remark 1.6 Let all assumptions of *Corollary 1.2* hold. It is then sufficient to check (1.12) on $\mathbb{B}_{x_0}^{R_k} \subset \mathbb{R}^n$, where $R_k \to \infty$ for $k \to \infty$.

Theorem 1.13 *Let X be as in Theorem 1.12. Then, both X and X^* are reflexive and separable.*

Proof We prove the statement for the space $\theta(\Omega)$. For the other spaces the proof can be carried out in the same manner. It is clear that, in order to prove separability of $\theta(\Omega)$, it suffices to show the separability of

$$\theta(\Omega \cap \mathbb{B}_{x_0}^R) = \left[W_{loc}^{1,2}(\Omega \cap \mathbb{B}_{x_0}^R)\right]^k \cap \left[L_{loc}^r(\Omega \cap \mathbb{B}_{x_0}^R)\right]^k$$

for each $\mathbb{B}_{x_0}^R \subset \mathbb{R}^n$. A proof of separability for $W^{1,2}$ and L^r can be found in [26]. Each space $\theta(\Omega \cap \mathbb{B}_{x_0}^R)$ is isometric isomorph to a closed subspace of the space $\mathcal{M} = \left[W^{1,2}(\Omega \cap \mathbb{B}_{x_0}^R)\right]^k \times \left[L^r(\Omega \cap \mathbb{B}_{x_0}^R)\right]^k$ which consists of pairs $\{z, z\}$, consequently, it is separable. This proves the separability of $\theta(\Omega)$.

Next, we prove reflexivity of $\theta(\Omega)$. According to *Lemmas 1.2 and 1.3*, it suffices to show that any bounded subset of $[\theta(\Omega)]^w$ is sequentially precompact. Thus, with *Corollary 1.2*, *Remark 1.6* and the Canter diagonal procedure, it is sufficient to show that $\theta(\Omega, \mathbb{B}_{x_0}^R)$ is reflexive for each $\mathbb{B}_{x_0}^R$. Let us prove this. Indeed, the space $\theta(\Omega \cap \mathbb{B}_{x_0}^R)$ is reflexive as a closed subspace of the reflexive B-space $\mathcal{M}$ [86], and the space $\theta(\Omega, \mathbb{B}_{x_0}^R)$ is reflexive as a closed subspace of the reflexive B-space $\theta(\Omega \cap \mathbb{B}_{x_0}^R)$. Consequently, $\theta(\Omega)$ is reflexive.

The space $[\theta(\Omega, \mathbb{B}_{x_0}^R)]^*$ is separable as the dual space to a reflexive B-space. Hence, due to *Theorem 1.12*, $[\theta(\Omega, \mathbb{B}_{x_0}^R)]^*$ is separable as well. This proves *Theorem 1.13*.   □

Corollary 1.3 *Let X be the same as in Theorem 1.13. Then, any bounded subset $\mathbb{B} \subset X$ is a metrisable precompact in the space X^w. In particular, any precompact in X^w is metrisable.*

1.2   Linear Elliptic Boundary Value Problems

Notation Let Ω be a bounded region in $\mathbb{R}^n$. For $\alpha = (\alpha_1, \cdots, \alpha_n)$ an n-tuple of nonnegative integers, recall that $D^\alpha = \prod_{i=1}^{n} \left(\dfrac{\partial}{\partial x_i}\right)^{\alpha_i}$, $|\alpha| = \sum_{i=1}^{n} \alpha_i$ and let $\xi^\alpha = \prod_{i=1}^{n}(\xi_i)^{\alpha_i}$ if $\xi \in C_1^n$.

Every linear differential operator L of order $2m$ ($m \in \mathbb{N}$) has the form

$$Lu = \sum_{|\alpha| \leqslant 2m} a_\alpha(x) \cdot D^\alpha u. \tag{1.13}$$

All coefficients $a_\alpha(x)$ are assumed to be real.
The partial differential operator

$$Lu = \sum_{|\alpha| \leqslant 2m} a_\alpha(x) \cdot D^\alpha u$$

is called elliptic of order $2m$ if its principal symbol,

$$p_0(x, \xi) = \sum_{|\alpha| = 2m} a_\alpha(x) \cdot \xi^\alpha$$

has the property that $p_0(x, \xi) \neq 0$ for all $x \in \Omega$, $\xi \in \mathbb{R}^n \setminus \{0\}$.

The differential operator L defined by (1.13) is called uniformly elliptic in Ω, if there is some $c > 0$, such that

$$(-1)^m \sum_{|\alpha| = 2m} a_\alpha(x)\xi^\alpha \geqslant C|\xi|^{2m} \text{ for every } x \in \Omega, \xi \in \mathbb{R}^n \setminus \{0\}. \tag{1.14}$$

Throughout we assume that $\partial\Omega$ is a smooth $(n-1)$-manifold.
Suppose now that L is elliptic and of order $2m$.

Let $\{m_i,\ 1 \leqslant i \leqslant m\}$ be distinct integers with $0 \leqslant m_i \leqslant 2m - 1$, and suppose that for $1 \leqslant i \leqslant m$ we prescribe a differential operator B_i of order m_i on $\partial\Omega$, by

$$B_i u(x) = \sum_{|\alpha| \leqslant m_i} b_{\alpha,i}(x) D^\alpha u(x), \quad i = 1, \cdots, m. \tag{1.15}$$

The family of boundary operators $B = \{B_1, \cdots, B_m\}$ is said to satisfy the Shapiro-Lopatinski covering condition with respect to L provided that the following algebraic condition is satisfied. For each $x \in \partial\Omega$, $\vec{N} \in \mathbb{R}^n \backslash \{0\}$ normal to $\partial\Omega$ at x and $\xi \in \mathbb{R}^n \backslash \{0\}$ with $\langle \xi, \vec{N} \rangle = 0$, consider the $(m + 1)$ polynomials of a single complex variable

$$\begin{aligned} \tau &\longmapsto p_0(x, \xi + \tau\vec{N}), \\ \tau &\longmapsto \sum_{|\alpha| = m_i} b_{\alpha,i}(x) \cdot (\xi + \tau\vec{N})^\alpha \equiv p_{0,i}(x, \xi, \tau), \quad 1 \leqslant i \leqslant m. \end{aligned} \tag{1.16}$$

Let $\tau_1^+, \cdots, \tau_m^+$ be the m complex zeros of $p_0(x, \xi + \tau\vec{N})$ which have positive imaginary part. Then $\{p_{0,i}(\tau)\}_{i=1}^m$ are assumed to be linearly independent modulo $\prod_{i=1}^m (\tau - \tau_i^+) = M^+(x, \xi, \vec{N}, \tau)$, i.e., after division by $M^+(x, \xi, N, \tau)$ all the various remainders are linearly independent.

In other words, let

$$p_{0,i}'(x, \xi, \vec{N}, \tau) = \sum_{k=0}^{m-1} b_{i,k}(x, \xi, \vec{N}) \cdot \tau^k, \quad i = 1, \cdots, m$$

be the remainders after division by $M^+(x, \xi, \vec{N}, \tau)$. Then the condition of the Shapiro-Lopatinski implies that

$$D(x, \xi, N) = \det \|b_{ik}(x, \xi, \vec{N})\| \neq 0 \tag{1.17}$$

for all $x \in \partial\Omega$, and for all $\vec{N} \in \mathbb{R}^n \backslash \{0\}$ normal to $\partial\Omega$ at x and $\xi \in \mathbb{R}^n \backslash \{0\}$ with $\langle \xi, \vec{N} \rangle = 0$.

Definition 1.7 We say that $(L, B_1, \cdots, B_m)$ defines an elliptic boundary value problem of order $(2m, m_1, \cdots, m_m)$ if L given by (1.13), is uniformly elliptic and of order $2m$, each B_i given by (1.15) has order m_i, $0 \leqslant m_i \leqslant 2m - 1$, the m_i's are distinct, $\partial\Omega$ is non characteristic to B_i at each point and $\{B_i\}_{i=1}^m$ satisfy the Shapiro-Lopatinski condition with respect to L (see [79]).

We have the following lemma (see [48, 63, 79]).

Lemma 1.7 Let $(L, B_1, \cdots, B_m)$ define an elliptic boundary value problem of order $(2m, m_1, \cdots, m_m)$. Then

$$(L \circ \triangle^l, B_1 \circ \tilde{\triangle}^l, \cdots, B_m \circ \tilde{\triangle}^l, L \circ \frac{\partial u}{\partial \vec{N}})$$

defines an elliptic boundary value problem of order $(2k + 2l, m_1 + 2l, \cdots, m_m + 2l, 2m + 1)$ *where* $\tilde{\triangle}$ *is the Laplace-Beltrami operator,* $l \in N$.

Proof The principal symbol of $L \circ \triangle^l$ is $|\xi|^{2l} \cdot p_0(x, \xi)$, so it is clear that $L \circ \triangle^l$ is uniformly elliptic.

Let $x \in \partial\Omega$ and $\xi, \vec{N} \in \mathbb{R}^n \backslash \{0\}$, with $\langle \xi, \vec{N} \rangle = 0$ and $\vec{N}$ normal to $\partial\Omega$ at x. It is obvious that the principal symbol operators $B_i \circ \tilde{\triangle}^l$ and $L \circ \frac{\partial}{\partial \vec{N}}$ at $\xi + \tau \vec{N}$ are $\psi_l(\xi) \cdot p_{0i}(x, \xi + \tau \vec{N})$ and $\tau p_0(x, \xi + \tau \vec{N})$ respectively, where $\psi_l(\xi) \neq 0$.

If $\tau_1^+, \cdots, \tau_m^+$ are the m roots of $p_0(x, \xi + \tau \vec{N}) = 0$ having positive imaginary part, then $\tau_1^+, \cdots, \tau_m^+, i \cdot \frac{|\vec{N}|}{|\xi|}$ constitute the $m+1$ roots of $|\xi + \tau \vec{N}|^2 \cdot p_0(x, \xi + \tau \vec{N}) = 0$ with positive imaginary part.

We must show that if $\lambda_1, \cdots, \lambda_{m+1} \in C_2$ ang $h(\tau)$ is a polynomial with

$$\begin{aligned} \psi_l(\xi) \sum_{i=1}^{m} \lambda_i \cdot p_{0i}(x, \xi + \tau \vec{N}) + \lambda_{m+1} \tau p_0(x, \xi + \tau \vec{N}) = \\ h(\tau) \cdot \left(\tau - \frac{i|\vec{N}|}{|\xi|}\right) \cdot \prod_{i=1}^{m}(\tau - \tau_i^+) \end{aligned} \tag{1.18}$$

then $\lambda_i = 0$, $1 \leq i \leq m + 1$ and $h(\tau) \equiv 0$. Due to the assumption that $(B_1, \cdots, B_m)$ satisfy the covering condition it is not difficult to see that $\lambda_1 = \cdots = \lambda_m = 0$. But then the right-hand side of (1.18) has more roots with positive imaginary part than does the left-hand side, so that $\lambda_{m+1} = 0$ and $h(\tau) \equiv 0$.

With appropriate smoothness conditions on the coefficients (see *Lemma 1.8* below), elliptic boundary value problems induce linear Fredholm operators in Sobolev spaces. Here the spaces $W^{2m+k-m_i-1/p,p}(\partial\Omega)$ with the fractional differentiation order $2m+k-m_i-\frac{1}{p}$ play a decisive role. Before giving a precise definition we wish to point out *a priori* the most important property of these spaces, i.e. the surjective boundary operator

$$T : C^\infty(\bar{\Omega}) \to C^\infty(\partial\Omega)$$

which assigns to each function $u \in C^\infty(\bar{\Omega})$ its classical boundary value Tu on $\partial\Omega$, can be extended uniquely to a continuous linear surjective operator

$$T : W^{2m+k,p}(\Omega) \to W^{2m+k-m_i-1/p,p}(\partial\Omega).$$

Here $k \geq 0$ and $m \geq 1$ are integers, and $1 < p < \infty$ (we are mainly interested in the case $p = 2$, $W^{2m,2}(\Omega) = H^{2m}(\Omega)$). Then Tu is described naturally as the generalized boundary value of $u \in W^{2m+k,p}(\Omega)$. These functions u have generalized derivatives $D^\alpha u$ up to order $2m+k$ on Ω. The functions $D^\alpha u$ with $|\alpha| \leq m_i$ have generalized boundary values which all lie in $W^{2m+k-m_i-1/p,p}(\partial\Omega)$, since $m_i < 2m$. Consequently, $B_i u \in W^{2m+k-m_i-1/p,p}(\partial\Omega)$ also. The differential operators L and the boundary operator B_i are thus to be understood in the space of generalized derivatives on Ω and as generalized boundary values respectively.

Definition of the Space $W^{m-1/p,p}(\partial\Omega)$.

Let Ω be an open subset of $\mathbb{R}^n$ with sufficiently smooth boundary and $\{U_i\}_{i=1}^l$ be an open covering of $\bar{\Omega}$ with diffeomorphisms $\varphi_i : U_i \to \mathbb{R}^n$, $\varphi_i \in C^m(U_i)$, such that $\varphi_i(U_i) = V_1 = \{y \in \mathbb{R}^n \mid |y| < 1\}$ if $U_i \subset \Omega$, and

$$\varphi_i(U_i \cap \bar{\Omega}) = V_1^+ = \{y \in \mathbb{R}^n \mid |y| < 1,\, y_n \geqslant 0\},$$

$$\varphi_i(U_i \cap \partial\Omega) = \tilde{V}_1 = \{y \in \mathbb{R}^n \mid |y| < 1,\, y_n = 0\} \text{ if } U_i \cap \partial\Omega \neq \varnothing.$$

Let $\chi_i(x)$ be a partition of unity subordinated to $\{U_i\}_{i=1}^l$ and let $\lambda_i(y) := \chi_i(\varphi_i^{-1}(y))$.

For each $u(x) \in C^m(\partial\Omega)$, $0 < \delta < 1$, $p > 1$ we define the norm:

$$\|u\|'_{m-\delta,p,\partial\Omega} = \left\{ \sum_{i\in I'} \left[\sideset{}{'}\sum_{|\alpha|\leqslant m-1} \int_{\tilde{V}_1} |D_y^\alpha(\lambda_i(y)\cdot u_i(y))|^p \, dy' + \right. \right.$$

$$\sideset{}{'}\sum_{|\alpha|=m-1} \int_{\tilde{V}_1}\int_{\tilde{V}_1} |D_y^\alpha(\lambda_i(y)\cdot u_i(y)) - D_z^\alpha(\lambda_i(z)\cdot u_i(z))|^p$$

$$\left. \left. \cdot \frac{dy'dz'}{|y'-z'|^{n+p-1-\delta p}} \right] \right\}^{\frac{1}{p}},$$

$$(1.19)$$

where $u_i(y) = u(\varphi_i^{-1}(y))$, $y' = (y_1, \cdots, y_{n-1})$, $I' \subset \{1, \cdots, l\}$ such that: $U_i \cap \partial\Omega \neq \varnothing$ and $\sum'$ implies that the sum is taken over those α for which $\alpha_n = 0$, $\alpha = (\alpha_1, \cdots, \alpha_n)$.

By definition, the norm in $W^{m-\frac{1}{p},p}(\partial\Omega)$, $p > 1$ is defined as the norm $\|\cdot\|'_{m-\frac{1}{p},p,\partial\Omega}$. For more details see [84, 96, 101].

Let us return to the discussion of elliptic boundary value problems. We first recall some results regarding linear Fredholm operators. Let X and Y be real Banach spaces. By $L(X, Y)$ we denote the Banach space of bounded linear operators from X to Y. An operator T in $L(X, Y)$ is called Fredholm if the $Ker\, T = \{x \in X | Tx = 0\}$ has finite dimension and the image of T, $R(T)$ is of finite codimension in Y, that is codim $R(T) = \dim Y/R(T) < \infty$. For a Fredholm operator $T : X \to Y$, the numerical Fredholm index of T, $ind(T)$ is defined by

$$ind(T) = \dim Ker\, T - \text{codim}(R(T)).$$

Lemma 1.8 *Let $\Omega \subset \mathbb{R}^n$ be open and bounded with $\partial\Omega$ smooth. Suppose that $s > n/2$, $a_\alpha \in H^s(\Omega)$ if $|\alpha| \leqslant 2m$, while $b_{\alpha,i} \in H^{s+2m-m_i}(\partial\Omega)$ and $i = 1, \cdots m$. Then the following three assertions are equivalent:*

(i) *The operator $A = (L, B_1, \cdots, B_m)$*

$$A : H^{s+2m}(\Omega) \longrightarrow H^s(\Omega) \times \prod_{i=1}^{m} H^{s+2m-m_i}(\partial\Omega) \tag{1.20}$$

   is an elliptic boundary value problem of order $(2m, m_1, \cdots, m_m)$
(ii) *The operator $A = (L, B_1, \cdots B_m)$ is Fredholm*
(iii) *There is some $c > 0$, such that if $u \in H^{s+2m}(\Omega)$, then*

$$\|u\|_{2m+s} \leqslant c \left[\|Lu\|_s + \sum_{i=1}^{m} \|B_i(x, D)u\|_{2m+s-m_i-\frac{1}{2}} + \|u\|_s \right]. \tag{1.21}$$

Proof If each $a_\alpha \in C^s(\Omega)$ and each $b_{\alpha,i} \in C^{2k+s-m_i}(\Omega)$, then a priori estimate (1.21) is contained in [3]. It is not difficult to see that (1.21) also holds
under the present smoothness conditions. Thus, in fact a priori estimate (1.21) and
equivalence (i) and (iii) follows from [3]. Equivalence (i) and (iii) to (ii) can be
proved analogously to [4].

Remark 1.7 Of course, the Fredholm index of $(L, B_1, \cdots, B_m)$ need not be equal
to 0. If L is uniformly elliptic and $B_i u(x) = \left(\frac{\partial}{\partial N} \right)^{i-1} u(x)$ for $1 \leqslant i \leqslant m$, then the
index

$$A = (L, B_1, \cdots, B_m) : H^{2m+s}(\Omega) \to H^s(\Omega) \times \prod_{i=1}^{m} H^{2m+s-m_i-\frac{1}{2}}(\partial\Omega)$$

is 0. (see [78]).

Remark 1.8 (C^γ-theory). The a priori estimates (1.21) remain valid if we choose
the following B- spaces for $0 < \gamma < 1$:

$$X = C^{2m+s,\gamma}(\bar\Omega), \ Y = C^{s,\gamma}(\bar\Omega), \ Z = C(\bar\Omega), Y_j = C^{2m+s-m_i,\gamma}$$

i.e.

$$\|u\|_X \leqslant \text{constant}(\|Lu\|_Y + \sum_{j=1}^{m} \|B_j u\|_{Y_j} + \|u\|_Z). \tag{1.22}$$

Remark 1.9 The important fact is that the index of corresponding operators is the
same in both theories.

Remark 1.10 As shown in [3, 4] the terms $\|u\|_s$ and $\|u\|_Z$ in (1.21), (1.22) disappear if dim $Ker A = \{0\}$, where $Au = (Lu, B_1 u, \cdots, B_m u)$.

1.3 Nemytskii Operator

The investigation of nonlinear equations in the following chapters relies on properties of mappings of the form $u \mapsto f(u)$ in the spaces $C^\alpha(\bar{\Omega})$ and $L^p(\Omega)$, $H^l(\Omega)$.

Definition 1.8 Let $\Omega \subset \mathbb{R}^n$ be a domain. We say that a function

$$\Omega \times \mathbb{R}^m \ni (x, u) \longmapsto f(x, u) \in \mathbb{R}$$

satisfies the Carathéodory conditions if

$$u \longmapsto f(x, u) \text{ is continuous for almost every } x \in \Omega$$

and

$$x \longmapsto f(x, u) \text{ is measurable for every } u \in \Omega.$$

Given any f satisfying the Carathéodory conditions and a function $u : \Omega \to \mathbb{R}^m$, we can define another function by composition

$$F(u)(x) := f(x, u(x)). \tag{1.23}$$

The composed operator F is called a Nemytskii operator. In this section we state some important results on the composition of $C^\alpha(\bar{\Omega})$, $L^p(\Omega)$, $H^l(\Omega)$ with nonlinear functions (some of them without proof [74, 106]).

Proposition 1.4 *Let $\Omega \subset \mathbb{R}^n$ be a bounded domain and*

$$\Omega \times \mathbb{R}^m \ni (x, u) \longmapsto f(x, u) \in \mathbb{R}$$

satisfy the Carathéodory conditions. In addition, let

$$|f(x, u)| \leqslant f_0(x) + c(1 + |u|)^r, \tag{1.24}$$

where $f_0 \in L^{p_0}(\Omega)$, $p_0 \geqslant 1$, and $r p_0 \leqslant p_1$. Then the Nemytskii operator a priori estimates F defined by (1.23) is bounded from $L^{p_1}(\Omega)$ into $L^{p_0}(\Omega)$, and

$$\|F(u)\|_{0,p_0} \leqslant C_1(1 + \|u\|_{p_1}^r). \tag{1.25}$$

Proof By (1.24) and (1.2)

$$\|F(u)\|_{o,p_0} \leqslant \|f_0(x)\|_{o,p_0} + C\|1\|_{o,p_0} + C\||u|^r\|_{o,p_0} \tag{1.26}$$

$$\leqslant C' + C\left(\int_\Omega |u|^{rp_0}dx\right)^{\frac{1}{p_0}} \tag{1.27}$$

$$= C' + \|u\|^r_{0,p_0r}. \tag{1.28}$$

Since Ω is bounded, then by Hölder's inequality

$$\|v\|_{o,q} \leqslant C(\Omega)\|v\|_{o,p} \quad \text{when } 1 \leqslant q \leqslant p, \ v \in L^p(\Omega), \tag{1.29}$$

where $C(\Omega) = (\text{mes}(\Omega))^{\frac{1}{q}-\frac{1}{p}}$. Inequalities (1.26) and (1.29) with $q = rp_0$ and $p = p_1$ imply (1.25). $\qquad\square$

It is well-known that the notions of continuity and boundedness of a nonlinear operator are independent of one another [74]. It turns out that the following is valid.

Theorem 1.14 *Let $\Omega \subset \mathbb{R}^n$ be a bounded domain and let*

$$\Omega \times \mathbb{R}^m \ni (x,u) \longmapsto f(x,u) \in \mathbb{R}$$

satisfy the Carathéodory conditions. In addition, let $p \in (1,\infty)$ and $g \in L^q(\Omega)$ (where $\frac{1}{p} + \frac{1}{q} = 1$) be given, and let f satisfy

$$|f(x,u)| \leqslant C|u|^{p-1} + g(x).$$

Then the Nemytskii operator F defined by (1.23) is a bounded and continuous map from $L^p(\Omega)$ to $L^q(\Omega)$.

For a more detailed treatment, the reader could consult [74, 106].

Theorem 1.15 *Let Ω be a bounded domain in $\mathbb{R}^n$ with smooth boundary and let*

$$\Omega \times \mathbb{R} \ni (x,u) \mapsto f(x,u) \in \mathbb{R}$$

satisfy the Carathéodory conditions. Then f induces

1) a continuous mapping from $H^s(\Omega)$ into $H^s(\Omega)$ if $f \in C^s$,
2) a continuously differentiable mapping from $H^s(\Omega)$ into $H^s(\Omega)$ if $f \in C^{s+1}$, where in both cases $s > n/2$.

Proof First we consider the simplest case, that is $f = f(u)$ is independent of x. By the Sobolev embedding theorem, we have $H^s(\Omega) \subset C(\bar\Omega)$. Hence we have $f(u) \in C(\bar\Omega)$ for every $u \in H^s(\Omega)$. Moreover, if u is in $C^{(s)}(\bar\Omega)$, we can obtain the derivatives of $f(u)$ by the chain rule, and in the general case, we can use approximation by smooth functions. Note that all derivatives of $f(u)$ have the form of a product involving a derivative of f and derivatives of u. The first factor is in $C(\bar{\Omega})$, while any l-th derivative of u lies in $H^{s-l}(\Omega)$, which imbeds into $L^{2n/(n-2(s-l))}(\Omega)$ if $s - l < \frac{n}{2}$.

We can use this fact and Hölder's inequality to show that all derivatives of $f(u)$ up to order s are in $L^2(\Omega)$; moreover, it is clear from this argument that f is actually continuous from $H^s(\Omega)$ into $H^s(\Omega)$. A proof of the differentiability in this special case is that $f = f(u)$ is based on the relation

$$f(u) - f(v) = \int_0^1 f_u'(v + \theta(u - v))(u - v)d\theta$$

and the same arguments as before.

Let us now consider the general case, that is $f = f(x, u)$. Let $|\alpha| \leqslant s$. We must show that

$$u \longmapsto D^\alpha F(u) \tag{1.30}$$

defines a continuous map of $H^s(\Omega)$ into $L^2(\Omega)$. It is not difficult to see that (1.29) is the finite sum of operators of the form

$$u(x) \longmapsto g(x, u(x)) \cdot D^\gamma u(x), \tag{1.31}$$

where $|\gamma| = \gamma_1 + \cdots + \gamma_n \leqslant s$, while g is a partial derivative of f order at most s. It is obvious that D^γ is continuous from $H^s(\Omega)$ into $L^2(\Omega)$ for $|\gamma| \leqslant s$. On the other hand, the continuous embedding of $H^s(\Omega)$ in $C(\bar{\Omega})$ implies that

$$u(x) \longmapsto g(x, u(x))$$

is continuous from $H^s(\Omega)$ into $C(\bar{\Omega})$. Thus

$$u(x) \longmapsto g(x, u(x)) \cdot D^\gamma u(x)$$

defines a continuous map of $H^s(\Omega)$ into $L^2(\Omega)$ and hence so does $u \longmapsto D^\alpha F(u)$.

For $p \in \mathbb{N}$, let $\tilde{p}$ be the number of multi-indices α with $|\alpha| \leqslant p$.

Corollary 1.4 *An analogous result is valid for a continuity of the operator*

$$F(u)(x) = f(x, u(x), \cdots, D^p u(x)) : H^{s+p}(\Omega) \to H^s(\Omega),$$

where $p, s \in \mathbb{N}$ with $s > \frac{n}{2}$ and $f : \Omega \times \mathbb{R}^{\tilde{p}} \to \mathbb{R}$ is C^s.

Corollary 1.5 *Let $p, s \in \mathbb{N}$ with $s > \frac{n}{2}$ and*

$$f : \Omega \times \mathbb{R}^{\tilde{p}} \to \mathbb{R} \text{ be } C^{s+1}.$$

Then the operator $F : H^{s+p}(\Omega) \to H^s(\Omega)$ *defined by*

$$F(u)(x) = f(x, u(x), \cdots, D^p u(x))$$

is Fréchet differentiable from $H^{s+p}(\Omega)$ *into* $H^s(\Omega)$.

We have the following continuity and C^1-differentiability results for a nonlinear differential operator of the form $Au(x) = f(x, u(x), ..D^{2p}u(x))$ in the Hölder spaces. They are based on *Theorems 1.16* and *1.17*.

Let $p \in \mathbb{N}$ and $\tilde{p}$ denote as before the number of multi-indices with $|\alpha| \leqslant p$. Let Ω be a bounded domain in $\mathbb{R}^n$.

Theorem 1.16 *Let the function* $f(x, y) = f(x, y_1, \cdots, y_{\tilde{p}})$ *be defined on* $\bar{\Omega} \times \mathbb{R}^{\tilde{p}}$ *which satisfies the following conditions:*

1) $f(x, 0) = 0$
2) For any $R > 0$, $\displaystyle\sup_{|y| \leqslant R} \left| \frac{\partial^2 f}{\partial y_i \partial y_j} \right| \leqslant C(R)$, $\displaystyle\sup_{|y| \leqslant R} \|f\|_{C^{1,\alpha}(\bar{\Omega})} \leqslant C(R)$, *where* $C(R)$
is constant depending on R.

Let $u_1(x), \cdots, u_{\tilde{p}}(x) \in C^\alpha(\bar{\Omega})$, $0 < \alpha < 1$, $\|u_i\|_{C^\alpha(\bar{\Omega})} \leqslant R$, $i = 1, \ldots, \tilde{p}$. *Then*

$$\|f(x, u_1(x), \cdots, u_{\tilde{p}}(x))\|_{C^\alpha(\bar{\Omega})} \leqslant C_1(R) \cdot \sum_{i=1}^{\tilde{p}} \|u_i\|_{C^\alpha(\bar{\Omega})}. \tag{1.32}$$

Proof Obviously,

$$\begin{aligned}
f(x, y, \ldots, y_{\tilde{p}}) &= \int_0^1 \frac{d}{dt} f(x, ty_1, \ldots, ty_{\tilde{p}}) dt \\
&= \sum_{j=1}^{\tilde{p}} y_j \int_0^1 \frac{\partial f(x, ty_1, \ldots, ty_{\tilde{p}})}{\partial y_j} dt \\
&= \sum_{j=1}^{\tilde{p}} \varphi_j(x, y_1, \ldots, y_{\tilde{p}}) \cdot y_j,
\end{aligned}$$

where

$$\varphi_j(x, y_1, \ldots, y_{\tilde{p}}) = \int_0^1 \frac{\partial f(x, ty_1, \ldots, ty_{\tilde{p}})}{\partial y_j} dt.$$

Hence

$$f(x, u_1(x), \cdots, u_{\tilde{p}}(x)) = \sum_{j=1}^{\tilde{p}} \varphi_j(x, u_1(x), \cdots, u_{\tilde{p}}(x)) \cdot u_j(x).$$

Since $C^\alpha(\bar{\Omega}), 0 < \alpha < 1$ is a Banach algebra, we have

$$\|f(x, u_1(x), \cdots u_{\tilde{p}}(x))\|_{C^\alpha} \leqslant \sum_{j=1}^{\tilde{p}} \|\varphi_j(x, u_1(x), \cdots u_{\tilde{p}}(x))\|_{C^\alpha} \cdot \|u_j\|_{C^\alpha}.$$

Hence we have to prove that

$$\sup_{|y| \leqslant R} \|\varphi_j(x, u_1(x), \cdots, u_{\tilde{p}}(x))\|_{C^\alpha} \leqslant C_1(R).$$

Indeed

$$|\varphi_j(x + \xi, u_1(x + \xi), \cdots, u_{\tilde{p}}(x + \xi)) - \varphi_j(x, u_1(x), \cdots, u_{\tilde{p}}(x))| \qquad (1.33)$$

$$\leqslant |\varphi_j(x + \xi, u_1(x + \xi), \cdots, u_{\tilde{p}}(x + \xi)) - \varphi_j(x, u_1(x + \xi), \cdots, u_{\tilde{p}}(x + \xi))|$$
$$\qquad (1.34)$$

$$+ |\varphi_j(x, u_1(x + \xi), \cdots, u_{\tilde{p}}(x + \xi)) - \varphi_j(x, u_1(x), \cdots, u_{\tilde{p}}(x))|. \qquad (1.35)$$

The first term on the right-hand side of (1.35) is bounded by $C(R)|\xi|^\alpha$. The second term is bounded by

$$\sup_{|y| \leqslant R} |\frac{\partial \varphi_j}{\partial y_k}| \|\varphi_j(x, u_1(x + \xi), \cdots, u_{\tilde{p}}(x + \xi)) - \varphi_j(x, u_1(x), s, u_{\tilde{p}}(x))|$$

$$\leqslant C(R) R |\xi|^\alpha. \qquad (1.36)$$

The estimates (1.35) and (1.36) yield (1.32).

Theorem 1.17 *Let the function $f(x, y) = f(x, y_1, \cdots, y_{\tilde{p}})$ be defined on $\bar{\Omega} \times R^{\tilde{p}}$ satisfy the following conditions:*

1) $f(x, 0) = 0$, $grad_y f(x, 0) = 0$

2) For any $R > 0$, $\sup_{|y| \leqslant R} \|f(x, y)\|_{C^{2,\alpha}(\bar{\Omega})} \leqslant C(R)$ and $\sup_{|y| \leqslant R} |\frac{\partial^3 f}{\partial y_i \partial y_j \partial y_k}| \leqslant C(R)$,

where $C(R)$ is constant depending on R. Let as before, $u_1(x), \cdots, u_{\tilde{p}}(x) \in C^\alpha(\bar{\Omega})$ with $\|u_i\|_{C^\alpha(\bar{\Omega})} \leqslant R, i = 1, \cdots, \tilde{p}$.
Then the following estimate holds.

$$\|f(x, u_1(x), \cdots, u_{\tilde{p}}(x))\|_{C^\alpha(\bar{\Omega})} \leqslant C_2(R) \cdot \sum_{i=1}^{\tilde{p}} \|u_i\|_{C^\alpha}^2. \qquad (1.37)$$

Proof Obviously we have

$$f(x, y_1, \cdots, u_{\tilde{p}}) = \sum_{i,j=1}^{\tilde{p}} g_{ij}(x, y_1, \cdots, y_{\tilde{p}}) \cdot y_i \cdot y_j,$$

so we can write

$$f(x, u_1(x), \cdots, u_{\tilde{p}}(x)) = \sum_{i,j=1}^{\tilde{p}} g_{ij}(x, u_1(x), \ldots, u_{\tilde{p}}(x)) \cdot u_i(x) \cdot u_j(x)$$

and we have

$$\|f(x, u_1(x), \cdots, u_{\tilde{p}}(x)\|_{C^\alpha(\bar{\Omega})}$$

$$\leqslant \sum_{i,j=1}^{\tilde{p}} \|g_{ij}(x, u_1(x), \cdots, u_{\tilde{p}}(x)\|_{C^\alpha(\bar{\Omega})} \cdot \|u_i\|_{C^\alpha} \cdot \|u_j\|_{C^\alpha}. \tag{1.38}$$

Due to *Theorem 1.16*, we obtain

$$\|g_{ij}(x, u_1(x), \cdots, u_{\tilde{p}}(x)\|_{C^\alpha(\bar{\Omega})} \leqslant C_0(R). \tag{1.39}$$

Hence the estimates (1.38) and (1.39) yield (1.37)

$$\|f(x, u_1(x), \cdots, u_{\tilde{p}}(x)\|_{C^\alpha(\bar{\Omega})} \leqslant C_2(R) \cdot \sum_{i=1}^{\tilde{p}} \|u_i\|_{C^\alpha}^2.$$

We apply *Theorems 1.16* and *1.17* to the operator

$$Au(x) = f(x, u(x), \cdots, D^{2p}u(x)),$$

where the function $f(x, y_1, \cdots, y_{\tilde{p}})$ satisfy conditions of *Theorems 1.16* and *1.17*, respectively. Hence we have

$$\|Au\|_{C^{2p,\alpha}} \leqslant C(R) \cdot \|u\|_{C^\alpha}.$$

Moreover, as it follows from *Theorem 1.17*, $A \in C^1$, $A'(0) = 0$ and

$$\|A'(u+h) - A'(u)\|_{L(C^{2p,\alpha}, C^\alpha)} \leqslant C \cdot \|h\|_{C^{2p,\alpha}(\bar{\Omega})}.$$

Remark 1.11 As shown in the proofs of Theorems *1.16* and *1.17*, continuity and differentiability of the operator $Au(x) = f(x, u(x), \cdots, D^{2p}u(x))$ between $C^{2p,\alpha}(\bar{\Omega})$ and $C^\alpha(\bar{\Omega})$ remains valid under slightly weaker conditions on a given function $f(x, y_1, \cdots, y_{\tilde{p}})$. We leave these as exercises for the reader.

Below we present the properties of the Nemytskii operators in the spaces $H^s(S^1)$ or $C^{p,\alpha}(S^1)$, where S^1 is the unit circle. We recall some of the properties which will be used often in the sequel. The norm in $C^\alpha(M)$ is given by

$$\|f\|_{C^\alpha(M)} = \|f\|_C + \sup_{x \neq y} \frac{|f(x) - f(y)|}{|x - y|^\alpha}, \quad M = S^1.$$

As before, by $C^{k,\alpha}(M)$ we denote the space of Hölder continuous functions, which have derivatives up to order k, with $D^k f \in C^\alpha(M)$. Let F be a superposition operator defined by

$$F(u)(x) = f(x, u(x)), \quad x \in M.$$

The following theorems are not hard to prove (although not obvious).

Theorem 1.18 *Let $k \in \mathbb{R}+$. Then the superposition operator $F : E_1 \to E_2$ defined by $F(u)(x) = f(x, u(x))$ acts as a bounded operator in each of the following cases (see also [48]).*

> *1)* $f \in C(S^1 \times \mathbb{R}, \mathbb{R})$, $E_1 = C(S^1)$, $E_2 = C(S^1)$
> *2)* $f \in C^1(S^1 \times \mathbb{R}, \mathbb{R})$, $E_1 = C^\alpha(S^1)$, $E_2 = C^\alpha(S^1)$, $0 < \alpha < 1$.

Theorem 1.19 *Let $k \in \mathbb{R}+, 0 < \alpha < 1$. Then the superposition operator $F : E_1 \to E_2$ defined by $F(u)(x) = f(x, u(x))$ is m times continuously differentiable if one of the following cases*

> *1)* $D^{0,j} f \in C^k(S^1 \times \mathbb{R}, \mathbb{R})$, $E_1 = C^k(S^1)$, $E_2 = C^k(S^1)$
> *2)* $D^{0,j} f \in C^{k+1}(S^1 \times \mathbb{R}, \mathbb{R})$, $E_1 = C^{k,\alpha}(S^1)$, $E_2 = C^{k,\alpha}(S^1)$,

The j-th derivative of F is given by

$$D^{0,j} F(x, u(x)) h_1(x) \ldots h_j(x) = D^j F(f)(h_1, \ldots h_j)(x).$$

Analogous results are valid in Sobolev spaces:

Theorem 1.20 *Let $X = Y = H^s(S^1)(s \geqslant 1)$ be the Sobolev space of real functions $x(\tau)$ on the circumference of a circle, where $0 \leqslant \tau < 2\pi$; $f(\tau, x)$ is a smooth real function, $x \in \mathbb{R}, 0 \leqslant \tau < 2\pi$. Then the operator $F : H^s(S^1) \to H^s(S^1)$ defined by $Fx(\tau) = f(\tau, x(\tau))$ is continuous.*

Proof It is not difficult to see, that

$$\left(\frac{d}{d\tau}\right)^k f(\tau, x(\tau)) = \sup_{\substack{p+q \leqslant k \\ r_1 + \ldots + r_q = k-p \\ r_j \geqslant 0}} C_{p,q,r_1 \ldots r_q} \frac{\partial^{p+q} f(\tau, x(\tau))}{\partial \tau^p \ldots \partial x^q} x^{(r_1)}(\tau) \ldots x^{(r_q)}(\tau),$$

where $C_{p,q,r_1\ldots r_q}$ are some constants. If $x(\tau) \in H^s$, then it follows that the derivatives $\left\{\frac{d^l x(\tau)}{d\tau^l} \big| 0 \leqslant l \leqslant s - 1\right\}$ are continuous. Therefore in $\frac{d^s}{d\tau^s} f(\tau, x(\tau))$ all terms without ones are continuous. The last term is equal to $\frac{d^s x(\tau)}{d\tau^s} \times Q(\tau)$ where $Q(\tau)$ is a continuous function, hence also square integrable. As a consequence of these arguments we obtain continuity.

Remark 1.12 An analogous result holds for vector functions, and also in the multidimensional case, for functions on arbitrary smooth compact manifold with boundary.

The following Lemma on the smoothness relations between u and $f(u)$ plays a decisive role in many applications (see [47]).

Lemma 1.9 *Let the function* $f \in C^2(\mathbb{R}, \mathbb{R})$ *satisfies* $C_1 |u|^{p-1} \leqslant f'(u) \leqslant C_1 |u|^{p-1}$, $p > 1$, *with* C_1 *and* C_2 *some positive constants. Then, for every* $s \in (0, 1)$ *and* $1 < q \leqslant \infty$, *we have*

$$\|u\|_{W^{s/p, pq}(\Omega)} \leqslant C_p \|f(u)\|_{W^{s,q}(\Omega)}^{1/p}$$

where the constant C_p *is independent of* u.

Proof Indeed, let f^{-1} be the inverse function to f. Then, due to conditions on f, the function $G(v) := \operatorname{sgn}(v)|f^{-1}(v)|^p$ is nondegenerate and satisfies

$$C_2 \leqslant G'(v) \leqslant C_1,$$

for some positive constants C_1 and C_2. Therefore, we have

$$|f^{-1}(v_1) - f^{-1}(v_2)|^p \leqslant C_p |G(v_1) - G(v_2)| \leqslant C_p' |v_1 - v_2|,$$

for all $v_1, v_2 \in \mathbb{R}$. Finally, according to the definition of the fractional Sobolev spaces (see e.g. [96, 101]),

$$\|f^{-1}(v)\|_{W^{s/p, qp}(\Omega)}^{pq} := \|f^{-1}(v)\|_{L^{pq}(\Omega)}^{pq} + \int_\Omega \int_\Omega \frac{|f^{-1}(v(x)) - f^{-1}(v(y))|^{pq}}{|x - y|^{n+sq}} \, dx \, dy$$

$$\leqslant C \|v\|_{L^q(\Omega)}^q + C_p' \int_\Omega \int_\Omega \frac{|v(x) - v(y)|^q}{|x - y|^{n+sq}} \, dx \, dy$$

$$= C_p'' \|v\|_{W^{s,q}(\Omega)}^q,$$

where we have implicitly used that $f^{-1}(v) \sim \operatorname{sgn}(v)|v|^{1/p}$. *Lemma 1.9* is proved.

$\square$

1.4 Maximum Principles and Their Applications

The maximum principle is one of the most useful and best known tools employed in the study of partial differential equations. Indeed, the maximum principle enables us to obtain information about solutions of differential equations and inequalities without any explicit knowledge of the solutions themselves, and thus can be a valuable tool in scientific research. The intention of *Sects. 1.4–1.7* is, on one hand, to survey some extension and applications of the classical maximum principles for elliptic operators and, on the other hand, to prove some new explicit and uniform bounds that ensure the boundedness and asymptotic dissipation of the solutions of semilinear elliptic equations in bounded (unbounded) domains based on the maximum principles. These type of explicit and uniform bounds for the solutions are crucial to the construction of the attractive basin of the trajectories in the dynamical systems approach developed in the subsequent chapters. In *Sects. 1.4, 1.6, and 1.7* we mainly follow [97].

We consider first the one-dimensional case. If

$$u'' \geqslant 0$$

on (a, b) and u is continuous on $[a, b]$, then

(a) u attains its largest value either at a or at b.
(b) If u attains its largest value at $c \in (a, b)$, then u is constant on $[a, b]$.
(c) If a non-constant u attains its maximum at b, then $u'_-(b) > 0$, if at a, then $u'_+(a) < 0$.
(d) u is convex on $[a, b]$.

The one-side derivatives in (c) are assumed to exist. If, for example, $u'_+(a)$ fails to exist, then, instead of $u'_+(a) < 0$, one has $D^+ u(a) = \limsup_{x \downarrow a} < 0$.

It is common to refer to all or any of the statements (a)–(d) as maximum principle. We shall call (a), (b), (c), (d) the weak maximum principle, the strong maximum principle, the boundary point lemma and the convexity theorem, respectively.

Each of the statements (a)–(d) can be easily verified directly. Let us check (a). Assume, for a direct proof, that there is a point $c \in (a, b)$ such that $u(c) > u(a)$ and $u(c) > u(b)$. Choose an $\varepsilon > 0$ sufficiently small so that $u(c) > u(a) + \varepsilon(c - a)^2$ and $u(c) > u(b) + \varepsilon(b - a)^2$ and define a function v by $v(x) = u(x) + \varepsilon(x - a)^2$. By the Weierstrass theorem, v attains its maximum in $[a, b]$, and, since $v(c) > v(a)$ $v(c) > v(b)$, v attains its maximum at some $\xi \in (a, b)$. Therefore, $v'(\xi) = 0$ and $v''(\xi) \leqslant 0$. This, however, is a contradiction to $v''(\xi) = u''(\xi) + 2\varepsilon > 0$, and (a) is proved.

There is another way to prove (a)–(d). First, (d) is a well-known result from calculus. Hence, its suffices to prove the implication sequence $(d) \Rightarrow (c) \Rightarrow (b) \Rightarrow (a)$. Let us check $(d) \Rightarrow (c)$. Without loss of generality, we may assume that

$u'_-(b) = 0$. Then, by convexity, the graph of u lies above the tangent at b, which means that $u(x) > u(b)$ and u must be constant.

It is interesting to note that, in some sense, $(a) \Rightarrow (d)$, namely: if for every γ the function $x \rightarrow u(x) - \gamma x$ satisfies the weak maximum principle on every interval $[\alpha, \beta] \subset [a, b]$, then u is convex on $[a, b]$. As we will see below, the maximum principles also hold for a wide class of general second order elliptic partial differential equations. Indeed, let Ω be an open connected set in $\mathbb{R}^n$ with boundary $\partial \Omega = \bar{\Omega} \cap (\mathbb{R}^n \backslash \Omega)$ Let L be the second order differential operator:

$$L = \sum_{i,j=1}^{n} a_{ij}(x)D_{ij} + \sum_{i=1}^{n} b_i(x)D_i + c(x)$$

with $a_{ij} \in L_{loc}^{\infty}(\Omega)$ and $b_i, c \in L^{\infty}(\Omega)$. Here we have used $D_i = \frac{\partial}{\partial x_i}$ and $D_{ij} = \frac{\partial}{\partial x_i} \frac{\partial}{\partial x_j}$. Without loss of generality one assumes $a_{ij} = a_{ji}$.

Definition 1.9 We will fix the following notions.

- The operator L is called elliptic on Ω if for every $x \in \Omega$ there is $\lambda(x) > 0$ such that

$$\sum_{i,j=1}^{n} a_{ij}(x)\xi_i\xi_j \geqslant \lambda(x)|\xi|^2 \text{ for all } \xi \in \mathbb{R}^n.$$

- The operator L is called strictly elliptic on Ω if there is $\lambda > 0$ such that

$$\sum_{i,j=1}^{n} a_{ij}(x)\xi_i\xi_j \geqslant \lambda|\xi|^2 \text{ for all } \xi \in \mathbb{R}^n \text{ and } x \in \Omega.$$

- The operator L is called uniformly elliptic on Ω if there are $\Lambda, \lambda > 0$ such that

$$\lambda|\xi|^2 \leqslant \sum_{i,j=1}^{n} a_{ij}(x)\xi_i\xi_j \leqslant \Lambda|\xi|^2 \text{ for all } \xi \in \mathbb{R}^n \text{ and } x \in \Omega.$$

Remark 1.13 These definitions are not uniform throughout the literature. However, if the a_{ij} are bounded on $\bar{\Omega}$ then strictly elliptic implies uniformly elliptic and most references then agree (see *Sect. 1.2*).

Remark 1.14 The assumption $a_{ij} \in L_{loc}^{\infty}(\Omega)$ is too weak to expect even solutions of $Lu = f \in C^{\infty}(\bar{\Omega})$ to satisfy $u \in C^2(\Omega)$ and for that reason one usually assumes a_{ij} to be more regular. The maximum principle however does not need a_{ij} to be continuous.

Some notations that we will use are as follows. For $r > 0$ and $y \in \mathbb{R}^n$ we will write an open ball by

$$B_r(y) = \{x \in \mathbb{R}^n : |x - y| < r\}.$$

For a function u we will use u^+, u^- which are defined by

$$u^+(x) = \max(0, u(x)),$$
$$u^-(x) = \max(0, -u(x)).$$

It is obvious that $u = u^+ - u^-$ and $|u| = u^+ + u^-$.

1.4.1 Classical Maximum Principles

Lemma 1.10 *Suppose that L is elliptic and that $c \leq 0$. If $u \in C^2(\Omega)$ and $Lu > 0$ in Ω, then u cannot attain a nonnegative maximum in Ω.*

A proof can be done with a contradiction argument. We leave it to the reader.

Theorem 1.21 (Weak Maximum Principle) *Suppose that Ω is bounded and that L is strictly elliptic with $c \leq 0$. If $u \in C^2(\Omega) \cap C(\overline{\Omega})$ and $Lu \geq 0$ in Ω, then a nonnegative maximum is attained at the boundary.*

Proof Suppose that $\Omega \subset \{|x_1| < d\}$. Consider $w(x) = u(x) + \varepsilon e^{\alpha x_1}$ with $\varepsilon > 0$. Then

$$Lw = Lu + \varepsilon(\alpha^2 a_{11}(x) + \alpha b_1(x) + c(x))e^{\alpha x_1}$$
$$\geq \varepsilon(\alpha^2 \lambda + \alpha \|b_1\|_\infty + \|c\|_\infty)e^{\alpha x_1}$$

One chooses α large enough to find $Lw > 0$. By the previous lemma w cannot have a nonnegative maximum in Ω. Hence

$$\sup_\Omega u \leq \sup_\Omega w \leq \sup_\Omega w^+ = \sup_{\partial\Omega} w^+ \leq \sup_{\partial\Omega} u^+ + \varepsilon e^{\alpha d}$$

if $\Omega \subset \{|x| < d\}$. The result follows for $\varepsilon \to 0$. □

The proof of this maximum principle uses local arguments. If we skip the assumption that Ω is bounded we obtain:

Corollary 1.6 *Suppose that L is strictly elliptic with $c \leq 0$. If $u \in C^2(\Omega) \cap C(\overline{\Omega})$ and $Lu \geq 0$ in Ω, then u cannot attain a strict[1] nonnegative maximum in Ω.*

Theorem 1.22 (Strong Maximum Principle) *Suppose that L is strictly elliptic and that $c \leq 0$. If $u \in C^2(\Omega) \cap C(\overline{\Omega})$ and $Lu \geq 0$ in Ω, then either $u \equiv \sup_\Omega u$ or u does not attain a nonnegative maximum in Ω.*

Proof Let $m = \sup_\Omega u$. and set $\Sigma = \{x \in \Omega; u(x) = m\}$. We are done if $\Sigma \in \{\Omega, \emptyset\}$. Arguing by contradiction we assume that Σ and $\Omega \backslash \Sigma$ are non-empty.

The argument proceeds in three steps. First one fixes an appropriate open ball and in the next step an auxiliary function is defined that is positive on and only on this ball. For the sum of u and this auxiliary function one obtains a contradiction on a second ball by the weak maximum principle. For more details we refer to [97].   □

Clearly, the weak maximum principle is a consequence of the strong one.

Corollary 1.7 (Positivity Preserving Property) *Let Ω be bounded and suppose that L is strictly elliptic with $c \leqslant 0$. If $u \in C^2(\Omega) \cap C(\overline{\Omega})$ satisfies*

$$\begin{cases} -Lu \geqslant 0 & \text{in } \Omega, \\ u \geqslant 0 & \text{on } \partial\Omega, \end{cases} \tag{1.40}$$

then either $u(x) > 0$ for $x \in \Omega$ or $u \equiv 0$.

Remark 1.15 Let $f \in C(\overline{\Omega})$ and $\varphi \in C(\partial\Omega)$. A function $w \in C^2(\Omega) \cap C(\overline{\Omega})$ satisfying

$$\begin{cases} -Lw \geqslant f & \text{in } \Omega, \\ w \geqslant \varphi & \text{on } \partial\Omega, \end{cases} \tag{1.41}$$

is called a supersolution for

$$\begin{cases} -Lu = f & \text{in } \Omega, \\ u = \varphi & \text{on } \partial\Omega. \end{cases} \tag{1.42}$$

A much more useful concept of supersolutions assumes that $w \in C(\Omega)$ and replaces $-Lw \geqslant f$ by

$$\int_\Omega [(-L\varphi)w - \varphi f]\,dx \geqslant 0 \text{ for all } \varphi \in C_0^\infty(\Omega) \text{ with } \varphi \geqslant 0.$$

If the maximum principle holds then one finds that a supersolution for (1.42) lies above a solution for (1.42); $w \geqslant u$. In particular, since solutions are also supersolutions, if there are two solutions u_1 and u_2 then both $u_1 \geqslant u_2$ and $u_2 \geqslant u_1$ hold true. In other words, (1.42) has at most one solution in $C^2(\Omega) \cap C(\overline{\Omega})$.

Assuming more for $\partial\Omega$ one obtains an even stronger conclusion, that is, E. Hopf's result in 1952.

Theorem 1.23 (Hopf's Boundary Point Lemma) *Suppose that Ω satisfies the interior sphere condition, that is, there is a ball $B \subset \Omega$ with $x_0 \in \partial B$ at $x_0 \in \partial\Omega$. Let L be strictly elliptic with $c \leqslant 0$: If $u \in C^2(\Omega) \cap C(\overline{\Omega})$ satisfies $Lu \geqslant 0$ and $\max_{\overline{\Omega}} u(x) = u(x_0)$. Then either $u \equiv u(x_0)$ on Ω or*

$$\liminf_{t\downarrow 0} \frac{u(x_0) - u(x_0 + tv)}{t} > 0 \ (possibly \ +\infty)$$

for every direction v pointing into an interior sphere.
If $u \in C^1(\Omega \cup \{x_0\})$, then either $\frac{\partial u(x_0)}{\partial v} < 0$.

There are two directions in order to weaken the restriction on c. Skipping the sign condition for c but adding one for u one obtains the next result.

Theorem 1.24 (Maximum Principle for Nonpositive Functions) *Let Ω be bounded. Suppose that L is strictly elliptic (no sign assumption on c). If $u \in C^2(\Omega) \cap C(\overline{\Omega})$ satisfies $Lu \geq 0$ and in Ω and $u \leq 0$ on $\overline{\Omega}$, then either $u(x) < 0$ for all $x \in \Omega$, or $u \equiv 0$. Moreover, if Ω satisfies an interior sphere condition at $x_0 \in \partial\Omega$ and $u \in C^1(\Omega \cup \{x_0\})$ with $u < u(x_0) = 0$ in Ω, then $\frac{\partial u(x_0)}{\partial v} < 0$ for every direction v pointing into an interior sphere.*

Proof Writing $c(x) = c^+(x) - c^-(x)$ with c^+, $c^- \geq 0$ one finds that $L - c^+$ satisfies the condition of the S.M.P. and moreover from $u \leq 0$ it follows that

$$(L - c^+)u \geq -c^+u \geq 0.$$

The conclusion for the derivative follows from *Theorem 1.23*. $\qquad\qquad\square$

Theorem 1.25 (Maximum Principle When a Positive Supersolution Exists) *Let Ω be bounded. Suppose that L is strictly elliptic (no sign assumption on c) and that there exists $w \in C^2(\overline{\Omega})$ with $w > 0$ and $-Lw \geq 0$ on $\overline{\Omega}$. If $u \in C^2(\Omega) \cap C(\overline{\Omega})$ satisfies $Lu \geq 0$ in Ω, then either there exists a constant $t \in \mathbb{R}$ such that $u \equiv tw$, or u/w does not attain a nonnegative maximum in Ω.*

Remark 1.16 One may rephrase this for supersolutions as follows. If there exists one function $w \in C^2(\overline{\Omega})$ with $-Lw \geq 0$ and $w > 0$ on $\overline{\Omega}$ then all functions $v \in C^2(\Omega) \cap C(\overline{\Omega})$ such that

$$\begin{cases} -Lv \geq 0 & \text{in } \Omega, \\ v \geq 0 & \text{on } \partial\Omega, \end{cases} \tag{1.43}$$

satisfy either $v \equiv 0$ or $v > 0$ in Ω. Apply the theorem to $-v$.

Remark 1.17 If $\overline{\Omega} \subset \Omega_1$ and if L is defined on Ω_1 and happens to have a positive eigenfunction φ with eigenvalue λ_{Ω_1} for the Dirichlet problem on Ω_1:

$$\begin{cases} -L\varphi = \lambda_{\Omega_1}\varphi & \text{in } \Omega_1, \\ \varphi = 0 & \text{on } \partial\Omega_1, \\ \varphi > 0 & \text{in } \Omega_1, \end{cases} \tag{1.44}$$

then this $w = \varphi$ may serve in the theorem above. It shows that a maximum principle holds on Ω for $c < \lambda_{\Omega_1}$.

Proof Set $v = u/w$. Then with $\tilde{b}_i = b_i + \sum_{j=1}^{n} \frac{2a_{ij}}{w} D_j w$ and $\tilde{c} = \frac{Lw}{w}$ one has

$$Lu = L(vw) = \left(\sum_{i,j=1}^{n} a_{ij} D_{ij} v + \sum_{i=1}^{n} \tilde{b}_i D_i v + \frac{Lw}{w} v \right) w. \qquad (1.45)$$

Then $Lu \leqslant 0$ implies $\tilde{L}v \leqslant 0$ with $\tilde{L}$ as in (1.45) and since this $\tilde{L}$ does satisfy the conditions of the Strong Maximum Principle, in particular the sign condition for $\tilde{c} = \frac{Lw}{w}$, one finds that either v is constant or that v does not attain a nonnegative maximum in Ω. $\qquad \square$

Theorem 1.26 (Maximum Principle for Narrow Domains) *Suppose that L is strictly elliptic (no sign assumption on c). Then there is $d > 0$ such that if $S \subset \{x \in \Omega; |x| < d\}$ and $u \in C^2(S) \cap C(\overline{S})$ satisfies $Lu \geqslant 0$ in S, then there exists $w \in C^2(S)$ with $w > 0$ and $-Lw \geqslant 0$ on $\overline{S}$.*

Proof For $w(x) = \cos(\alpha x_1)$ and $|x_1| \leqslant \frac{\pi}{4\alpha}$ we have $w(x) > 0$ and

$$-Lw = (\alpha^2 a_{11} - c) \cos(\alpha x_1) + \alpha b_1 \sin(\alpha x_1)$$

$$\geqslant (\alpha^2 \lambda - \alpha \|b_1\|_\infty - \|c\|_\infty) \frac{1}{2} \sqrt{2}.$$

The claim follows by taking α large enough and defining $d = \frac{\pi}{4\alpha}$. $\qquad \square$

1.5 Uniform Estimates and Boundedness of the Solutions of Semilinear Elliptic Equations

Here we obtain explicit and uniform bounds that ensure the boundedness and the asymptotic dissipation of the solutions of semilinear elliptic equations in bounded (unbounded domains). These estimates are crucial to construct the attractive basin of the trajectories in the dynamical system approach (see *Chaps. 3–5*).

First, we give a general boundary maximum principle for non-negative subsolutions:

Lemma 1.11 *Let Ω be an open subset of $\mathbb{R}^N$ and $v \in C^2(\Omega) \cap C^0(\overline{\Omega})$ be a solution of*

$$\begin{cases} \Delta v \geqslant A v^p & \text{in } \Omega, \\ v \geqslant 0 & \text{in } \Omega, \\ v = \bar{v} & \text{on } \partial\Omega, \end{cases}$$

with $A > 0$, $p > 1$ and $\bar{v} \in L^\infty(\partial\Omega)$. Then, for any $x \in \Omega$,

$$v(x) \leqslant \|\bar{v}\|_{L^\infty(\partial\Omega)}.$$

Proof Fix any

$$S > \|\bar{v}\|_{L^\infty(\partial\Omega)}. \tag{1.46}$$

Let $C > 0$ (to be conveniently chosen in what follows), $\alpha := 2/(p - 1)$ and $R := (S^{-1}A^{-1/(p-1)}C)^{1/\alpha}$.

For any $x \in B(0, R)$, we define

$$w(x) := \frac{A^{-1/(p-1)}CR^\alpha}{(R^2 - |x|^2)^\alpha}.$$

Of course,

$$S = A^{-1/(p-1)}CR^{-\alpha} = w(0) \leqslant w(x) \quad \text{for any } x \in B(0, R), \tag{1.47}$$

and

$$\lim_{|x| \to R^-} w(x) = +\infty. \tag{1.48}$$

Moreover, by a direct computation one sees that there exists $C_o(N, p) > 0$ such that, if $C \geqslant C_o(N, p)$,

$$-\Delta w + Aw^p \geqslant 0 \quad \text{in } B(0, R). \tag{1.49}$$

We claim that

$$v(x_0) \leqslant S \quad \text{for any } x_0 \in \Omega. \tag{1.50}$$

To prove this, we define $\vartheta_{x_0}(x) := w(x - x_0)$ and we compare v with ϑ_{x_0}, by distinguishing two cases: either $B(x_0, R) \subset \Omega$ or not.

If $B(x_0, R) \subset \Omega$, we have that $v \in C^0(\overline{B(x_0, R)}) \subset L^\infty(\overline{B(x_0, R)})$, and so, by (1.48), there exists $\bar{R} \in (0, R)$ such that

$$v(x) \leqslant \vartheta_{x_0}(x) \quad \text{for any } x \in \partial B(x_0, \bar{R}). \tag{1.51}$$

On the other hand, if we set

$$c(x) := Ap \int_0^1 \left(tv(x) + (1 - t)\vartheta_{x_0}(x)\right)^{p-1} dt$$

$$\text{and} \quad z(x) := v(x) - \vartheta_{x_0}(x), \tag{1.52}$$

we have that $z \in C^2(\overline{B(x_0, \bar{R})})$ and

$$\Delta z = \Delta(v - \vartheta_{x_0}) \geqslant A(v^p - \vartheta_{x_0}^p) = cz. \tag{1.53}$$

Also, $z \leqslant 0$ on $\partial B(x_0, \bar{R})$, thanks to (1.51). Hence, by the maxiumum principle, $z \leqslant 0$ in $B(x_0, \bar{R})$. In particular,

$$v(x_0) = z(x_0) + \vartheta_{x_0}(x_0) \leqslant 0 + w(0) = S. \tag{1.54}$$

This proves (1.50) when $B(x_0, R) \subset \Omega$.

On the other hand, in $B(x_0, R) \not\subset \Omega$, we set $K := \overline{B(x_0, R)} \cap \overline{\Omega}$. We have that $v \in C^0(K) \subset L^\infty(K)$, and so, by (1.48), there exists $\tilde{R} \in (0, R)$ such that

$$v(x) \leqslant \vartheta_{x_0}(x) \text{ for any } x \in \left(\partial B(x_0, \tilde{R})\right) \cap \Omega. \tag{1.55}$$

Moreover, if $x \in B(x_0, R) \cap (\partial\Omega)$, we have that

$$v(x) = \bar{v}(x) \leqslant \|\bar{v}\|_{L^\infty(\partial\Omega)} < S \leqslant w(x - x_0) = \vartheta_{x_0}(x),$$

thanks to (1.46) and (1.47). This observation and (1.55) give that $z \leqslant 0$ on ∂K, where z is defined in (1.52). Also, z is C^2 in the interior of K and continuous up to the boundary, and (1.53) holds true in the interior of K. Accordingly, the maximum principle yields that $z \leqslant 0$ in K, and therefore, by arguing as in (1.54), we conclude that (1.50) holds true also when $B(x_0, R) \not\subset \Omega$.

Since, by (1.46), S can be taken arbitrarily close to $\|\bar{v}\|_{L^\infty(\partial\Omega)}$, the desired result follows from (1.50). □

Corollary 1.8 *Let $r > 2$, $c_1 \geqslant 0$, $c_2 > 0$. Let Ω be an open subset of $\mathbb{R}^n$. Let $f : \mathbb{R} \to \mathbb{R}$ be a measurable function such that*

$$f(s)s \geqslant -c_1 + c_2 s^r \text{ for any } s \geqslant 0. \tag{1.56}$$

Let $u \in C^2(\Omega) \cap C^0(\overline{\Omega})$ be a solution of

$$\begin{cases} \Delta u \geqslant f(u) & \text{in } \Omega, \\ u \geqslant 0 & \text{in } \Omega, \\ u = \bar{u} & \text{on } \partial\Omega, \end{cases}$$

with $\bar{u} \in L^\infty(\partial\Omega)$.

Then $u \in L^\infty(\Omega)$ and, for any $x \in \overline{\Omega}$,

$$0 \leqslant u(x) \leqslant \max\left\{ \left(\frac{c_1}{c_2}\right)^{1/r}, \|\bar{u}\|_{L^\infty(\partial\Omega)} \right\}. \tag{1.57}$$

Proof Let $p := r/2 > 1$ and $v := u^2$. Then

$$\Delta v = 2u\,\Delta u + 2|\nabla u|^2 \geqslant 2uf(u) + 0$$
$$\geqslant -2c_1 + 2c_2 u^r = -2c_1 + 2c_2 v^p \quad \text{in } \Omega. \tag{1.58}$$

Now we fix $\epsilon \in (0, 2c_2)$, to be taken arbitrarily small in the sequel. Let

$$\Omega_\epsilon := \{x \in \Omega \text{ s.t. } 2c_2 v^p(x) > 2c_1 + \epsilon v^p(x)\}.$$

We claim that, for any $x \in \Omega$,

$$0 \leqslant u(x) \leqslant \max\left\{\left(\frac{2c_1}{2c_2 - \epsilon}\right)^{1/r}, \|\bar u\|_{L^\infty(\partial\Omega)}\right\}. \tag{1.59}$$

To establish it, we notice that

if $x \in \Omega \backslash \Omega_\epsilon$, then

$$u(x) = \sqrt{v(x)} \leqslant \left(\frac{2c_1}{2c_2 - \epsilon}\right)^{1/(2p)} = \left(\frac{2c_1}{2c_2 - \epsilon}\right)^{1/r}. \tag{1.60}$$

If $\Omega_\epsilon = \Omega$, then (1.59) follows from (1.60), so we may suppose that Ω_ϵ is a non-empty open subset of $\mathbb{R}^N$, with $v \in C^2(\Omega_\epsilon) \cap C^0(\overline{\Omega}_\epsilon)$. Also, by virtue of (1.58),

$$\Delta v \geqslant -2c_1 + 2c_2 v^p \geqslant \epsilon v^p \quad \text{in } \Omega_\epsilon. \tag{1.61}$$

Now, we claim that

$$\text{if } x \in \partial\Omega_\epsilon, \text{ then } 0 \leqslant v(x) \leqslant \max\left\{\left(\frac{2c_1}{2c_2 - \epsilon}\right)^{1/p}, \|\bar u\|^2_{L^\infty(\partial\Omega)}\right\}. \tag{1.62}$$

Indeed, if $x \in (\partial\Omega_\epsilon) \cap \Omega$, we have that

$$2c_2 v^p(x) = 2c_1 + \epsilon v^p(x),$$

which implies (1.62). If, on the other hand, $x \in (\partial\Omega_\epsilon) \backslash \Omega$, then $x \in \partial\Omega$ and so $v(x) = \bar u(x)^2$, which gives (1.62) in this case.

Exploiting (1.61) and Lemma 1.11, we conclude that v in Ω_ϵ is controlled by $\|v\|_{L^\infty(\partial\Omega_\epsilon)}$, and so, by (1.62),

$$v(x) \leqslant \max\left\{\left(\frac{2c_1}{2c_2 - \epsilon}\right)^{1/p}, \|\bar u\|^2_{L^\infty(\partial\Omega)}\right\}$$

for any $x \in \Omega_\epsilon$. This and (1.60) prove (1.59), which, in turn, proves (1.57) by taking ϵ as close to 0 as we wish.   □

We remark that the condition $r > 2$ in Corollary 1.8 cannot be removed. As a countarexample, take $\Omega := \mathbb{R}_+^N$, $f(r) = r$ and $u(x) = u(x_1, \ldots, x_N) := e^{x_1}$. Then $u = 1$ on $\partial\Omega$ and (1.56) holds true with $r = 2$ (instead of $r > 2$), but u is unbounded.

The forthcoming Theorems 1.27 and 1.28 give some a priori bounds on subsolutions: namely assuming that the solution has some bound, then the bound may be made explicit and universal.

Theorem 1.27 *Let $N \geqslant 2$, and $u \in C^2(\mathbb{R}_{++}^N) \cap C^0(\overline{\mathbb{R}_{++}^N})$ be a solution of*

$$\begin{cases} \Delta u \geqslant f(u) \ in \ \mathbb{R}_{++}^N, \\ u\Big|_{x_1=0} = u_0, \\ u\Big|_{x_N=0} = 0. \end{cases}$$

Suppose that $u_0 \in L^\infty(\overline{\mathbb{R}_{++}^N} \cap \{x_1 = 0\})$ and that f satisfies

$$f(s) \leqslant 0 \ for \ any \ s \in [0, \mu] \ and \ f(s) \geqslant 0 \ for \ any \ s \in [\mu, +\infty),$$

for a suitable $\mu > 0$.

Then, if u is bounded from above, we have that

$$u(x) \leqslant \max\left\{\mu, \|u_0\|_{L^\infty(\overline{\mathbb{R}_{++}^N} \cap \{x_1=0\})}\right\} \qquad for \ any \ x \in \mathbb{R}_{++}^N. \tag{1.63}$$

Proof Let

$$M := \max\left\{\mu, \|u_0\|_{L^\infty(\overline{\mathbb{R}_{++}^N} \cap \{x_1=0\})}\right\} \ and \ v := u - M.$$

We see that v solves

$$\begin{cases} \Delta v \geqslant f(u) \ in \ \mathbb{R}_{++}^N, \\ v\Big|_{x_1=0} = u_0 - M \leqslant 0, \\ v\Big|_{x_N=0} = -M \leqslant 0. \end{cases}$$

Also, v is bounded from above if so is u. We claim that

$$\{x \in \mathbb{R}_{++}^N \ s.t. \ v(x) > 0\} = \varnothing. \tag{1.64}$$

The proof of (1.64) is by contradiction: if not, let Ω_o be a connected component of the set in (1.64): then $\Delta v \geqslant f(u) \geqslant 0$ in Ω_o and $v\big|_{\partial\Omega_o} \leqslant 0$. The maximum principle then gives that $v \leqslant 0$ in Ω_o, while $\Omega_o \subseteq \{v > 0\}$. This is a contradiction and so (1.64) is established. In turn, (1.64) implies (1.63).   □

Theorem 1.28 *Let* $N \geqslant 2$, *and* $u \in C^2(\mathbb{R}_{++}^N) \cap C^0(\overline{\mathbb{R}_{++}^N})$ *be a solution of*

$$\begin{cases} \Delta u \geqslant f(u) \ \ in \ \mathbb{R}_{++}^N, \\ u\big|_{x_1=0} = u_0, \\ u\big|_{x_N=0} = 0. \end{cases}$$

Suppose that $u_0 \in L^\infty(\overline{\mathbb{R}_{++}^N} \cap \{x_1 = 0\})$ *and that* f *satisfies*

$$f(s)s \geqslant -c_1 + c_2 s^r \ \ for \ any \ s \geqslant 0$$

for suitable $c_1 \geqslant 0$, $c_2 > 0$ *and* $r \geqslant 2$, *and that* $u \geqslant 0$ *in* $\mathbb{R}_{++}^N$.
  Then, if $u \in L^\infty(\mathbb{R}_{++}^N)$, *we have that*

$$u(x) \leqslant \left(\frac{c_1}{c_2} \right)^{1/r} + \|u_0\|_{L^\infty(\overline{\mathbb{R}_{++}^N} \cap \{x_1=0\})} e^{-\gamma x_1} \qquad for \ any \ x \in \mathbb{R}_{++}^N, \qquad (1.65)$$

where

$$\gamma := \sqrt{ \frac{c_2 \|u_0\|_{L^\infty(\overline{\mathbb{R}_{++}^N} \cap \{x_1=0\})}^{r-2}}{2} }.$$

Proof We define

$$w := u^2 - \|u_0\|_{L^\infty(\overline{\mathbb{R}_{++}^N} \cap \{x_1=0\})}^2 e^{-2\gamma x_1} - \left(\frac{c_1}{c_2} \right)^{2/r}.$$

By a direct computation, one sees that

$$\begin{aligned} \Delta w &= 2u\Delta u + 2|\nabla u|^2 - 4\|u_0\|_{L^\infty(\overline{\mathbb{R}_{++}^N} \cap \{x_1=0\})}^2 \gamma^2 e^{-2\gamma x_1} \\ &\geqslant 2uf(u) - 4\|u_0\|_{L^\infty(\overline{\mathbb{R}_{++}^N} \cap \{x_1=0\})}^2 \gamma^2 e^{-2\gamma x_1} \\ &\geqslant 2c_2 u^r - 2c_1 - 4\|u_0\|_{L^\infty(\overline{\mathbb{R}_{++}^N} \cap \{x_1=0\})}^2 \gamma^2 e^{-2\gamma x_1} \qquad (1.66) \end{aligned}$$

in $\mathbb{R}^N_{++}$. Moreover

$$w\Big|_{x_1=0} = u_0^2 - \|u_0\|^2_{L^\infty(\overline{\mathbb{R}^N_{++}}\cap\{x_1=0\})} - \left(\frac{c_1}{c_2}\right)^{2/r} \leqslant 0$$

$$\text{and}\quad w\Big|_{x_N=0} = -\|u_0\|^2_{L^\infty(\overline{\mathbb{R}^N_{++}}\cap\{x_1=0\})} e^{-2\gamma x_1} - \left(\frac{c_1}{c_2}\right)^{2/r} \leqslant 0. \qquad (1.67)$$

We claim that

$$\{x \in \mathbb{R}^N_{++} \text{ s.t. } w(x) > 0\} = \varnothing. \qquad (1.68)$$

The proof is by contradiction. Indeed, if $\{w > 0\} \neq \varnothing$, for any point $x \in \{w > 0\}$ we have that

$$u^2(x) > \|u_0\|^2_{L^\infty(\overline{\mathbb{R}^N_{++}}\cap\{x_1=0\})} e^{-2\gamma x_1} + \left(\frac{c_1}{c_2}\right)^{2/r}$$

and so

$$u^r(x) > \left(\|u_0\|^2_{L^\infty(\overline{\mathbb{R}^N_{++}}\cap\{x_1=0\})} e^{-2\gamma x_1} + \left(\frac{c_1}{c_2}\right)^{2/r}\right)^{r/2}. \qquad (1.69)$$

Since $r/2 \geqslant 1$, this gives that

$$u^r(x) > \|u_0\|^r_{L^\infty(\overline{\mathbb{R}^N_{++}}\cap\{x_1=0\})} e^{-r\gamma x_1} + \frac{c_1}{c_2}. \qquad (1.70)$$

That is, for any $x \in \{w > 0\}$,

$$2c_2 u^r(x) - 2c_1 > 2c_2 \|u_0\|^r_{L^\infty(\overline{\mathbb{R}^N_{++}}\cap\{x_1=0\})} e^{-r\gamma x_1}$$

$$= 4\|u_0\|^2_{L^\infty(\overline{\mathbb{R}^N_{++}}\cap\{x_1=0\})} \gamma^2 e^{-r\gamma x_1}.$$

Thus, recalling (1.66), we infer that $\Delta w \geqslant 0$ in $\{w > 0\}$. Also, recalling (1.67), we see that $w \leqslant 0$ on $\partial\{w > 0\}$. Hence, since w is bounded, the maximum principle implies that $w \leqslant 0$ on $\{w > 0\}$. This is a contradiction and so (1.68) is established. The estimate in (1.65) then easily follows from (1.68). $\qquad\qquad\square$

We remark that the condition $r \geqslant 2$ was used, in the proof of Theorem 1.28, to pass from (1.69) to (1.70). On the other hand, the case $r \in (0, 2)$ may also be treated by using instead the formula

$$(a^{r/2} + b^{r/2}) \leqslant 2^{1-(r/2)}(a + b)^{r/2} \qquad \text{for any } r \in (0, 2) \text{ and } a, b \geqslant 0.$$

   Also, it is interesting to notice that Theorem 1.28 provides universal bounds for any solution, without any sign condition on f at the origin.

   The following Theorem *1.29* will be used in the study of the symmetry and monotonicity properties of positive solutions of a semilinear elliptic equations.

Theorem 1.29 (Maximum Principle for Small Domains) *Suppose that Ω is bounded and that L is strictly elliptic (without sign condition for c). Then there exists a constant δ, with $\delta = \delta(n, \mathrm{diam}(\Omega), \lambda, \|b\|_{L^n(\Omega)}, \|c^+\|_\infty)$, such that the following holds.*
If $|\Omega| < \delta$ and $u \in C^2(\Omega) \cap C(\overline{\Omega})$ satisfies $Lu \geq 0$ in Ω and $u \leq 0$ on $\partial\Omega$, then $u \leq 0$ in Ω.

Proof The operator $L - c^+$ satisfies the condition of Theorem 5.5 from [97] and from $-Lu \leq 0$ it follows that

$$(L - c^+)u \geq -c^+ u \geq -c^+ u^+$$

and hence, since $\sup_{\partial\Omega} u^+ = 0$,

$$\sup_\Omega u \leq \frac{C \,\mathrm{diam}\,\Omega}{\lambda} \|c^+ u^+\|_{L^n(\Omega)}$$

$$\leq \frac{C \,\mathrm{diam}\,\Omega}{\lambda} \|c^+\|_\infty |\Omega|^{\frac{1}{n}} \sup_\Omega u^+.$$

If $\delta = (\frac{C \,\mathrm{diam}\,\Omega}{\lambda} \|c^+\|_\infty)^{-n}$ one finds $\sup_\Omega u \leq 0$. Here we define

$$C = \exp\left(\frac{2^{n-2}}{\sigma_n n^n}\left(\left\|\frac{|b|}{\mathcal{D}^*}\right\|_{L^n(\Gamma^+ \cap \Omega^+)}\right) + 1\right), \quad \mathcal{D}^*(x) = (\det(a_{ij}(x)))^{1/n}, \qquad (1.71)$$

which depends on $\|b\|_{L^n(\Omega)}$.   □

1.6   The Sweeping Principle and the Moving Plane Method in a Bounded Domain

Using comparison principles for connected families of sub- and supersolutions one obtains a very powerful tool in deriving a priori estimates. One such result is the moving plane method used by Gidas, Ni and Nirenberg [65] to prove symmetry of positive solutions to

$$\begin{cases} -\Delta u = f(u) & \text{in } \Omega, \\ \qquad u = 0 & \text{on } \partial\Omega, \end{cases} \qquad (1.72)$$

on domains Ω satisfying some symmetry conditions. Another one is McNabb's sweeping principle [80] used to its full extend by Serrin. First let us recall the notion of supersolution for (1.72).

Definition 1.10 A function $v \in C(\overline{\Omega}) \cap C^2(\Omega)$ is called a supersolution for

$$\begin{cases} -\Delta u = f(u) & \text{in } \Omega, \\ \quad\ u = h & \text{on } \partial\Omega, \end{cases} \tag{1.73}$$

if $v \geqslant h$ on $\partial\Omega$ and $-\Delta v \geqslant f(v)$.

Remark 1.18 A much more useful concept of supersolution assumes $v \in C(\overline{\Omega})$ and replaces $-\Delta v \geqslant f(v)$ by

$$\int_\Omega ((-\Delta\phi)v - \phi f(v))\, dx \geqslant 0$$

for all $\phi \in C_0^\infty(\Omega)$ with $\phi \geqslant 0$.

The argument combining the strong maximum principle and continuous perturbations goes as follows. One needs:

- a continuous family of supersolutions v_t, say $t \in [0, 1]$, possibly on a subdomain or on an appropriate family of subdomains Ω_t;
- a strong maximum principle on each of the subdomains.

Roughly spoken the conclusion is that if $v_0 > u$ on Ω_t, then either $v_t > u$ for all $t \in [0, 1]$ or there is a $t_1 \in [0, 1]$ such that $v_{t_1} \equiv u$. For a precise statement we have to refine the notion of positivity.

For a fixed domain such a result is known as a 'sweeping principle'. A first reference to this result is a paper of McNabb from 1961, see [80]. In the following version we assume that $\partial\Omega = \Gamma_1 \cup \Gamma_2$, with Γ_1, Γ_2 closed and disjoint, possibly empty. We let $e \in C(\overline{\Omega}) \cap C^1(\Omega \cup \Gamma_2)$ be such that $e > 0$ on $\Omega \cup \Gamma_1$ and $\frac{\partial}{\partial n}e < 0$ on Γ_2, where n denotes the outward normal. Moreover, we define

$$C_e(\overline{\Omega}) = \{w \in C(\overline{\Omega});\ |w| < ce \text{ for some } c > 0\},$$

$$\|w\|_e = \left\| \frac{w}{e} \right\|_\infty.$$

Theorem 1.30 (Sweeping Principle) *Let Ω be bounded with $\partial\Omega \in C^2$ and $\partial\Omega = \Gamma_1 \cup \Gamma_2$ as above, $f \in C^1(\mathbb{R})$ and let $u \in C(\overline{\Omega}) \cap C^1(\Omega \cup \Gamma_2) \cap C^2(\Omega)$ be a solution of (1.73). Suppose $\{v_t; t \in [0, 1]\}$ is a family of supersolutions in $C(\overline{\Omega}) \cap C^1(\Omega \cup \Gamma_2) \cap C^2(\Omega)$ for (1.73) such that:*

1. *$t \mapsto (v_t - v_0) \in C_e(\overline{\Omega})$ is continuous (with respect to the $\|\cdot\|_e$-norm);*

2. *$v_t = u$ on Γ_2 and $v_t > u$ on Γ_1;*

3. *$v_0 \geqslant u$ in $\overline{\Omega}$;*

Then either $v_t \equiv u$ for some $t \in [0, 1]$, or there exists $c > 0$ such that $v_t \geqslant u + ce$ on $\overline{\Omega}$ for all $t \in [0, 1]$.

Proof Set $I = \{t \in [0, 1]; \ v_t \geqslant u \text{ in } \overline{\Omega}\}$ and assume that $v_t \not\equiv u$ for all $t \in [0, 1]$. By the way, notice that $v_t \equiv u$ for some $t \in [0, 1]$ can only occur when Γ_1 is empty. The set I is nonempty by assumption 3 and closed by assumption 1. We will show that I is open. Indeed, if t is such that $v_t \geqslant u$ in $\overline{\Omega}$ then $-\Delta(v_t - u) = f(v_t) - f(u)$ and setting g as (see also [97])

$$g(x) := \begin{cases} \frac{f(x, u_2(x)) - f(x, u_1(x))}{u_2(x) - u_1(x)} & \text{if } u_2(x) \neq u_1(x), \\ \frac{\partial f}{\partial u}(x, u_2(x)) & \text{if } u_2(x) = u_1(x), \end{cases}$$

one obtains

$$\begin{cases} -\Delta(v_t - u) + g^-(v_t - u) \geqslant g^+(v_t - u) \geqslant 0 & \text{in } \Omega, \\ v_t - u \geqslant 0 & \text{on } \partial\Omega, \end{cases}$$

and hence by the strong maximum principle $v_t - u > 0$ in Ω or $v_t \equiv u$. Moreover, for the first case Hopf's boundary point Lemma and the assumption that $u, v_t \in C^1(\Omega \cup \Gamma_2)$ imply that $-\frac{\partial}{\partial n}(v_t - u) > 0$ on Γ_2. Hence there is $c_t > 0$ such that $v_t - u \geqslant c_t e$ on $\overline{\Omega}$. By assumption 1 it follows that t lies in the interior of I. Hence I is open, implying $I = [0, 1]$. Moreover, since I is compact a uniform $c > 0$ exists. □

Example 1.1 Suppose that $f \in C^1$ is such that $f(u) > 0$ for $u < 1$ and $f(u) < 0$ for $u > 1$. Using the above version of the sweeping principle one may show that for $\lambda \gg 1$ every positive solution u_λ of

$$\begin{cases} -\Delta u = \lambda f(u) & \text{in } \Omega, \\ u = 0 & \text{on } \partial\Omega, \end{cases} \tag{1.74}$$

is near 1 in the interior of Ω. Indeed, let φ_1, μ_1 denote the first eigenfunction/eigenvalue of $-\Delta\varphi = \mu\varphi$ in B_1 and $\varphi = 0$ on ∂B_1, where B_1 is the unit ball in $\mathbb{R}^n$. This first eigenfunction is radially symmetric and we assume it to be normalized such that $\varphi(0) = 1$. Now let $\varepsilon > 0$ and take $\delta_\varepsilon > 0$ such that

$$\delta_\varepsilon u \leqslant f(u) \text{ for } u \in [0, 1 - \varepsilon].$$

Set $v_t(x) = t\varphi_1(\frac{\sqrt{\lambda\delta_\varepsilon}}{\sqrt{\mu}}(x - x^*))$ for any $x^* \in \Omega$ with distance to the boundary $d(x^*, \partial\Omega) > \frac{\sqrt{\mu}}{\sqrt{\lambda\delta_\varepsilon}} =: r_0$. On $B_{r_0}(x^*)$ one finds for $t \in [0, 1 - \varepsilon]$ that $-\Delta v_t = \lambda\delta_\varepsilon v_t \leqslant \lambda f(v_t)$. Since $v_t = 0 < u_\lambda$ on $\partial B_{r_0}(x^*)$ and $v_0 = 0 < u_\lambda$ on $\overline{B_{r_0}(x^*)}$ it is an appropriate family of subsolutions. From $v_{1-\varepsilon} < u_\lambda$ in $B_{r_0}(x^*)$ one concludes that $1 - \varepsilon = v_{1-\varepsilon}(x^*) < u_\lambda(x^*)$. Since the strong maximum principle implies that $\max u_\lambda < 1$ we may summarize:

$$1 - \varepsilon < u_\lambda(x) < 1 \text{ for } x \in \Omega \text{ with } d(x^*, \partial\Omega) > \frac{\sqrt{\mu}}{\sqrt{\lambda\delta_\varepsilon}}.$$

More intricate results not only consider a family of supersolutions but also simultaneously modify the domain. One such result is the result by Gidas, Ni and Nirenberg ([65] or [64]). The idea was used earlier by Serrin in [93].

Before stating the result let us first fix some notations. We will be moving planes in the x_1-direction but of course also any other direction will do.

Some sets that will be used are:

the moving plane:	$T_\lambda = \{x \in \mathbb{R}^n; x_1 = \lambda\}$,
the subdomain:	$\Sigma_\lambda = \{x \in \Omega; x_1 < \lambda\}$,
the reflected point:	$x_\lambda = (2\lambda - x_1, x_2, \ldots, x_n)$,
the reflected subdomain:	$\Sigma'_\lambda = \{x_\lambda; x \in \Sigma_\lambda\}$,
the starting value for λ :	$\lambda_0 = \inf\{x_1; x \in \Omega\}$,
the maximum value for λ :	$\lambda^* = \sup\{\lambda; \Sigma'_\mu \subset \Omega \text{ for all } \mu < \lambda\}$.

Theorem 1.31 (Moving Plane Argument) *Assume that f is Lipschitz, that Ω is bounded and that λ_0, λ^* and Σ_λ are as above. If $u \in C(\overline{\Omega}) \cap C^2(\Omega)$ satisfies (1.72) and $u > 0$ in Ω, then*

$$u(x) < u(x_\lambda) \quad \text{for all } \lambda < \lambda^* \text{ and } x \in \Sigma_\lambda,$$
$$\frac{\partial u}{\partial x_1}(x) > 0 \qquad \text{for all } x \in \Sigma_\lambda.$$

Remark 1.19 We will use the moving plane arguments, sweeping principle and sliding techniques (see subsequent sections) in *Chaps. 3–7* to study nonlinear elliptic boundary value problems in unbounded domains from the dynamical systems viewpoint.

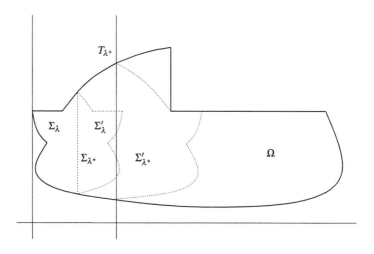

Proof First remark that if $u \equiv 0$ on some open set in Ω, then $f(0) = 0$. If $f(0) = 0$ then u satisfies $-\Delta u = c(x)u(x)$ with $c(x) = f(u(x))/u(x)$ a bounded function. By using the Maximum Principle for nonpositive functions it follows that $u \equiv 0$ in Ω. Hence $u \equiv 0$ on any open set in Ω.

We will consider $w_\lambda(x) = u(x_\lambda) - u(x)$ for $x \in \Sigma_\lambda$. Defining g_λ via

$$g_\lambda(x) = \begin{cases} \frac{f(u(x_\lambda)) - f(u(x))}{u(x_\lambda) - u(x)} & \text{if } u(x_\lambda) \neq u(x), \\ 0 & \text{if } u(x_\lambda) = u(x), \end{cases}$$

we find

$$-\Delta w_\lambda(x) = -\Delta(u(x_\lambda)) + \Delta u(x) = -(\Delta u)(x_\lambda) + \Delta u(x)$$
$$= f(u(x_\lambda)) - f(u(x)) = g_\lambda(x)w_\lambda(x) \text{ in } \Sigma_\lambda. \quad (1.75)$$

Moreover, for $\lambda < \lambda^*$ we have $u(x_\lambda) \geqslant 0 = u(x)$ on $\partial\Omega \cap \partial\Sigma_\lambda$ and $u(x_\lambda) = u(x)$ on $\Omega \cap \partial\Sigma_\lambda$. Hence

$$w_\lambda(x) \geqslant 0 \text{ on } \partial\Sigma_\lambda. \quad (1.76)$$

Also note that since u is bounded and f is Lipschitz the function g_λ is uniformly bounded on $\overline{\Omega}$ and hence that we may use the maximum principles for linear equations with a uniform bound for $\|c^+\|_\infty$.

The two basic ingredients in the proof are the maximum principle for small domains (*Theorem 1.29*) and again the strong maximum principle for nonpositive functions (*Theorem 1.24*).

Set $\lambda_* = \sup\{\lambda \in [\lambda_0, \lambda^*]; w_\mu(x) \geqslant 0 \text{ for all } x \in \Sigma_\mu \text{ and } \mu \in [0, \lambda]\}$. We will suppose that $\lambda_* < \lambda^*$ and arrive at a contradiction.

By the maximum principle for small domains (or even the one for narrow domains) one finds from (1.75) and (1.76) that there is $\lambda_1 > \lambda_0$ such that $w_\lambda(x) \geqslant 0$ on $\overline{\Sigma}_\lambda$ for all $\lambda \in (\lambda_0, \lambda_1]$. Hence $\lambda_* > 0$. By the strong maximum principle either $w_\lambda > 0$ in Σ_λ for all $\lambda \in (\lambda_0, \lambda_1]$ or $w_\lambda \equiv 0$ on $\overline{\Sigma}_\lambda$ for some $\lambda \in (\lambda_0, \lambda_1]$. We have found that $\lambda_* \geqslant \lambda_1$.

Next we will show that $w_\lambda \equiv 0$ on $\overline{\Sigma}_\lambda$ does not occur for $\lambda \in (\lambda_0, \lambda_1)$. If $w_\lambda \equiv 0$ on $\overline{\Sigma}_\lambda$ for some $\lambda \in (\lambda_0, \lambda^*)$ then $u(x) = 0$ for $x \in \partial(\Sigma'_\lambda \cup \overline{\Sigma}_\lambda)$ and moreover, since part of this boundary $\Omega \cap \partial(\Sigma'_\lambda \cup \overline{\Sigma}_\lambda)$ lies inside Ω it contradicts $u > 0$ in Ω.

For $\lambda = \lambda_*$ we have that $w_{\lambda_*}(x) \geqslant 0$ on $\overline{\Sigma}_\lambda$ and since $\Sigma'_{\lambda_*} \cup \overline{\Sigma}_{\lambda_*} \neq \Omega$ we just found that $w_{\lambda_*}(x) \not\equiv 0$ on $\overline{\Sigma}_\lambda$. By the strong maximum principle for signed functions (*Theorem 1.24*) it follows that $w_{\lambda_*}(x) > 0$ in Σ_{λ_*} and even that $\frac{\partial}{\partial x_1} w_{\lambda_*}(x) > 0$ for every $x \in \partial\Sigma_{\lambda_*} \cap \Omega$. In order to find a strict bound for w_λ away from 0 we restrict ourselves to a compact set as follows. Let δ be as in *Theorem 1.29* and let Γ_δ be an open neighborhood of $\partial\Omega \cap \partial\Sigma_{\lambda_*}$ such that $|\Omega \cap \Gamma_\delta| < \frac{1}{2}\delta$. By the previous estimate we have for some $c > 0$ that

$$w_{\lambda_*}(x) > c(\lambda_* - x_1) \text{ for } x \in \Sigma_{\lambda_*} \backslash \Gamma_\delta.$$

By continuity we may increase λ somewhat without loosing the positivity. Indeed, by the fact that $w_{\lambda_*} \in C^1(\Sigma_{\lambda_*})$ there exists $\varepsilon_0 > 0$ such that for all $\varepsilon \in [0, \varepsilon_0]$ one finds

$$w_{\lambda_*+\varepsilon}(x) > 0 \text{ for } x \in A_\varepsilon,$$

where $A_\varepsilon = \{x \in \Omega; (x_1 + \varepsilon, x_2, \ldots, x_n) \in \overline{\Sigma}_{\lambda_*} \backslash \Gamma_\delta\} \subset \Sigma_{\lambda_*+\varepsilon}$ is a shifted Σ_{λ_*}.

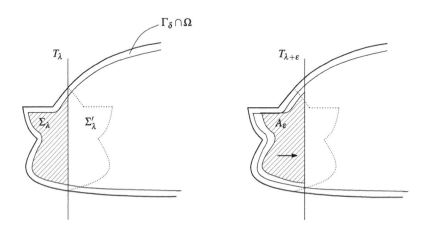

Since $\lambda \mapsto |\Sigma_\lambda|$ is continuous and $A_\varepsilon \subset \Sigma_{\lambda+\varepsilon}$ we may take $\varepsilon_1 \in (0, \varepsilon_0)$ such that

$$\left|\overline{\Sigma}_{\lambda_*+\varepsilon} \backslash (\Gamma_\delta \cap A_\varepsilon)\right| < \frac{1}{2}\delta \text{ for } \varepsilon \in [0, \varepsilon_1].$$

Finally we consider the maximum principle for small domains on the remaining subset of $\Sigma_{\lambda+\varepsilon}$ defined by $R_\varepsilon = (\Sigma_{\lambda_*+\varepsilon} \cap \Gamma_\delta) \cup (\Sigma_{\lambda_*+\varepsilon} \backslash A_\varepsilon)$ with $\varepsilon \in (0, \varepsilon_1)$. Since $w_{\lambda_*+\varepsilon} > 0$ on ∂R_ε and $-\Delta w_{\lambda_*+\varepsilon}(x) = g_{\lambda_*+\varepsilon}(x) w_{\lambda_*+\varepsilon}(x)$ in R_ε we find that $w_{\lambda_*+\varepsilon} \geqslant 0$ in R_ε and hence $w_\lambda \geqslant 0$ on $\overline{\Sigma}_\lambda$ for all $\lambda \in [\lambda_0, \lambda_* + \varepsilon_1]$, a contradiction.

The conclusion that $\frac{\partial}{\partial x_1} u(x) > 0$ for $x \in \Sigma_{\lambda^*}$ follows from the fact that $w_\lambda(x) > 0$ on Σ_λ for all $\lambda \in (\lambda_0, \lambda^*)$ and by Hopf's boundary point Lemma $\frac{\partial}{\partial x_1} u(x) = \frac{1}{2}\frac{\partial}{\partial x_1} w_\lambda(x) > 0$ on $T_\lambda \cap \Omega$.   □

Corollary 1.9 *Let $B_R(0) \subset \mathbb{R}^n$. If f is Lipschitz and if $u \in C^2(B) \cap C(\overline{B})$ satisfies*

$$\begin{cases} -\Delta u = f(u) & \text{in } B, \\ \quad\quad u = 0 & \text{on } \partial B, \\ \quad\quad u > 0 & \text{in } B, \end{cases} \tag{1.77}$$

then u is radially symmetric and $\frac{\partial}{\partial |x|} u(x) < 0$ for $0 < |x| < R$.

Proof From the previous theorem we find that $x_1 \frac{\partial}{\partial x_1} u(x) < 0$ for $x \in B_R(0)$ with $x_1 \neq 0$. Hence $\frac{\partial}{\partial x_1} u(x) < 0$ for $x_1 = 0$. Since Δ is radially invariant this holds for every direction and we find $\frac{\partial}{\partial \tau} u(x) = 0$ in $B_R(0)$ for any tangential direction. In other words, u is radially symmetric. Since $\frac{\partial}{\partial |x|} u(x) = \frac{\partial}{\partial \tau} u(r, 0, \ldots, 0)$ for $0 < r = |x| < R$ the second claim follows from $\frac{\partial u}{\partial x_1}(x) < 0$ for $x \in B_R(0)$ with $x_1 > 0$.   □

1.7   The Sliding and the Moving Plane Method in General Domains

In the sliding method introduced in [18] one compares translations of the function rather then reflections. As we will see below, the approach based on the maximal principle for 'narrow domains' yields not only a simpler but a more general result. To illustrate the main idea, we now state and prove a monotonicity result for equation

$$\begin{cases} \Delta u + f(u) = 0, & \text{in } \Omega := \{x : |x| < R\} \subset \mathbb{R}^N, \\ u = 0 & \text{on } \partial\Omega. \end{cases} \tag{1.78}$$

but with other boundary conditions, in an arbitrary domain and with general regularity assumptions on the solution. Here we follow [18].

Theorem 1.32 *Let Ω be an arbitrary bounded domain of $\mathbb{R}^n$ which is convex in the x_1-direction. Let $u \in W_{loc}^{2,n}(\Omega) \cap C(\overline{\Omega})$ be a solution of*

$$\Delta u + f(u) = 0 \text{ in } \Omega, \tag{1.79}$$

$$u = \varphi \text{ on } \partial\Omega. \tag{1.80}$$

The function f is supposed to be Lipschitz continuous. Here we assume that for any three points $x' = (x_1', y)$, $x = (x_1, y)$, $x'' = (x_1'', y)$ lying on a segment parallel to the x_1-axis, $x_1' < x_1 < x_1''$, with $x', x'' \in \partial\Omega$, the following hold

$$\varphi(x') < u(x) < \varphi(x'') \text{ if } x \in \Omega \tag{1.81}$$

and

$$\varphi(x') \leqslant \varphi(x) \leqslant \varphi(x'') \text{ if } x \in \partial\Omega. \tag{1.82}$$

Then, u is monotone with respect to x_1 in Ω:

$$u(x_1 + \tau, y) > u(x_1, y) \text{ for } (x_1, y), (x_1 + \tau, y) \in \Omega \text{ and } \tau > 0.$$

Furthermore, if f is differentiable, then $u_{x_1} > 0$ in Ω. Finally, u is the unique solution of (1.79), (1.80), in $W_{loc}^{2,n}(\Omega) \cap C(\overline{\Omega})$ satisfying (1.81).

Condition (1.82) requires monotonicity of φ on any segment parallel to the x_1-axis lying on $\partial\Omega$. It is obviously a necessary condition for the result to hold.

Proof Theorem 1.32 is proved by using the sliding technique. For $\tau \geqslant 0$, we let $u^\tau(x_1, y) = u(x_1 + \tau, y)$. The function u^τ is defined on the set $\Omega^\tau = \Omega - \tau e_1$ obtained from Ω by sliding it to the left a distance τ parallel to the x_1-axis. The main part of the proof consists in showing that

$$u^\tau > u \text{ in } \Omega^\tau \cap \Omega \text{ for any } \tau > 0. \tag{1.83}$$

Indeed, (1.83) means precisely that u is monotone increasing in the x_1 direction.

Set $w^\tau(x) = u^\tau(x) - u(x)$ i.e. $w^\tau(x_1, y) = u(x_1 + \tau, y) - u(x_1, y)$; w^τ is defined in $D^\tau = \Omega \cap \Omega^\tau$. As before, since u^τ satisfies the same Eq. (1.79) in Ω^τ as does u in Ω, we see that w^τ satisfies an equation

$$\begin{cases} \Delta w^\tau + c^\tau(x) w^\tau = 0 & \text{in } D^\tau, \\ w^\tau \geqslant 0 & \text{on } \partial D^\tau \end{cases} \tag{1.84}$$

where c^τ is the same L^∞ function satisfying $|c^\tau(x)| \leqslant b$, $\forall x \in D^\tau$, $\forall \tau$. The inequality on the boundary $\partial D^\tau \subset \partial\Omega \cap \partial\Omega^\tau$ follows the assumptions (1.81)–(1.82).

Let $\tau_0 = \sup\{\tau > 0; D^\tau \neq \emptyset\}$. For $0 < \tau_0 - \tau$ small, $|D^\tau|$ is small, that is D^τ is a 'narrow domain'. Therefore, from (1.84) it follows that for $0 < \tau_0 - \tau$ small, $w^\tau > 0$ in D^τ.

Next, let us start sliding Ω^τ back to the right, that is we decrease τ from τ_0 to a critical position $\tau \in [0, \tau_0)$: let (τ, τ_0) be a maximal interval, with $\tau \geqslant 0$, such that for all τ in $\tau < \tau' \leqslant \tau_0$, $w^{\tau'} \geqslant 0$ in $D^{\tau'}$. We want to prove that $\tau = 0$. We argue by contradiction, assuming $\tau > 0$.

By continuity, we have $w^\tau \geqslant 0$ in D^τ. Furthermore, we know by (1.81) that for any $x \in \Omega \cap \partial D^\tau$, $w^\tau > 0$. It follows easily that $w^\tau \not\equiv 0$ in every component of the open set D^τ. By the strong Maximal Principle it follows from (1.84) that $w^\tau > 0$ in D^τ. Indeed, if $x = (x_1, y)$ is any interior point of D^τ, then the half line $\{x_1 + t, y; t \geqslant 0\}$ hits ∂D^τ at a point $\bar{x} \in \partial\Omega^\tau \cap \partial\Omega$. This point $\bar{x}$ is on the boundary of that component of D^τ to which x belongs, and $u(\bar{x}) > 0$.

Now choose $\delta > 0$ and carve out of D^τ a closed set $K \subset D^\tau$ such that $|D^\tau \backslash K| < \delta/2$. We know that $w^\tau > 0$ in K. Hence, for small $\varepsilon > 0$, $w^{\tau - \varepsilon}$ is also positive on K. Moreover, for $\varepsilon > 0$ small, $|D^{\tau-\varepsilon} \backslash K| < \delta$. Since $\partial(D^{\tau-\varepsilon} \backslash K) \subset \partial D^{\tau-\varepsilon} \cup K$, we see that $w^{\tau-\varepsilon} \geqslant 0$ on $\partial(D^{\tau-\varepsilon} \backslash K)$. Thus, $w^{\tau-\varepsilon}$ satisfies

$$\begin{cases} \Delta w^{\tau-\varepsilon} + c^{\tau-\varepsilon}(x) w^{\tau-\varepsilon} = 0 & \text{in } D^{\tau-\varepsilon} \backslash K, \\ w^{\tau-\varepsilon} \geqslant 0 & \text{on } \partial(D^{\tau-\varepsilon} \backslash K). \end{cases} \tag{1.85}$$

It then follows from *Theorem 1.29* (see also [20]) that $w^{\tau-\varepsilon} \geqslant 0$ in $D^{\tau-\varepsilon} \backslash K$ and hence in all of $D^{\tau-\varepsilon}$. We have reached a contradiction and thus proved that u is monotone:

$$u^{\tau} > u \text{ in } D^{\tau}, \ \forall \tau > 0.$$

(Again this follows from the Eq. (1.84) since we know $w^{\tau} \geqslant 0$ and $w^{\tau} \not\equiv 0$ which implies $w^{\tau} > 0$.)

If, further more, f is differentiable, u_{x_1} satisfies a linear equation in Ω, by differentiation (1.79):

$$\Delta u_{x_1} + f'(u)u_{x_1} = 0 \text{ in } \Omega.$$

Since we already know that $u_{x_1} \geqslant 0$, $u_{x_1} \not\equiv 0$, we infer from this equation that $u_{x_1} > 0$ in Ω.

To prove the last assertion of the theorem suppose v is another solution. We argue exactly as before but instead of $w^{\tau} = u^{\tau} - u$ we now take $w^{\tau} = v^{\tau} - u$. The same proof shows that $v^{\tau} \geqslant u \ \forall \tau \geqslant 0$. Hence $v \geqslant u$, and by symmetry we have $v = u$.

$\square$

Next, we discuss an extension of the moving plane method to the case of an unbounded domain. Here we follow [14]. Let $\Omega = \{x \in \mathbb{R}^n \,|\, x_n > \varphi(x_1, \ldots, x_{n-1})\}$, where $\varphi : \mathbb{R}^{n-1} \to \mathbb{R}$ is a coercive Lipschitz graph, that is, they assume that

$$\lim_{\substack{|x| \to \infty \\ x \in \mathbb{R}^{n-1}}} \varphi(x) = +\infty. \tag{1.86}$$

Theorem 1.33 *If Ω is bounded by a coercive continuous graph (in the above understanding), then, for any locally Lipschitz function f and given any solution u of (1.86) -be it bounded or not-, it satisfies*

$$\frac{\partial u}{\partial x_n} > 0 \text{ in } \Omega.$$

Notice that for this result the graph φ needs not be globally Lipschitz. Note that, Esteban and Lions get this result for a smooth graph φ (see [53] and references therein). But using the improved version of the moving plane method in [17], it was possible to extend this result to the case of merely continuous graph φ. This result rests on the method of reflection in moving planes.

For $\lambda \in \mathbb{R}$, $T_\lambda = \{x_n = \lambda\}$, as before, is the *moving hyperplane*, $\lambda_0 = \inf\{\lambda \in \mathbb{R}; T_\lambda \cap \Omega \neq \varnothing\}$, (this number λ_0 is finite because of (1.86))

$$\Omega_\lambda = \{x \in \Omega_\lambda; x_n < \lambda\}$$

is the lower cap of Ω cut out by T_λ.

For $x \in \Omega_\lambda$, $x = (x_1, \ldots, x_n)$, let $x^\lambda = (x_1, \ldots, x_{n-1}, 2\lambda - x_n)$ denote its reflection in Ω,

$$v^\lambda(x) = u(x^\lambda),$$

$$w^\lambda(x) = u(x^\lambda) - u(x).$$

These two functions are defined in Ω_λ.

Notice that since Ω is defined as an epigraph, the cap Ω_λ is always induced in Ω and is not empty for all $\lambda > \lambda_0$. The main consequence of the coercivity assumption (1.86) is that Ω_λ is always *bounded*, for all values of λ. Actually, (1.86) is equivalent to the boundedness of Ω_λ for all λ. This is the reason why the moving planes method for bounded domains applies as such to epigraphs when (1.86) is satisfied.

Indeed, as usual, essentially, one wants to prove that $w^\lambda > 0$ in Ω_λ for all $\lambda > \lambda_0$. For this shows that u is nondecreasing with respect to x_n, hence that $\frac{\partial u}{\partial x_n} \geqslant 0$ in Ω. Since $\frac{\partial u}{\partial x_n}$ satisfies some linear equation, viz. $\Delta(\frac{\partial u}{\partial x_n}) + f'(u)\frac{\partial u}{\partial x_n} = 0$, one infers from the strong maximum principle that $\frac{\partial u}{\partial x_n} > 0$ in Ω.

To prove that $w^\lambda > 0$ in Ω_λ, one uses the Eq. (1.86): v^λ satisfies the same equation, $\Delta v^\lambda + f(v^\lambda) = 0$ in Ω_λ. Hence, the function w^λ satisfies some linear equation

$$\Delta w^\lambda + c_\lambda(x)w^\lambda = 0 \text{ in } \Omega_\lambda, \tag{1.87}$$

where

$$c_\lambda(x) = \frac{f(v^\lambda(x)) - f(u(x))}{v^\lambda(x) - u(x)}.$$

The function c_λ is bounded in $L^\infty(\Omega_\lambda)$, uniformly with respect to λ in bounded intervals. On the contrary, w^λ satisfies

$$w^\lambda \geqslant 0, \; w^\lambda \not\equiv 0 \text{ on } \partial\Omega_\lambda. \tag{1.88}$$

In fact, $w^\lambda = 0$ on T_λ whereas $w^\lambda > 0$ on the remaining part of the boundary, that is $\partial\Omega\backslash T_\lambda = \{x \in \partial\Omega, \; x_n < \lambda\}$.

We now use the method of [17]. It relies on the following version of the maximum principle in domains of small value (see *Theorem 1.29* and [20]).

Consider an elliptic operator

$$L = a_{ij}(x)\frac{\partial^2}{\partial x_i \partial x_j} + b_i(x)\frac{\partial}{\partial x_i} + c(x) \tag{1.89}$$

in a domain Ω. Assume that $a_{ij} \in C(\overline{\Omega})$, b_i, $c \in L^\infty(\Omega)$, that the operator is uniformly elliptic, that is:

$$a_{ij}(x)\xi_i\xi_j \geqslant C_0|\xi|^2 \quad \forall \xi \in \mathbb{R}^n, \ \forall x \in \Omega.$$

Let $b \geqslant \|b_i\|_{L^\infty(\Omega)}$, $\|c\|_{L^\infty(\Omega)}$. Let us recall that L satisfies the maximum principle in Ω if for any function w which satisfies

$$Lw \leqslant 0 \text{ in } \Omega$$

$$w \geqslant 0 \text{ on } \partial\Omega$$

one can infer that $w \geqslant 0$ in Ω.

As we discussed in *Sect. 1.4*, a sufficient condition for L to satisfy the maximum principle is that $c(x) \leqslant 0$.

In general, when $c(x)$ is allowed to be positive, the maximum principle does not hold. However, if the domain is sufficiently small, it still holds, see *Theorem 1.29*.

For a systematic study of conditions under which the maximum principle holds and for some general results related to it we refer also to [20].

Step 1 $w^\lambda > 0$ in Ω_λ for $0 < \lambda - \lambda_0$ small. Indeed, then, Ω_λ has small measure and since $\|c_\lambda\|_{L^\infty(\Omega_\lambda)}$ is bounded uniformly on compact sets of λ, the maximum principle applies (by *theorem 1.29* above). This could also be seen by using the fact that Ω_λ is *narrow* in the x_n direction, if $0 < \lambda - \lambda_0$ is small (see *Theorem 1.26* and [20]). Therefore, $w^\lambda > 0$ in Ω_λ if $0 < \lambda - \lambda_0$ is small.

Step 2 Suppose $w^\lambda > 0$ in Ω_λ for all values of λ in some interval (λ_0, μ). Then, by continuity, $w^\mu \geqslant 0$ in Ω_μ and since $w^\mu \not\equiv 0$, we infer from (1.87) that $w^\mu > 0$ in Ω_μ. Let K be a compact set included in Ω_μ such that the measure of $\Omega_\mu \setminus K$ is small, namely,

$$\text{meas}(\Omega_\mu \setminus K) < \frac{\delta}{2}$$

for some small enough δ. Then since $\min_{x \in K} w^\mu(x) = \eta > 0$, if $\varepsilon > 0$ is sufficiently small, by continuity, we see that for all values of $\lambda \in [\mu, \mu + \varepsilon]$,

$$\min_{x \in K} w^\lambda(x) \geqslant \frac{\eta}{2} > 0.$$

Hence, $w^\lambda > 0$ on K and, furthermore,

$$\text{meas}(\Omega_\lambda \setminus K) < \delta.$$

By the maximum principle for domains with small value (see *Theorem 1.29* and [17, 20]), if $\delta > 0$ is chosen sufficiently small, then the operator $\Delta + c_\lambda(x)$ satisfies the maximum principle in $\Omega_\lambda \setminus K$. Therefore, since w^λ satisfies

$$\Delta w^\lambda + c_\lambda(x)w^\lambda = 0 \quad \text{in} \quad \Omega_\lambda \backslash K$$
$$w^\lambda \geqslant 0, \; w^\lambda \not\equiv 0 \quad \text{on} \quad \partial\Omega$$
$$w^\lambda > 0 \quad \text{on} \quad K$$

we see that w^λ in $\Omega_\lambda \backslash K$, hence in all of Ω_λ.

We have thus shown that $w^\lambda > 0$ for all λ, not only in (λ_0, μ), but also in $(\lambda_0, \mu + \varepsilon]$.

We may now conclude. The above two steps show that $w^\lambda > 0$ for all $\lambda > \lambda_0$ and this implies that $\frac{\partial u}{\partial x_n} > 0$ in Ω.

As said earlier, this proof does not extend to more general epigraph domains when coercivity condition (1.86) is not satisfied. There are two difficulties. Firstly, λ_0 is not necessarily finite and, second, Ω_λ is not bounded any more for all values of λ and the problem becomes much more involved.

Next, observe that *Theorem 1.33* implies, in the case of a half space, a *symmetry* and monotonicity result. Indeed, consider the case of a half space $\Omega = \{x \in \mathbb{R}^n; x_n > 0\}$ which corresponds to a flat graph $\varphi \equiv 0$. Then, for any direction $\xi \in \mathbb{R}^n \backslash \{0\}$ transversal to Ω and entering Ω, that is, such that $\xi_n > 0$, for any system of coordinates $y_1, \ldots, y_n$, there is a $\eta_n = \xi_n$ and one can read the half space as defined by $\Omega = \{y_n > \psi(x_1, \ldots, x_n)$ with $\psi : \mathbb{R}^{n-1} \to \mathbb{R}$ Lipschitz$\}$.

Therefore, if f satisfies the following conditions

1. there exists a $\mu > 0$, such that $f > 0$ in $(0, \mu)$, $f(0) = f(\mu) = 0$;
2. $f(s) \geqslant \eta s$ for $s \in [0, \delta]$ for some $\eta, \delta > 0$;
3. f is nonincreasing in $[\mu - \delta, \mu]$, $\delta > 0$;
4. f is Lipschitz continuous on $[0, \mu]$;

and Ω is a half space, for any such direction ξ, *Theorem 1.33* implies

$$\frac{\partial u}{\partial \xi} > 0 \text{ in } \Omega.$$

This implies symmetry. Indeed, let $x, x' \in \Omega$ be two points at equal distance of $\partial\Omega$, i.e. $x_n = x'_n$.

Let $e_n = (0, 0, \ldots, 1)$. Choosing appropriate directions ξ, the previous monotonicity property yields

$$u(x) < u(x' + \varepsilon e_n)$$
$$u(x') < u(x + \varepsilon e_n)$$

hence, by letting $\varepsilon \to 0$,

$$u(x) = u(x').$$

1.8   Variational Solutions of Elliptic Equations

In this section, we formulate auxiliary results concerning the regularity of solutions to linear elliptic equations of the form (2.161), which we will use in *Chap. 2*.

Let $\omega \subset \mathbb{R}^n$ be a bounded polyhedral domain. Then $H_Q^2(\omega)$ denotes the space of all variational solutions $v \in H^1(\omega)$ to the problem

$$(1 - \Delta)v = g, \; \partial_n v|_{\partial\omega} = 0, \tag{1.90}$$

where the right-hand side g belongs to $L^2(\omega)$. $H_Q^2(\omega)$ is a Hilbert space in a natural manner.

For the moment, let $\mathcal{G}_0$ be the space of all variational solutions to the problem (1.90), with ω replaced by $\Omega_{T_1-1, T_2+1} = (T_1 - 1, T_2 + 1) \times \omega$. When speaking on Ω_{T_1, T_2}, we shall not consider Ω_{T_1, T_2} as a bounded polyhedral domain itself, but as piece of the cylinder $\mathbb{R} \times \omega$ over the bounded polyhedral domain ω. In slight abuse of notation we define:

Definition 1.11 $H_Q^2(\Omega_{T_1, T_2})$ is the space of all restrictions of functions from $\mathcal{G}_0$ to Ω_{T_1, T_2} with the norm

$$\|v, \Omega_{T_1, T_2}\|_{2,Q} \equiv \inf\{\|u\|_{\mathcal{G}_0} : u \in \mathcal{G}_0, \; u|_{\Omega_{T_1, T_2}} = v\}.$$

Let $H_Q^{3/2}(\omega)$ denote the space of traces on the set $\{t = 0\}$ of functions belonging to $H_Q^2(\Omega_0)$ with the norm

$$\|u_0, \omega\|_{3/2,Q} = \inf\{\|u, \Omega_0\|_{2,Q} : u|_{t=0} = u_0\}.$$

Definition 1.12 $H_{Q,\mathrm{loc}}^2(\Omega_+)$ denotes the Fréchet space of all distributions u on Ω_+ such that $u|_{\Omega_T}$ belongs to $H_Q^2(\Omega_T)$ for every $T \geqslant 0$.

$H_{Q,b}^2(\Omega_+)$ denotes the Banach space of all functions from $H_{Q,\mathrm{loc}}^2(\Omega_+)$ for which the following norm is finite:

$$\|u\|_b = \sup_{T \geqslant 0} \|u, \Omega_T\|_{2,Q}.$$

Set $V_0 = [H_Q^{3/2}(\omega)]^k$, $\Theta_0^+ = [H_{Q,\mathrm{loc}}^2(\Omega_+)]^k$, $F_0^+ = [H_{Q,b}^2(\Omega_+)]^k$, where $\Omega_+ := \mathbb{R}_+ \times \omega$.

Lemma 1.12 *Let u be a variational solution to the problem*

$$\begin{cases} \partial_t^2 u + \Delta u = g, \; \partial_n u|_{\partial\omega} = 0, \\ u|_{t=T_1} = u_1, \; u|_{t=T_2} = u_2, \end{cases}$$

where u_1, $u_2 \in V_0$ and $g \in L^2(\Omega_{T_1,T_2})$. Then $u \in H^2_Q(\Omega_{T_1,T_2})$. Moreover, the following estimate holds:

$$\|u, \Omega_{T_1,T_2}\|_{2,Q} \leqslant C(\|u_1, \omega\|_{3/2,Q} + \|u_2, \omega\|_{3/2,Q} + \|g, \Omega_{T_1,T_2}\|_{0,2}).$$

Proof By definition, there is a function $v \in H^2_Q(\Omega_{T_1,T_2})$ with $v|_{t=T_1} = u_1$, $v|_{t=T_2} = u_2$ such that

$$\|v, \Omega_{T_1,T_2}\|_{2,Q} \leqslant C(\|u_1, \omega\|_{3/2,Q} + \|u_2, \omega\|_{3/2,Q}).$$

We show that $w = u - v \in H^2_Q(\Omega_{T_1,T_2})$. The function w satisfies the equation

$$\begin{cases} \partial_t^2 w + \Delta w = g_1 \equiv g - (\partial_t^2 v + \Delta v) \in L^2(\Omega_{T_1,T_2}), \\ w|_{t=T_1} = w|_{t=T_2} = 0, \partial_n w|_{\partial\omega} = 0. \end{cases}$$

Take a cut-off function $\phi \in C_0^\infty(\mathbb{R})$ with $\phi(t) = 1$ for $t \in (T_1, T_2)$ and $\phi(t) = 0$ for $t \notin (T_1 - \varepsilon, T_2 + \varepsilon)$, where $0 < \varepsilon < T_2 - T_1$. It is readily seen that the function

$$W(t) = \phi(t)\tilde{w}(t) \equiv \phi(t) \begin{cases} -w(2T_1 - t) \text{ for } t \in (T_1 - 1, T_1), \\ w(t) \text{ for } t \in (T_1, T_2), \\ -w(2T_2 - t) \text{ for } t \in (T_2, T_2 + 1) \end{cases}$$

belongs to $\mathcal{G}_0$. Indeed,

$$\partial_t^2 W + \Delta W = \phi(t)\tilde{g}_1(t) + 2\phi'(t)\partial_t \tilde{w}(t) + \phi''(t)\tilde{w}(t) \in L^2(\Omega_{T_1-1,T_2+1}),$$

where $\tilde{g}_1$ is defined similarly to $\tilde{w}$. In addition, W satisfies the appropriate boundary conditions. Hence, according to *Definition 1.11*, w belongs to $H^2_Q(\Omega_{T_1,T_2})$. □

Theorem 1.34 *The space $H^2_Q(\Omega_{T_1,T_2}) \cap L^\infty(\Omega_{T_1,T_2})$ is dense in $H^2_Q(\Omega_{T_1,T_2})$.*

Proof It suffices to prove that $\mathcal{G}_0 \cap L^\infty(\Omega_{T_1-1,T_2+1})$ is dense in $\mathcal{G}_0$. Take a function $u \in \mathcal{G}_0$ and a function $g \in L^2(\Omega_{T_1-1,T_2+1})$ which satisfy (1.90). Let $\{g_m\} \subset L^\infty(\Omega_{T_1-1,T_2+1})$ be a sequence such that

$$g_m \to g \text{ in } L^2(\Omega_{T_1-1,T_2+1}) \text{ as } m \to \infty. \tag{1.91}$$

Let $u_m \in \mathcal{G}_0$ be the variational solution to (1.90) with right-hand side g_m. It follows from (1.91) and the definition of the space $\mathcal{G}_0$ that

$$u_m \to u \text{ in } \mathcal{G}_0 \text{ as } m \to \infty.$$

Hence *Theorem 1.34* will be proved when we have shown that $u_m \in L^\infty(\Omega_{T_1-1,T_2+1})$.

We use the following maximum principle.

Lemma 1.13 *Let $u_i \in H^1(\Omega_{T_1-1,T_2+1})$ for $i = 1, 2$ be variational solutions to (1.90) with right-hand sides $g_i \in H^1(\Omega_{T_1-1,T_2+1})^*$. Assume, in addition, that*

$$\langle g_1, \Phi \rangle \geq \langle g_2, \Phi \rangle \text{ for all } \Phi \in H^1(\Omega_{T_1-1,T_2+1}), \quad \Phi \geq 0.$$

Then

$$u_1(t,x) \geq u_2(t,x) \text{ for } (t,x) \in \Omega_{T_1-1,T_2+1} \text{ a.e.}.$$

Proof Consider the function $u = u_2 - u_1$. Then

$$\langle \partial_t u, \partial_t \Phi \rangle + \langle \nabla u, \partial_t \Phi \rangle + \langle u, \Phi \rangle \geq 0 \text{ for all } \Phi \in H^1(\Omega_{T_1-1,T_2+1}), \quad \Phi \geq 0. \tag{1.92}$$

Now introduce $u_+(t,x) = \max\{u, 0\}$, $u_-(t,x) = \max\{-u, 0\}$. Then $u = u_+ - u_-$. It is known that u_+, $u_- \in H^1(\Omega_{T_1-1,T_2+1})$ and

$$\langle u_+, u_- \rangle = 0, \quad \langle \partial_t u_+, \partial_t u_- \rangle + \langle \nabla u_+, \nabla u_- \rangle = 0$$

(see [104]). Upon replacing Φ in (1.92) with u_-, we obtain

$$-\langle \partial_t u_-, \partial_t u_- \rangle - \langle \nabla u_-, \nabla u_- \rangle - \langle u_-, u_- \rangle = 0.$$

This implies that $\langle u_-, u_- \rangle = 0$ or $u_-(t,x) = 0$ for $(t,x) \in \Omega_{T_1-1,T_2+1}$ a.e. The lemma is proved. □

Corollary 1.10 *Let $u \in H^1(\Omega_{T_1-1,T_2+1})$ be a variational solution to (1.90) with right-hand side $g \in L^\infty(\Omega_{T_1-1,T_2+1})$. Then $u \in L^\infty(\Omega_{T_1-1,T_2+1})$.*

This completes the proof of *Theorem 1.34*. □

Theorem 1.35 *The embedding*

$$H_Q^2(\Omega_{T_1,T_2}) \subset L^q(\Omega_{T_1,T_2}) \tag{1.93}$$

holds for

$$q \leq q_0 = 2\frac{n+1}{n-3}.$$

If $q < q_0$, then this embedding is compact. Furthermore, if $u \in H_Q^2(\Omega_{T_1,T_2})$, then $u|u|^{(q_0-2)/2} \in H^1(\Omega_{T_1,T_2})$ and

$$\big\||u|u|^{(q_0-2)/2}, \Omega_{T_1,T_2}\big\|_{1,2} \leqslant C\|u, \Omega_{T_1,T_2}\|_{2,Q}^{q_0/2}. \tag{1.94}$$

Proof Let $u \in H_Q^2(\Omega_{T_1,T_2})$. By definition, there is a function $\tilde{u} \in H_Q^2(\Omega_{T_1-1,T_2+1})$, $\tilde{u}|_{\Omega_{T_1-1,T_2+1}} = u$, such that

$$\langle \partial_t \tilde{u}, \partial_t \Phi \rangle + \langle \nabla \tilde{u}, \partial_t \Phi \rangle + \langle \tilde{u}, \Phi \rangle = \langle \tilde{g}, \Phi \rangle \text{ for all } \Phi \in H^1(\Omega_{T_1-1,T_2+1}) \tag{1.95}$$

for a certain right-hand side $\tilde{g} \in L^2(\Omega_{T_1-1,T_2+1})$ and

$$\|u, \Omega_{T_1-1,T_2+1}\|_{2,Q} \leqslant C\|\tilde{g}, \Omega_{T_1-1,T_2+1}\|_{0,2}.$$

Approximate $\tilde{g} \in L^2(\Omega_{T_1-1,T_2+1})$ by a sequence $\{\tilde{g}_m\} \subset L^\infty(\Omega_{T_1-1,T_2+1})$, $\tilde{g}_m \to \tilde{g}$ in $L^2(\Omega_{T_1-1,T_2+1})$ as $m \to \infty$. Let $\tilde{u}_m$ be the solution to the variational problem (1.95) with the right-hand side $\tilde{g}$ replaced with $\tilde{g}_m$. Then $\tilde{u}_m \in L^\infty(\Omega_{T_1-1,T_2+1})$ by *Corollary 1.10*. Hence the function $\tilde{u}_m|\tilde{u}_m|^{l-2}$ belongs to $H^1(\Omega_{T_1-1,T_2+1})$, where $l_0 - 2 = 4/(n-3) = (q_0 - 2)/2$. Replacing $\tilde{u}$ by $\tilde{u}_m$ and Φ by $\tilde{u}_m|\tilde{u}_m|^{l_0-2}$ in (1.95) and arguing as in the proof of estimate (2.180) we obtain the inequality

$$\big\|\tilde{u}_m|\tilde{u}_m|^{(l_0-2)/2}, \Omega_{T_1-1,T_2+1}\big\|_{1,2}^2 \leqslant C(1 + |\langle \tilde{g}_m, \tilde{u}_m|\tilde{u}_m|^{l_0-2}\rangle|). \tag{1.96}$$

By Sobolev's embedding theorem, $H^1(\Omega_{T_1-1,T_2+1}) \subset L^{2n/(n-2)}(\Omega_{T_1-1,T_2+1})$. It follows that

$$\|\tilde{u}_m, \Omega_{T_1-1,T_2+1}\|_{0,q_0}^{l_0} = \big\|\tilde{u}_m|\tilde{u}_m|^{(l_0-2)/2}, \Omega_{T_1-1,T_2+1}\big\|_{0,2n/(n-2)}^2$$

$$= C\big\|\tilde{u}_m|\tilde{u}_m|^{(l_0-2)/2}, \Omega_{T_1-1,T_2+1}\big\|_{1,2}^2.$$

By Hölder's inequality, we further obtain

$$|\langle \tilde{g}_m, \tilde{u}_m|\tilde{u}_m|^{l_0-2}\rangle| \leqslant \|g, \Omega_{T_1-1,T_2+1}\|_{0,2}\|\tilde{u}_m, \Omega_{T_1-1,T_2+1}\|_{0,q_0}^{l_0-1}$$

$$\leqslant \mu\|\tilde{u}_m, \Omega_{T_1-1,T_2+1}\|_{0,q_0}^{l_0} + C_\mu\|g_m, \Omega_{T_1-1,T_2+1}\|_{0,2}^{l_0},$$

where $\mu > 0$ is arbitrary. Applying these estimates to (1.96) for a sufficiently small $\mu > 0$, we get

$$\|\tilde{u}_m, \Omega_{T_1-1,T_2+1}\|_{0,q_0}^{l_0} + \big\|\tilde{u}_m|\tilde{u}_m|^{(l_0-2)/2}, \Omega_{T_1-1,T_2+1}\big\|_{1,2}^2 \leqslant C\|g_m, \Omega_{T_1-1,T_2+1}\|_{0,2}^{l_0}.$$

We know that $\tilde{g}_m \to \tilde{g}$ in $L^2(\Omega_{T_1-1,T_2+1})$, hence the sequence $\{\tilde{u}_m\}$ is bounded in the space $L^{q_0}(\Omega_{T_1-1,T_2+1})$. Without loss of generality, we may assume that $\tilde{u}_m \to \tilde{u}$ weakly in $L^{q_0}(\Omega_{T_1-1,T_2+1})$. Thus $\tilde{u} \in L^{q_0}(\Omega_{T_1-1,T_2+1})$ and

$$\|u, \Omega_{T_1,T_2}\|_{0,q_0} \leqslant \|\tilde{u}, \Omega_{T_1-1,T_2+1}\|_{0,q_0} \leqslant C\|\tilde{g}, \Omega_{T_1-1,T_2+1}\|_{0,2} \leqslant C_1\|u, \Omega_{T_1,T_2}\|_{2,Q}.$$

The statement $u|u|^{l_0-2} \in H^1(\Omega_{T_1,T_2})$ is shown analogously.

We finally prove the compactness of embedding (1.93) for $q < q_0$. By the interpolation inequality between H_1 and L^{q_0},

$$H_Q^2(\Omega_{T_1,T_2}) \subset H^{\varepsilon,q_0}(\Omega_{T_1,T_2})$$

for some $\varepsilon > 0$. The assertion follows, since the embedding $H^{\varepsilon,q_0} \subset L^q$ is compact.

The proof of *Theorem 1.35* is complete. □

Corollary 1.11 *The embedding*

$$H_Q^2(\Omega_{T_1,T_2}) \subset C([T_1, T_2], L^{p_0}(\omega))$$

holds. Here $p_0 = 2l_0 = 2 + 4/(n-3)$ is the maximum value of p in (2.162).

In fact, the estimate (1.94) together with Sobolev's embedding theorem imply that $u|u|^{(l_0-2)/2} \in C([T_1, T_2], L^2(\omega))$ if $u \in H_Q^2(\Omega_{T_1,T_2})$. Moreover, we infer from the embedding $H_Q^2 \subset H^1$ that $u \in C([T_1, T_2], L^2(\omega))$. Now, employing standard arguments, we finally open that $u \in C([T_1, T_2], L^{p_0}(\omega))$.

Theorem 1.36 *Let $u \in H_Q^2(\Omega_{T_1,T_2})$. Then $\partial_t^2 u \in L^2(\Omega_{T_1,T_2})$, $\partial_t \nabla u \in L^2(\Omega_{T_1,T_2})$. Moreover, the following estimate holds:*

$$\|\partial_t^2 u, \Omega_{T_1,T_2}\|_{0,2} + \|\partial_t \nabla u, \Omega_{T_1,T_2}\|_{0,2} \leqslant C\|u, \Omega_{T_1,T_2}\|_{2,Q}. \tag{1.97}$$

Proof By definition, there exists a function $\tilde{u} \in \mathcal{G}_0$ that satisfies $\tilde{u}|_{\Omega_{T_1,T_2}} = u$ and

$$\begin{cases} \partial_t^2 \tilde{u} + \Delta \tilde{u} - \tilde{u} = g(x), \quad \partial_n \tilde{u}|_{\partial\omega} = 0, \\ \tilde{u}|_{t=T_1-1} = 0, \ \tilde{u}|_{t=T_2+1} = 0 \end{cases} \tag{1.98}$$

for some function $g \in L^2(\Omega_{T_1-1,T_2+1})$. $\|g, \Omega_{T_1,T_2}\|_{0,2} \leqslant C\|u, \Omega_{T_1,T_2}\|_{2,Q}$. Below we only give the formal arguments for deriving estimate (1.97). A rigorous proof can be supplied by exploiting the Galerkin approximation method.

Multiplying Eq. (1.98) by $\partial_t^2 \tilde{u}$ and integrating over Ω_{T_1-1,T_2+1}, we obtain after an integration by parts

$$\langle |\partial_t^2 \tilde{u}|^2, 1\rangle + \langle |\partial_t \nabla \tilde{u}|^2, 1\rangle + \langle |\partial_t \tilde{u}|^2, 1\rangle = \langle g, \partial_t^2 \tilde{u}\rangle. \tag{1.99}$$

Applying Hölder's inequality

$$\langle g, \partial_t^2 \tilde{u}\rangle \leqslant \frac{1}{2}\langle |g|^2, 1\rangle + \frac{1}{2}\langle |\partial_t^2 \tilde{u}|^2, 1\rangle$$

to the right-hand side in (1.99), we then find (1.97). The proof is finished. □

From *Theorem 1.36* we conclude:

Corollary 1.12 *We have*

$$H_Q^2(\Omega_{T_1,T_2}) \subset H^1((T_1, T_2), H^1(\omega)) \cap H^2((T_1, T_2), L^2(\omega)).$$

In particular, the functions $t \mapsto \|u(t)\|_{1,2}$ and $t \mapsto \|\partial_t u(t)\|_{0,2}$ are defined and continuous for every $u \in H_Q^2(\Omega_{T_1,T_2})$.

Corollary 1.13 *Furthermore,*

$$H_Q^2(\Omega_{T_1,T_2}) = H^2((T_1, T_2), L^2(\omega)) \cap L^2((T_1, T_2), H_Q^2(\omega)). \tag{1.100}$$

Remark 1.20 For smooth domains $\omega \subset \mathbb{R}^n$, all previous results of this section are consequences of L^2-elliptic regularity for the Laplace operator (e.g., [101]), especially the fact

$$H_Q^2(\Omega_{T_1,T_2}) = \{u \in H^2(\Omega_{T_1,T_2}) : \partial_n u|_{\partial\omega} = 0\} \tag{1.101}$$

and Sobolev's embedding theorem. For polyhedral domains ω, however, (1.101) is in general not fulfilled.

The following deep result may be found in [40].

Theorem 1.37 *Let $\omega \subset \mathbb{R}^n$ be a bounded polyhedral domain. Then there exists an ε satisfying $0 < \varepsilon \leqslant 1/2$ such that*

$$H_Q^2(\omega) \subseteq H^{3/2+\varepsilon}(\omega). \tag{1.102}$$

Corollary 1.14 *Let $\omega \subset \mathbb{R}^n$ be a bounded polyhedral domain. Then*

$$H_Q^2(\Omega_{T_1,T_2}) \subseteq H^{3/2+\varepsilon}(\Omega_{T_1,T_2}), \tag{1.103}$$

where ε is the same as in Theorem 1.37.

(1.103) is actually a consequence of (1.100), (1.102).

Corollary 1.15 *Let $u \in H_Q^2(\Omega_{T_1,T_2})$. Then*

$$\partial_n u|_{\partial\omega} \in H^\varepsilon((T_1, T_2) \times \partial\omega).$$

This follows from (1.103) and Sobolev's embedding theorem. In particular, using Green's formula (see [78]) we obtain that $\partial_n u|_{\partial\omega} = 0$ for $u \in H_Q^2(\Omega_{T_1,T_2})$. Thus solutions u to the problem (2.161) which belong to Θ_0^+ satisfy the homogeneous Neumann boundary condition in the proper sense.

1.9 Elliptic Regularity for the Neumann Problem for the Laplace Operator on an Infinite Edge

In this section, we discuss elliptic regularity for the Neumann problem for the Laplace operator on an infinite cone $\Gamma \subset \mathbb{R}^2$ and the infinite wedge $\mathbb{R} \times \Gamma \subset \mathbb{R}^3$. We will use the result of this section in *Chap. 2* (see *Sect. 2.13*) while studying the trajectory attractor for an elliptic system in the half cylinder $\Omega_+ = \mathbb{R}_+ \times \omega$, where ω is a bounded polyhedral in $\mathbb{R}^n$. In this section we mainly follow [91], preserving the notations of the authors for the convenience of the reader. It is worth to note that the results of this section are of independent interest. Let $\Gamma \subset \mathbb{R}^2$ be an open cone with angle α. Throughout we shall suppose that $\Gamma = \{(r, \theta); \ 0 < \theta < \alpha\}$. Here (r, θ) denote polar coordinates in $\mathbb{R}^2$. We further suppose that $\alpha > \pi$ (see Remark 1.22 (a)).

Since the model cone Γ arises from flatting out the boundary of ω near a fixed conical point of $\partial \omega$, we shall consider operators $\mathrm{Id} - \Delta_y - M(y, \partial_y)$ on Γ, where $y = (y_1, y_2)$ are Euclidian coordinates in $\mathbb{R}^2$ and $M(y, \partial_y) = \sum_{|\gamma| \leq 2} b_\gamma(y) \partial_y^\gamma$ is a second-order partial differential operator subject to the following conditions: For $\gamma \in \mathbb{N}^2$, $|\gamma| = 2$, $b_\gamma \in C^\infty(\overline{\Gamma})$ and

(a1) $\|b_\gamma\|_{L^\infty(\Gamma)} \leq \delta$, $b_\gamma(0) = 0$;
(a2) $\|\nabla_y b_\gamma\|_{L^\infty(B_{\delta/K} \cap \Gamma)} \leq K$,
    where $B_\varrho = \{y \in \mathbb{R}^2; \ |y| \leq \varrho\}$. For $\gamma \in \mathbb{N}^2$, $|\gamma| \leq 1$, $b_\gamma = b_{\gamma 1} + b_{\gamma 2}$, where $b_{\gamma 1}, b_{\gamma 2} \in C^\infty(\overline{\Gamma})$ and
(b1) $\mathrm{supp} b_{\gamma 1} \subseteq B_\delta \cap \overline{\Gamma}$, $\|b_{\gamma 1}\|_{L^\infty(\Gamma)} \leq K$;
(b2) $\|b_{\gamma 2}\|_{L^\infty(\Gamma)} \leq \delta$

for some constant $K > 1$ and a certain $\delta = \delta(K)$ sufficiently small, $0 < \delta < 1$.

Let $H_\varrho^2(\Gamma)$ denote the space of all variational solutions v to the problem

$$(\mathrm{Id} - \Delta_y - M(y, \partial_y))v = g, \quad \partial_n v|_{\partial \Gamma} = 0 \tag{1.104}$$

with right-hand side $g \in L^2(\Gamma)$ (see *Sect. 1.8*).

Remark 1.21 It is a well-known fact that the space $H_\varrho^2(\Gamma)$ is actually independent of the choice of the operator $M(y, \partial_y)$ satisfying (a), (b) provided that $\delta > 0$ is small enough (see [83, 89]).

Moreover, from *Theorem 1.37* and its corresponding version for a model cone it follows that a solution to (1.104) belongs to $H^{3/2+\varepsilon}(\Gamma)$ for a certain $\varepsilon > 0$.

For the special case $M(y, \partial_y) \equiv 0$, it is readily seen that

$$H_\varrho^2(\Gamma) = H_N^2(\Gamma) \oplus \mathrm{span}\{S\}, \quad S(y) = \psi(r) r^{\pi/\alpha} \cos(\pi \theta / \alpha), \tag{1.105}$$

where $H_N^2(\Gamma) = \{v \in H^2(\Gamma); \ \partial_n v|_{\partial \Gamma} = 0\}$ (see [40, 83]). Here $\psi \in C_0^\infty(\overline{\Gamma})$ is some fixed cut-off function, depending only on the radial coordinate r, such that

$\psi(r) = 1$ in a neighbourhood of 0 and ψ is supported sufficiently close to 0. Notice that $S \in H^{1+\pi/\alpha-\varepsilon}(\Gamma)$ for any $\varepsilon > 0$, but $S \notin H^{1+\pi/\alpha}(\Gamma)$.

Lemma 1.14 *For $\delta > 0$ sufficiently small (depending on K), the differential operator*

$$\mathrm{Id} - \Delta_y - M(y, \partial_y) : H_Q^2(\Gamma) \to L^2(\Gamma) \tag{1.106}$$

induces an isomorphism, where $H_Q^2(\Gamma)$ is the space given in (1.105). Moreover, we have the estimate

$$\|v\|_{H_Q^2(\Gamma)} \leqslant C \|(\mathrm{Id} - \Delta_y - M(y, \partial_y))v\|_{L^2(\Gamma)} \tag{1.107}$$

for $v \in H_Q^2(\Gamma)$, where the constant $C > 0$ only depends on δ, K.

Proof It is known that $\mathrm{Id} - \Delta$ is an isomorphism from $H_Q^2(\Gamma)$ onto $L^2(\Gamma)$. Furthermore it is seen that $M(y, \partial_y)$ maps $H_Q^2(\Gamma)$ into $L^2(\Gamma)$, and it can be shown that

$$\|M(y, \partial_y)\|_{H_Q^2(\Gamma) \to L^2(\Gamma)} \leqslant C(K)\delta^{1/2} \tag{1.108}$$

with some constant $C(K) > 0$. To prove (1.108) it suffices to observe that

$$\left\| \sum_{|\gamma|=2} b_\gamma \partial_y^\gamma S \right\|_{L^2(\Gamma)} \leqslant C K^{1-\pi/\alpha} \delta^{\pi/\alpha}$$

and

$$\left\| \sum_{|\gamma| \leqslant 1} b_{\gamma 1} \partial_y^\gamma v \right\|_{L^2(\Gamma)} \leqslant \sum_{|\gamma| \leqslant 1} \|b_{\gamma 1}\|_{L^4(\Gamma)} \|v\|_{H^{1,4}(\Gamma)} \leqslant C K \delta^{1/2} \|v\|_{H_Q^2(\Gamma)}.$$

In fact, for $|\gamma| = 2$, we have $b_\gamma(y) = r\tilde{b}_\gamma(y)$ with $\tilde{b}_\gamma(y) = \int_0^1 (y/r) \cdot \nabla_y b_\gamma(sy)\, ds$ such that $\|\tilde{b}_\gamma\|_{L^\infty(B_{\delta/K} \cap \Gamma)} \leqslant K$ and

$$\int_{B_1 \cap \Gamma} |b_\gamma(y) \partial_y^\gamma S(y)|^2\, dy \leqslant C \int_0^{\delta/K} \int_0^\alpha |\tilde{b}_\gamma(y)|^2 r^{2\pi/\alpha-1}\, d\theta dr$$

$$+ C \int_{\delta/K}^1 \int_0^\alpha |b_\gamma(y)|^2 r^{2\pi/\alpha-3}\, d\theta dr$$

$$\leqslant C \left(K^2 (\delta/K)^{2\pi/\alpha} + \delta^2 (\delta/K)^{2(\pi/\alpha-1)} \right)$$

$$= C K^{2(1-\pi/\alpha)} \delta^{2\pi/\alpha}.$$

Moreover, $H_Q^2(\Gamma) \subset H^{1,4}(\Gamma)$ follows from the explicit description of $H_Q^2(\Gamma)$, $H^{1,4}(\Gamma)$ and Sobolev's embedding theorem.

Now choose $\delta > 0$ dependent on K so small that

$$\|M(y, \partial_y)\|_{H_Q^2(\Gamma) \to L^2(\Gamma)} < \|(\mathrm{Id} - \Delta)^{-1}\|_{L^2(\Gamma) \to H_Q^2(\Gamma)},$$

where $(\mathrm{Id} - \Delta)^{-1}$ stands for the inverse to $\mathrm{Id} - \Delta : H_Q^2(\Gamma) \to L^2(\Gamma)$. Then the differential expression $\mathrm{Id} - \Delta - M(y, \partial_y)$ in (1.106) induces an isomorphism.

The estimate (1.107) immediately follows. □

Remark 1.22

(a) The same argumentation yields that $H_Q^2(\Gamma) = H_N^2(\Gamma)$ when $\alpha < \pi$. In subsequent discussion we always assume that $\alpha > \pi$.
(b) From (1.105) it follows that each $v \in H_Q^2(\Gamma)$ can uniquely be represented in the form

$$v = v_0 + dS, \tag{1.109}$$

where $v_0 \in H_N^2(\Gamma), d \in \mathbb{C}$. It is important to observe that the coefficient d in (1.109) is independent of the particular cut-off function ψ, i.e., choosing another cut-off function possessing the same properties as ψ we obtain d as before.

Next we discuss the space $H_Q^2(\mathbb{R} \times \Gamma)$ of variational solutions v to

$$(\mathrm{Id} - \partial_t^2 - \Delta y - M(y, \partial_y))v = g, \quad \partial_n v|_{\mathbb{R} \times \partial \Gamma} = 0$$

with right-hand side $g \in L^2(\mathbb{R} \times \Gamma)$, where $M(y, \partial_y)$ is a second-order partial differential operator as above, but satisfying the additional conditions

$$\operatorname{supp} b_{\gamma 1} \subseteq B_{\delta 2/K} \cap \overline{\Gamma}, \quad b_{\gamma 2}(y) \equiv 0 \text{ for } |\gamma| \leq 1. \tag{1.110}$$

Again it turns out that the space $H_Q^2(\mathbb{R} \times \Gamma)$ is independent of the operator $M(y, \partial_y)$ provided that $\delta > 0$ is small enough.

We need the following result in the cases $s = 2, s = 0$. For a proof, see [40, 89].

Lemma 1.15 *Let $\Gamma \subset \mathbb{R}^2$ be an open cone, $s \in \mathbb{R}$. Then an equivalent norm on $H^s(\mathbb{R} \times \Gamma)$ is given by*

$$\|u\|_{H^s(\mathbb{R} \times \Gamma)} = \left\{ \int_{-\infty}^{\infty} \langle \tau \rangle^{2s} \|\kappa(\tau)^{-1} \hat{u}(\tau)\|_{H^s(\Gamma)}^2 \, d\tau \right\}^{1/2},$$

where $\hat{u}(\tau) = F_{t \to \tau} u(\tau), \kappa(\tau) = \kappa_{\langle \tau \rangle}, \langle \tau \rangle = (1 + |\tau|^2)^{1/2}$ and

$$\kappa_\lambda u(y) = \lambda u(\lambda y), \ \lambda > 0, \ y \in \Gamma,$$

for $u \in H^s(\Gamma)$.

Notice that $\{\kappa_\lambda\}_{\lambda>0}$ is a strongly continuous group on $H^s(\Gamma)$. It consists of isometries when $s = 0$.

Lemma 1.16 *Let $\Gamma \subset \mathbb{R}^2$ be an open cone as above. Then we have*

$$H^2_Q(\mathbb{R} \times \Gamma)$$
$$= H^2_N(\mathbb{R} \times \Gamma) \oplus \{F^{-1}_{\tau \to t}\{\langle\tau\rangle\psi(r\langle\tau\rangle)(r\langle\tau\rangle)^{\pi/\alpha}\cos(\pi\theta/\alpha)\hat{d}(\tau)\}; \ d \in H^2(\mathbb{R})\}, \tag{1.111}$$

where $H^2_N(\mathbb{R} \times \Gamma) = \{v \in H^2(\mathbb{R} \times \Gamma); \ \partial_n v|_{\mathbb{R} \times \partial\Gamma} = 0\}$.

Proof Let v be a solution to (1.109) with right-hand side $g \in L^2(\mathbb{R} \times \Gamma)$. Upon applying the Fourier transformation $F_{t \to \tau}$ and afterwards the group action $\kappa(\tau)^{-1}$ we obtain the equation

$$\begin{cases} (\mathrm{Id} - \Delta - M_\tau(y, \partial_y))\kappa(\tau)^{-1}\hat{v}(\tau) = \langle\tau\rangle^{-2}\kappa(\tau)^{-1}\hat{g}(\tau) \text{ in } \Gamma, \\ \partial_n u(\kappa(\tau)^{-1}\hat{v}(\tau))|_{\partial\Gamma} = 0 \end{cases} \tag{1.112}$$

with parameter $\tau \in \mathbb{R}$, where $M_\tau(y, \partial_y) = \langle\tau\rangle^{-2}M(\langle\tau\rangle^{-1}y, \langle\tau\rangle\partial_y)$. Now it is seen that the operator $M_\tau(y, \partial_y) = \sum_{1 \le |\gamma| \le 2}\langle\tau\rangle^{-2+|\gamma|}b_\gamma(\langle\tau\rangle^{-1}y)\partial_y^\gamma$ satisfies requirements (a), (b) with the same $\delta > 0$, $K > 1$ as before. Indeed, for $|\gamma| \le 1$, we put $b_{\gamma1,\tau}(y) = \langle\tau\rangle^{-1}b_\gamma(\langle\tau\rangle^{-1}y)$, $b_{\gamma2,\tau}(y) = 0$ if $\langle\tau\rangle \le K/\delta$ and $b_{\gamma1,\tau}(y) = 0$, $b_{\gamma2,\tau}(y) = \langle\tau\rangle^{-1}b_\gamma(\langle\tau\rangle^{-1}y)$ if $\langle\tau\rangle > K/\delta$.

Hence we conclude from Eq. (1.112) together with (1.104), (1.109) that

$$\kappa(\tau)^{-1}\hat{v}(\tau) = \kappa(\tau)^{-1}\hat{v}_0(\tau) + \hat{d}(\tau)S(y), \ S(y) = \psi(r)r^{\pi/\alpha}\cos(\pi\theta/\alpha). \tag{1.113}$$

Moreover, from (1.107) we derive the estimate

$$\|\kappa(\tau)^{-1}\hat{v}_0(\tau)\|^2_{H^2(\Gamma)} + |\hat{d}(\tau)|^2 \le C\langle\tau\rangle^{-4}\|\kappa(\tau)^{-1}\hat{g}(\tau)\|^2_{L^2(\Gamma)}.$$

Therefore, we get

$$\int_{-\infty}^\infty \langle\tau\rangle^4\|\kappa(\tau)^{-1}\hat{v}_0(\tau)\|^2_{H^2(\Gamma)}\,d\tau + \int_{-\infty}^\infty \langle\tau\rangle^4|\hat{d}(\tau)|^2\,d\tau$$
$$\le C\int_{-\infty}^\infty \|\kappa(\tau)^{-1}\hat{g}(\tau)\|^2_{L^2(\Gamma)}\,d\tau = C\|g\|^2_{L^2(\mathbb{R}\times\Gamma)}$$

showing that $v_0 \in H^2(\mathbb{R} \times \Gamma)$, $d \in H^2(\mathbb{R})$ by *Lemma 1.15*. From (1.113) we finally get

$$v = v_0 + F_{\tau \to t}^{-1}\{\hat{d}(\tau)(\kappa(\tau)S)(y)\}$$

which gives us the decomposition (1.111), observing that the sum on the right-hand side of (1.111) is direct and is obviously contained in $H_Q^2(\mathbb{R} \times \Gamma)$. $\qquad\square$

Remark 1.23 The proof of *Lemma 1.16* shows that

$$\|u\|'_{H_Q^2(\mathbb{R} \times \Gamma)} = \left\{ \int_{-\infty}^{\infty} \langle \tau \rangle^4 \|\kappa(\tau)^{-1}\hat{u}(\tau)\|^2_{H_Q^2(\Gamma)} \, d\tau \right\}^{1/2}$$

is an equivalent norm on $H_Q^2(\mathbb{R} \times \Gamma)$. Since $H_Q^2(\Gamma)$ is a cone Sobolev space of functions possessing asymptotics of a certain discrete asymptotic type near $y = 0$, $H_Q^2(\mathbb{R} \times \Gamma)$ is in fact a wedge Sobolev space (see [88–90]).

After these preparations, we are now in position to discuss elliptic regularity and asymptotics for the cylinder $\mathbb{R} \times \omega$ and the half-cylinder $\Omega_+ = \mathbb{R}_+ \times \omega$. Let ω be a bounded, polyhedral domain in $\mathbb{R}^2$. The boundary $\partial \omega$ is in particular smooth except for a finite number of conical points. Only the conical points with an obtuse angle deserve further interest, for H^2-regularity holds up to conical points with an acute angle (see Remark 1.22 (a)).

Let $\{b_1, \ldots, b_\kappa\}$ denote the set of these conical points. Let α_j be the angle at b_j, $\alpha_j > \pi$. For every j, $1 \le j \le \kappa$, we choose an open cone $\Gamma_j \subset \mathbb{R}^2$, open subsets U_j, V_j in $\mathbb{R}^2$ with $U_j \ni b_j$, $V_j \ni 0$, and a diffeomorphism $\chi_j : U_j \to V_j$ such that $\chi_j(b_j) = 0$ and $\chi_j(\overline{\omega} \cap U_j) = \overline{\Gamma}_j \cap V_j$. We assume that $\Gamma_j = \{(r, \theta); 0 < \theta < \alpha_j\}$. Furthermore, we suppose that the diffeomorphism χ_j are chosen in such a manner that $\chi_j'(x)^t \chi_j'(x)$ is a positive scalar multiple of the identity for $x \in \partial \omega \cap U_j$. Moreover, $\chi_j'(b_j) \in SO(2; \mathbb{R})$. It can be shown that such a choice is possible, if U_j is sufficiently small. Then the homogeneous Neumann boundary condition is preserved under the diffeomorphisms χ_j.

Notice that the assumption implies that

$$(\chi_j)_*\Delta = \Delta + M_j(y, \partial_y)$$

close to $y = 0$, where $M_j(y, \partial_y)$ is a second-order differential operator (without zero-order terms). By shrinking U_j, if necessary, we may suppose that $M_j(y, \partial_y)$ satisfies, for $\Gamma = \Gamma_j$, $K = 1$ and $\delta > 0$ sufficiently small, the assumptions (a), (b) previous to *Lemma 1.14* as well as condition (1.110).

Let $U_0 \subset \mathbb{R}^2$ be an open set not meeting $\{b_1, \ldots, b_\kappa\}$ such that $\{U_0\} \cup \{U_j\}_{j=1}^\kappa$ forms an open covering of $\overline{\omega}$. Let $\{\phi_0\} \cup \{\phi_j\}_{j=1}^\kappa$ be a subordinate partition of unity, $\phi_0 + \sum_{j=1}^\kappa \phi_j = 1$ on $\overline{\omega}$, $\phi_j = 1$ in a neighbourhood of b_j for all j, $1 \le j \le \kappa$. Eventually we assume that, for $1 \le j \le \kappa$, $\psi_j = (\chi_j)_*\phi_j$ only depends on the radial variable r, i.e., $\psi_j = \psi_j(r)$.

Remark 1.24 Before we proceed we give an intrinsic interpretation of (1.107): There is a short split exact sequence

$$0 \longrightarrow H_N^2(\omega) \longrightarrow H_Q^2(\omega) \longrightarrow \Pi_{j=1}^{\kappa}\mathbb{C} \longrightarrow 0 \qquad (1.114)$$

with the surjection assigning to each function $u \in H_Q^2(\omega)$ its sequence $(d_1, \ldots, d_\kappa)$ of singular coefficients. Thereby, d_j is explained as the coefficient appearing in the front of S for $v = (\chi_j)_*(\phi_j u)$, $\Gamma = \Gamma_j$.

To see that (1.114) is correctly defined observe that the coefficient d_j is not only independent of the choice of the cut-off function ψ_j (see Remark 1.22 (b)), but also independent of the choice of the diffeomorphism χ_j meeting all of the assumptions above. A splitting of (1.114) is obtained via (1.105) after having fixed the diffeomorphisms χ_j and the cut-off functions ψ_j. More precisely, we may write

$$u = u_0 + \sum_{j=1}^{\kappa} d_j(\chi_j)^*(\psi_j(r)r^{\pi/\alpha_j}\cos(\pi\theta/\alpha_j))$$

for $u \in H_Q^2(\omega)$, where $u_0 \in H_N^2(\omega)$, $d_j \in \mathbb{C}$ are uniquely determined. The coefficients d_j can be calculated using the formula

$$d_j = \lim_{r \to 0+} \beta_j^{-2}\big(r^{-\pi/\alpha_j}((\chi_j)_*(\phi_j u)(r,\theta) - u(b_j)), \cos(\pi\theta/\alpha_j))\big)_{L^2(0,\alpha_j)},$$
$$(1.115)$$

where $(\cdot, \cdot)_{L^2(0,\alpha_j)}$ denotes the scalar product in $L^2(0, \alpha_j)$, $u(b_j)$ is the value of u at b_j, and $\beta_j = \{\int_0^{\alpha_j} |\cos(\pi\theta/\alpha_j)|^2 \, d\theta\}^{1/2}$. The value $u(b_j) = (\chi_j)_*(\phi_j u)(0)$ is well-defined by *Theorem 1.37*.

Notice that an equivalent norm on $H_Q^2(\mathbb{R} \times \omega)$ is given by

$$\|u\|_{H_Q^2(\mathbb{R}\times\omega)} = \left\{ \|\phi_0 u\|_{H^2(\mathbb{R}\times\omega)}^2 + \sum_{j=1}^{\kappa} \|(\chi_j)_*(\phi_j u)\|_{H_Q^2(\mathbb{R}\times\Gamma_j)}^2 \right\}^{1/2}. \qquad (1.116)$$

This follows from the fact that $u \in H_Q^2(\mathbb{R} \times \omega)$ if and only if $\phi_l u \in H_Q^2(\mathbb{R} \times \omega)$ for all $l, 0 \leqslant l \leqslant \kappa$, and obviously $\phi_0 u \in H_Q^2(\mathbb{R} \times \omega)$ if and only if $\phi_0 u \in H^2(\mathbb{R} \times \omega)$, while, for $1 \leqslant j \leqslant \kappa$, $\phi_j u \in H_Q^2(\mathbb{R} \times \omega)$ if and only if $(\chi_j)_*(\phi_j u) \in H_Q^2(\mathbb{R} \times \Gamma_j)$.

From *Lemma 1.16* and (1.116) we conclude that

$$H_Q^2(\mathbb{R} \times \omega) = H_N^2(\mathbb{R} \times \omega) \oplus \left\{ \sum_{j=1}^{\kappa}(\chi_j)^*(F_{\tau \to t}^{-1}\{\langle\tau\rangle\psi_j(r\langle\tau\rangle)(r\langle\tau\rangle)^{\pi/\alpha_j} \right.$$

$$\left. \times \cos(\pi\theta/\alpha_j)\hat{d}_j(\tau)\}); d_j \in H^2(\mathbb{R}), \ 1 \leqslant j \leqslant \kappa \right\}.$$
$$(1.117)$$

Analogously to (1.114) we have the following lemma.

Lemma 1.17 *For $\omega \subset \mathbb{R}^2$ being a bounded, polyhedral domain as above, there is a short split exact sequence*

$$0 \longrightarrow H_N^2(\mathbb{R} \times \omega) \longrightarrow H_Q^2(\mathbb{R} \times \omega) \xrightarrow{(\tau_1, \ldots, \tau_\kappa)} \Pi_{j=1}^\kappa H^{1-\pi/\alpha_j}(\mathbb{R}) \longrightarrow 0,$$
$$(1.118)$$

where the operators τ_j are given by

$$\tau_j u(t) = \lim_{r \to 0^+} \beta_j^{-2} \left(r^{-\pi/\alpha_j} ((\chi_j)_* (\phi_j u)(t, r, \theta) - u(t, b_j)), \cos(\pi\theta/\alpha_j)) \right)_{L^2(0,\alpha_j)}.$$
$$(1.119)$$

Moreover, a splitting of (1.118) is given by the mapping

$$(d_{11}, \ldots, d_{\kappa 1}) \longmapsto \sum_{j=1}^\kappa (\chi_j)^* (F_{\tau \to t}^{-1} \{\psi_j(r\langle\tau\rangle) \hat{d}_{j1}(\tau)\} r^{\pi/\alpha_j} \cos(\pi\theta/\alpha_j)).$$
$$(1.120)$$

Proof According to (1.112) and the short exact sequence (1.114), the functions $d_j \in H^2(\mathbb{R})$ appearing in the representation of $u \in H_Q^2(\mathbb{R} \times \omega)$ as

$$u = u_0 + \sum_{j=1}^\kappa (\chi_j)^* (F_{\tau \to t}^{-1} \{\langle\tau\rangle \psi_j(r\langle\tau\rangle)(r\langle\tau\rangle)^{\pi/\alpha_j} \cos(\pi\theta/\alpha_j) \hat{d}_j(\tau)\})$$

$$= u_0 + \sum_{j=1}^\kappa (\chi_j)^* (F_{\tau \to t}^{-1} \{\psi_j(r\langle\tau\rangle) \hat{d}_{j1}(\tau)\}^{\pi/\alpha_j} \cos(\pi\theta/\alpha_j)),$$

where $u_0 \in H_N^2(\mathbb{R} \times \omega)$, are uniquely determined, independently of the choice of the diffeomorphisms χ_j and the cut-off function ψ_j. Likewise, the same is then true for the functions $d_{j1} = \langle D \rangle^{1+\pi/\alpha_j} d_j \in H^{1-\pi/\alpha_j}(\mathbb{R})$. Therefore, the surjection in (1.118) is well-defined. Moreover, it becomes clear that (1.118) is exact and a splitting of it is provided by (1.120).

Thus it remains to show (1.119). From (1.115), applied to $\Gamma = \Gamma_j$, $v = (\chi_j)_*(\phi_j u)$, and Eq. (1.112), in which $d = d_j$, we conclude that

$$\hat{d}_j(\tau)$$
$$= \lim_{r \to 0^+} \beta_j^{-2} \left(r^{-\pi/\alpha_j} (\langle\tau\rangle^{-1} \hat{v}(\tau, r\langle\tau\rangle^{-1}, \theta) - \langle\tau\rangle^{-1} \hat{v}(\tau, 0)), \cos(\pi\theta/\alpha_j)) \right)_{L^2(0,\alpha_j)}$$
$$= \lim_{r \to 0^+} \beta_j^{-2} \left((r\langle\tau\rangle)^{-\pi/\alpha_j} \langle\tau\rangle^{-1} (\hat{v}(\tau, r, \theta) - \hat{v}(\tau, 0)), \cos(\pi\theta/\alpha_j)) \right)_{L^2(0,\alpha_j)},$$

the latter line upon replacing r with $r\langle\tau\rangle$, i.e.,

$$\hat{d}_{j1}(\tau) = \langle\tau\rangle^{1+\pi/\alpha_j}\hat{d}_j(\tau)$$

$$= \lim_{r\to 0^+}\beta_j^{-2}\big(r^{-\pi/\alpha_j}(\hat{v}(\tau,r,\theta)-\hat{v}(\tau,0)),\cos(\pi\theta/\alpha_j)\big)_{L^2(0,\alpha_j)},$$

$$d_{j1}(t) = \lim_{r\to 0^+}\beta_j^{-2}\big(r^{-\pi/\alpha_j}((\chi_j)_*(\phi_j u)(t,r,\theta)-u(t,b_j)),\cos(\pi\theta/\alpha_j)\big)_{L^2(0,\alpha_j)}.$$

This proves *Lemma 1.17* completely.   □

From (1.119) we obtain in particular that the trace operation on an edge is local.

Corollary 1.16 *For $u \in H_Q^2(\mathbb{R}\times\omega)$, we have* $\mathrm{supp}(\tau_j u) \subseteq \mathrm{supp}(u) \cap (\mathbb{R}\times\{b_j\})$.

Remark 1.25

(a) To interpret the functions $d_{j1} \in H^{1+\pi/\alpha_j}(\mathbb{R})$, $1 \le j \le \kappa$, as coefficients in the asymptotic expansion of $u \in H_Q^2(\mathbb{R}\times\omega)$ close to the edge $\mathbb{R}\times\{b_j\}$, we observe that $F_{\tau\to t}^{-1}\{\psi_j(r\tau)\hat{d}_{j1}(\tau)\} = d_{j1}(t)$ when $r = 0$.
(b) It can be shown that

$$\beta_j^{-2}(r^{-\pi/\alpha_j}((\chi_j)_*(\phi_j u)(t,r,\theta)-u(t,b_j)),\cos(\pi\theta/\alpha_j))_{L^2(0,\alpha_j)} \in H^1(\mathbb{R})$$

for $u \in H_Q^2(\mathbb{R}\times\omega)$, and convergence in (1.119) takes place in $H^{1-\pi/\alpha_j}(\mathbb{R})$.

The final goal in this section is to conclude the form of asymptotics when going over from $H_Q^2(\mathbb{R}\times\omega)$ to its factor space $H_Q^2(\mathbb{R}_+\times\omega)$. This is achieved by constructing a suitable splitting of (1.118) in terms of a continuous projection Π_2 in $H_Q^2(\mathbb{R}\times\omega)$ by means of a reformulation of the asymptotic information.

Theorem 1.38 *Let $\omega \subset \mathbb{R}^2$ be a bounded, polyhedral domain as above. Then there exists a continuous projection Π_2 in $H_Q^2(\mathbb{R}\times\omega)$ obeying the following properties:*

(a) $\ker\Pi_2 = H_N^2(\mathbb{R}\times\omega)$;
(b) $T_s\Pi_2 = \Pi_2 T_s$ *for all* $s \in \mathbb{R}$;
(c) $\mathrm{supp}\,u \subseteq \overline{\mathbb{R}}_-$ *implies* $\mathrm{supp}\,\Pi_2 u \subseteq \overline{\mathbb{R}}_-$;
(d) Π_2 *is* $(H_{Q,b}^2(\mathbb{R}\times\omega), H_{Q,b}^2(\mathbb{R}\times\omega))$*-continuous;*
(e) Π_2 *is* $(H_{Q,\mathrm{loc}}^2(\mathbb{R}\times\omega), H_{Q,\mathrm{loc}}^2(\mathbb{R}\times\omega))$*-continuous.*

In the proof of *Theorem 1.38*, we shall make use of the following result.

Lemma 1.18 *Let $\Gamma \subset \mathbb{R}^2$ be an open cone. Further let $\psi \in \mathcal{S}(\mathbb{R})$, $\psi_1 \in \mathcal{S}(\overline{\mathbb{R}})$, $d_1 \in H^{1-\pi/\alpha}(\mathbb{R})$. Then*

$$\psi_1(r)F_{\tau\to t}^{-1}\{(\psi(r\langle\tau\rangle)-\psi(r\tau))\hat{d}_1(\tau)\}r^{\pi/\alpha}\cos(\pi\theta/\alpha) \in H_N^2(\mathbb{R}\times\Gamma).$$

Proof Let $u(t, r) = \psi_1(r) F_{\tau \to t}^{-1} \{ (\psi(r\langle \tau \rangle) - \psi(r\tau)) \hat{d}_1(\tau) \} r^{\pi/\alpha} \cos(\pi\theta/\alpha)$. Then we have

$$\|u\|_{H_N^2(\mathbb{R} \times \Gamma)}$$

$$= \left\{ \int_{-\infty}^{\infty} \langle \tau \rangle^2 \|\kappa(\tau)^{-1} (\psi_1(r) (\psi(r\langle \tau \rangle) - \psi(r\tau)) \hat{d}_1(\tau) r^{\pi/\alpha} \cos(\pi\theta/\alpha))\|_{H_N^2(\Gamma)}^2 \, d\tau \right\}^{1/2}$$

$$= \left\{ \int_{-\infty}^{\infty} \langle \tau \rangle^2 |\hat{d}(\tau)|^2 \|\psi_1(r\langle \tau \rangle^{-1}) (\psi(r) - \psi(r\tau/\langle \tau \rangle)) r^{\pi/\alpha} \cos(\pi\theta/\alpha))\|_{H_N^2(\Gamma)}^2 \, d\tau \right\}^{1/2}$$

$$\leqslant C \left\{ \int_{-\infty}^{\infty} \langle \tau \rangle^2 |\hat{d}(\tau)|^2 \, d\tau \right\}^{1/2}, \tag{1.121}$$

where $d = \langle D \rangle^{-1 - \pi/\alpha} d_1 \in H^2(\mathbb{R})$. Thereby,

$$\|\psi_1(r\langle \tau \rangle^{-1}) (\psi(r) - \psi(r\tau/\langle \tau \rangle)) r^{\pi/\alpha} \cos(\pi\theta/\alpha)\|_{H_N^2(\Gamma)} \leqslant C$$

for a certain constant $C > 0$ independent of τ is seen from the fact that $\psi_2(r) \mapsto \psi_2(r) r^{\pi/\alpha} \cos(\pi\theta/\alpha)$ constitutes a bounded map from $\{\psi_2 \in \mathcal{S}(\overline{\mathbb{R}}_+); \ \psi_2(0) = 0\}$ into $H_N^2(\Gamma)$, while $\{\psi_1(r\langle \tau \rangle^{-1}) (\psi(r) - \psi(r\tau/\langle \tau \rangle)); \ \tau \in \mathbb{R}\}$ for $\psi \in \mathcal{S}(\mathbb{R})$, $\psi_1 \in \mathcal{S}(\overline{\mathbb{R}}_+)$ is bounded in $\{\psi_2 \in \mathcal{S}(\overline{\mathbb{R}}_+); \ \psi_2(0) = 0\}$. Hence the right-hand side in (1.121) is finite proving that $u \in H_N^2(\mathbb{R} \times \Gamma)$. $\quad \square$

Proof of Theorem 1.38 *Lemma 1.18* allows to replace

$$F_{\tau \to t}^{-1} \{\langle \tau \rangle \psi_j(r\langle \tau \rangle) (r\langle \tau \rangle)^{\pi/\alpha_j} \cos(\pi\theta/\alpha_j) \hat{d}_j(\tau)\}$$

in (1.117) by

$$\psi_{j1} F_{\tau \to t}^{-1} \{\psi_j(r\tau) \hat{d}_{j1}(\tau)\} r^{\pi/\alpha_j} \cos(\pi\theta/\alpha_j),$$

i.e., we have

$$H_Q^2(\mathbb{R} \times \omega) = H_N^2(\mathbb{R} \times \omega) \oplus$$

$$\left\{ \sum_{j=1}^{\kappa} (\chi_j)^* (\psi_{j1}(r) F_{\tau \to t}^{-1} \{\psi_j(r\tau) \hat{d}_{j1}(\tau)\} r^{\pi/\alpha_j} \cos(\pi\theta/\alpha_j)); \ d_{j1} \in H^{1 - \pi/\alpha_j}(\mathbb{R}) \right\},$$

where, for each j, $1 \leqslant j \leqslant \kappa$, $\psi_j \in \mathcal{S}(\mathbb{R})$, $\psi_{j1} \in C_0^\infty(\overline{\mathbb{R}}_+)$, $\psi_j(0) = \psi_{j1}(0) = 1$, and ψ_{j1} is supported in V_j when considered as a function on Γ_j. If especially the ψ_j are chosen in a way such that supp $F^{-1} \psi_j \subseteq \overline{\mathbb{R}}_-$ holds for all j, then

$$\Pi_2 u = \sum_{j=1}^{\kappa} (\chi_j)^* (\psi_{j1}(r) F_{\tau \to t}^{-1} \{\psi_j(r\tau)(\tau_j u)\widehat{}(\tau)\} r^{\pi/\alpha_j} \cos(\pi\theta/\alpha_j)) \qquad (1.122)$$

for $u \in H_Q^2(\mathbb{R} \times \omega)$ is a projection in $H_Q^2(\mathbb{R} \times \omega)$ meeting all the requirements (a)–(e). That Π_2 is a projection follows from the fact that $\tau_j \Pi_2 u = \tau_j u$ holds for $u \in H_Q^2(\mathbb{R} \times \omega)$, (a), (c) are immediate, (b) is the locality of the trace operator τ_j (see *Corollary 1.16*) and the translation invariance of the pseudo-differential operator $d_1 \mapsto F_{\tau \to t}^{-1}(\psi_j(r\tau)\widehat{d}_1(\tau))$, where $r > 0$ is regarded as a parameter, and (d), (e) come from the observation that $\psi_{j1}(r) F_{\tau \to t}^{-1} \{\psi_j(r\tau)(\tau_j u)\widehat{}(\tau) r^{\pi/\alpha_j} \cos(\pi\theta/\alpha_j)\}$ belongs to $H_{Q,b}^2(\mathbb{R} \times \Gamma_j)$ and $H_{Q,\mathrm{loc}}^2(\mathbb{R} \times \Gamma_j)$, respectively, for u belonging to $H_{Q,b}^2(\mathbb{R} \times \Gamma_j)$ and $H_{Q,\mathrm{loc}}^2(\mathbb{R} \times \Gamma_j)$, as an easy calculation reveals.   □

The following consequences of *Theorem 1.38* supply the projection Π_2^+ in $H_{Q,b}^2(\mathbb{R} \times \omega)$ onto its closed subspace comprising the asymptotic information as well as the short exact sequences used in *Chap. 2*.

Theorem 1.39 *Let $\omega \subset \mathbb{R}^2$ be a bounded, polyhedral domain as above. Then there exists a continuous projection Π_2^+ in $H_{Q,b}^2(\mathbb{R}_+ \times \omega)$ obeying the following properties:*

(a) $\ker \Pi_2^+ = H_{N,b}^2(\mathbb{R}_+ \times \omega)$;
(b) $T_s \Pi_2^+ = \Pi_2^+ T_s$ *for all $s \geq 0$.*

Moreover, Π_2^+ is $(H_{Q,\mathrm{loc}}^2(\overline{\mathbb{R}}_+ \times \omega), H_{Q,\mathrm{loc}}^2(\overline{\mathbb{R}}_+ \times \omega))$-continuous.

Proof The theorem follows from *Theorem 1.38 (a)–(e)* by continuous extension of the projection Π_2 to $H_{Q,b}^2(\mathbb{R} \times \omega)$ and its subsequent factorization to $H_{Q,b}^2(\mathbb{R}_+ \times \omega)$.   □

Notice that a projection Π_2^+ satisfying the requirements of *Theorem 1.39* is

$$\Pi_2^+ u = \sum_{j=1}^{\kappa} (\chi_j)^* (\psi_{j1}(r) F_{\tau \to t}^{-1} \{\psi_j(\tau r)((\tau_j^+ u)_{\mathrm{ext}})\widehat{}(\tau)\} r^{\pi/\alpha_k} \cos(\pi\theta/\alpha_j)),$$

$$(1.123)$$

$u \in H_{Q,b}^2(\mathbb{R}_+ \times \omega)$, where ψ, ψ_1 are as in (1.122). Here $(\tau_j^+ u)_{\mathrm{ext}}$ means an arbitrary extension of $\tau_j^+ u \in H_b^{1-\pi/\alpha_j}(\mathbb{R}_+)$ to a function in $H_b^{1-\pi/\alpha_j}(\mathbb{R})$.

Corollary 1.17 *The short exact sequence* (1.118) *extends by continuity and factors subsequently to a short split exact sequence*

$$0 \longrightarrow H_{N,b}^2(\mathbb{R}_+ \times \omega) \longrightarrow H_{Q,b}^2(\mathbb{R}_+ \times \omega) \xrightarrow{(\tau_1,\ldots,\tau_\kappa)} \Pi_{j=1}^{\kappa} H_b^{1-\pi/\alpha_j}(\mathbb{R}_+) \longrightarrow 0,$$

where $(\tau_1, \ldots, \tau_\kappa)$ is the vector of trace operators. A splitting is obtained from (1.123) *by replacing $\tau_j^+ u$ by $d_{1j} \in H_b^{1-\pi/\alpha_j}(\mathbb{R}_+)$.*

Chapter 2
Trajectory Dynamical Systems and Their Attractors

2.1 Kolmogorov ε-Entropy and Its Asymptotics in Functional Spaces

We start with the definition of Kolmogorov ε-entropy, via which we define fractal dimension of the compact set in the metric space. We will use these two concepts in the sequel.

Definition 2.1 Let K be a (pre)compact set in a metric space M. Then, due to Hausdorff's criteria, it can be covered by a finite number of ε-balls in M. Let $N_\varepsilon(K, M)$ be the minimal number of ε-balls that cover K. Then, we can call Kolmogorov's ε-entropy of K the logarithm of this number;

$$\mathbb{H}_\varepsilon(K, M) := \log_2 N_\varepsilon(K, M).$$

We now give several examples of typical asymptotics for the ε-entropy.

Example 2.1 We assume that $K = [0, 1]^n$ and $M = \mathbb{R}^n$ (more generally, K is an n-dimensional compact Lipschitz manifold of the metric space M). Then

$$\mathbb{H}_\varepsilon(K, M) = (n + \bar{\bar{o}}(1)) \log_2 \frac{1}{\varepsilon} \text{ as } \varepsilon \to 0.$$

This example justifies the definition of the fractal dimension.

Definition 2.2 The fractal dimension $\dim_F(K, M)$ is defined as

$$\dim_F(K, M) := \limsup_{\varepsilon \to 0} \frac{\mathbb{H}_\varepsilon(K, M)}{\log_2 1/\varepsilon}.$$

© Springer Nature Switzerland AG 2018
M. Efendiev, *Symmetrization and Stabilization of Solutions of Nonlinear Elliptic Equations*, Fields Institute Monographs 36,
https://doi.org/10.1007/978-3-319-98407-0_2

Hence, for a compact n-dimensional Lipschitz manifold K in a metric space M, $\dim_F(K, M) = n$.

The following example shows that, for sets that are not manifolds, the fractal dimension may be a non-integer.

Example 2.2 Let K be a standard ternary Cantor set in $M = [0, 1]$. Then $\dim_F(K, M) = \frac{\ln 2}{\ln 3} < 1$.

Proof Let K be the Cantor set obtained from the segment $[0,1]$ by the sequential removal of the centre thirds. First we remove all the points between $1/3$ and $2/3$. Then the centre thirds $(1/9, 2/9)$ and $(7/9, 8/9)$ of the remaining segments $[0, 1/3]$ and $[2/3, 1]$ are deleted. After that the centre parts $(1/27, 2/27)$, $(7/27, 8/27)$, $(19/27, 20/27)$ and $(25/27, 26/27)$ of the four remaining segments $[0, 1/9]$, $[2/9, 1/3]$, $[2/3, 7/9]$ and $[8/9, 1]$, respectively, are deleted. If we continue this process to infinity, it will lead to the standard Cantor set K. Next we calculate its fractal dimension. We emphasize that $K = \bigcap_{M=0}^{\infty} \theta_m$, where $\theta_0 = [0, 1]$, $\theta_1 = [0, 1/3] \cup [2/3, 1]$, $\theta_2 = [0, 1/9] \cup [2/9, 1/3] \cup [2/3, 7/9] \cup [8/9, 1]$ and so on. Each of the sets θ_m can be considered as a union of 2^m segments of length 3^{-m}. In particular, the cardinality of the covering of the set K with segments of length 3^{-m} is equal to 2^m. Consequently
$\dim_F(K, [0, 1]) = \lim_{m \to \infty} \frac{\ln 2^m}{\ln(3^m)} = \frac{\ln 2}{\ln 3}$ It is not difficult to show that

(1) if $K_1 \subseteq K_2$, then $\dim_F(K_1, M) \leqslant \dim_F(K_2, M)$
(2) $\dim_F(K_1 \cup K_2, M) \leqslant \max \{\dim_F(K_1, M); \dim_F(K_2, M)\}$
(3) $\dim_F(K_1 \times K_2, M \times M) \leqslant \dim_F(K_1, M) + \dim_F(K_2, M)$
(4) let g be a Lipschitzian mapping of one metric space M_1 into another M_2. Then $\dim_F(g(K), M_2) \leqslant \dim_F(K, M_1)$.

The next example gives the typical behavior of the entropy in classes of functions with finite smoothness.

Example 2.3 Let V be a smooth bounded domain of $\mathbb{R}^n$ and let K be the unit ball in the Sobolev space $W^{l_1, p_1}(V)$ and M be another Sobolev space $W^{l_2, p_2}(V)$ such that the embedding $W^{l_1, p_1} \subset W^{l_2, p_2}$ is compact, i.e.

$$l_1 > l_2 \geqslant 0, \quad \frac{l_1}{n} - \frac{1}{p_1} > \frac{l_2}{n} - \frac{1}{p_2}.$$

Then, the entropy $\mathbb{H}_\varepsilon(K, M)$ has the following asymptotics (see [101]):

$$C_1 \left(\frac{1}{\varepsilon}\right)^{n/(l_1 - l_2)} \leqslant \mathbb{H}_\varepsilon(K, M) \leqslant C_2 \left(\frac{1}{\varepsilon}\right)^{n/(l_1 - l_2)}.$$

Finally, the last example shows the typical behavior of the entropy in classes of analytic functions.

Example 2.4 Let $V_1 \subset V_2$ be two bounded domains of $\mathbb{C}_3^n$. We assume that K is the set of all analytic functions ϕ in V_2 such that $\|\phi\|_{C(V_2)} \leqslant 1$ and that $M = C(V_1)$. Then

$$C_1 \left(\log_2 1/\varepsilon\right)^{n+1} \leqslant \mathbb{H}_\varepsilon (K|_{V_1}, M) \leqslant C_2 \left(\log_2 1/\varepsilon\right)^{n+1},$$

(see [73]).

2.2 Global Attractors and Finite-Dimensional Reduction

It is well-known that, one of the main concepts of the modern theory of DS in infinite dimensions is that of the *global attractor*. We give below its definition for an abstract semigroup $S(t)$ acting on a metric space Φ, although, without loss of generality, the reader may think that $(S(t), \Phi)$ is just a DS associated with one of the PDEs described in the introduction.

To this end, we first recall that a subset K of the phase space Φ is an attracting set of the semigroup $S(t)$ if it attracts the images of all the *bounded* subsets of Φ, i.e., for every bounded set B and every $\varepsilon > 0$, there exists a time T (depending in general on B and ε) such that the image $S(t)B$ belongs to the ε-neighborhood of K if $t \geqslant T$. This property can be rewritten in the equivalent form

$$\lim_{t\to\infty} \mathrm{dist}_H (S(t)B, K) = 0,$$

where $\mathrm{dist}_H (X, Y) := \sup_{x\in X} \inf_{y\in Y} d(x, y)$ is the nonsymmetric Hausdorff distance between subsets of Φ.

We now give the definition of a global attractor, following Babin-Vishik (see [1, 35, 48, 98]).

Definition 2.3 A set $\mathcal{A} \subset \Phi$ is a global attractor for the semigroup $S(t)$ if

1) $\mathcal{A}$ is *compact* in Φ;
2) $\mathcal{A}$ is *strictly invariant*: $S(t)\mathcal{A} = \mathcal{A}$, for all $t \geqslant 0$;
3) $\mathcal{A}$ is an *attracting* set for the semigroup $S(t)$.

Thus, the second and third properties guarantee that a global attractor, if it exists, is unique and that the DS reduced to the attractor contains all the nontrivial dynamics of the initial system. Furthermore, the first property indicates that the reduced phase space $\mathcal{A}$ is indeed "thinner" than the initial phase space Φ (we recall that, in infinite dimensions, a compact set cannot contain, e.g., balls and should thus be nowhere dense).

In most applications, one can use the following attractor's existence theorem.

Theorem 2.1 *Let a DS $(S(t), \Phi)$ possess a compact attracting set and the operators $S(t) : \Phi \to \Phi$ be continuous for every fixed t. Then, this system possesses the global attractor $\mathcal{A}$ which is generated by all the trajectories of $S(t)$ which are defined for all $t \in \mathbb{R}$ and are globally bounded.*

The strategy for applying this theorem to concrete equations of mathematical physics is the following. In a first step, one verifies a so-called *dissipative* estimate which has usually the form

$$\|S(t)u_0\|_\Phi \leqslant Q(\|u_0\|_\Phi)e^{-\alpha t} + C_*, \quad u_0 \in \Phi, \tag{2.1}$$

where $\|\cdot\|_\Phi$ is a norm in the function space Φ and the positive constants α and C_* and the monotonic function Q are independent of t and $u_0 \in \Phi$ (usually, this estimate follows from energy estimates and is sometimes even used in order to "define" a dissipative system). This estimate obviously gives the existence of an attracting set for $S(t)$ (e.g., the ball of radius $2C_*$ in Φ), which is however *noncompact* in Φ. In order to overcome this problem, one usually derives, in a second step, a *smoothing* property for the solutions, which can be formulated as follows:

$$\|S(1)u_0\|_{\Phi_1} \leqslant Q_1(\|u_0\|_\Phi), \quad u_0 \in \Phi, \tag{2.2}$$

where Φ_1 is another function space which is *compactly* embedded into Φ. In applications, Φ is usually the space $L^2(\Omega)$ of square integrable functions, Φ_1 is the Sobolev space $H^1(\Omega)$ of the functions u such that u and $\nabla_x u$ belong to $L^2(\Omega)$ and estimate (2.2) is a classical smoothing property for solutions of parabolic equations (for parabolic equations in unbounded domains and for hyperbolic equations, a slightly more complicated *asymptotic* smoothing property should be used instead of (2.2), see the Sect. 3.2 of the monograph [48] and the references therein.

Since the continuity of the operators $S(t)$ usually poses no difficulty (if the uniqueness is proven), then the above scheme gives indeed the existence of the global attractor for most of the PDE of mathematical physics in bounded domains.

Remark 2.1 As was shown in [1] the assumption that $S(t) : \Phi \rightarrow \Phi$ be continuous for every fixed t can be replaced by the closedness of the graph $\{(u_0, S(t)u_0), u_0 \in \Phi\}$.

Remark 2.2 Although the global attractor has usually a very complicated geometric structure, there exists one exceptional class of DS for which the global attractor has a relatively simple structure which is completely understood, namely the DS having a global Lyapunov function. We recall that a continuous function $\mathcal{L} : \Phi \rightarrow \mathbb{R}$ is a global Lyapunov function if

1) $\mathcal{L}$ is non-increasing along the trajectories, i.e. $\mathcal{L}(S(t)u_0) \leqslant \mathcal{L}(u_0)$, for all $t \geqslant 0$;
2) $\mathcal{L}$ is *strictly* decreasing along all non-equilibrium solutions, i.e. $\mathcal{L}(S(t)u_0) = \mathcal{L}(u_0)$ for some $t > 0$ and u_0 implies that u_0 is an equilibrium of $S(t)$.

It is well known that, if a DS possesses a global Lyapunov function, then, at least under the generic assumption that the set $\mathcal{R}$ of equilibria is finite, every trajectory $u(t)$ *stabilizes* to one of these equilibria as $t \rightarrow +\infty$. Moreover, every complete bounded trajectory $u(t)$, $t \in \mathbb{R}$, belonging to the attractor is a heteroclinic orbit joining two equilibria. Thus, the global attractor $\mathcal{A}$ can be described as follows [1, 48, 98]:

$$\mathcal{A} = \bigcup_{u_0 \in \mathcal{R}} \mathcal{M}^+(u_0),$$

where $\mathcal{M}^+(u_0)$ is the so-called unstable set of the equilibrium u_0 (which is generated by all heteroclinic orbits of the DS which start from the given equilibrium $u_0 \in \mathcal{A}$). It is also known that, if the equilibrium u_0 is *hyperbolic* (generic assumption [1]), then the set $\mathcal{M}^+(u_0)$ is a κ-dimensional submanifold of Φ, where κ is the instability index of u_0. Thus, under the generic hyperbolicity assumption on the equilibria, the attractor $\mathcal{A}$ of a DS having a global Lyapunov function is a finite union of smooth finite-dimensional submanifolds of the phase space Φ. These attractors are called *regular* (following Babin-Vishik (see [1]).

It is also worth emphasizing that, in contrast to general global attractors, regular attractors are robust under perturbations. Moreover, in some cases, it is also possible to verify the so-called *transversality* conditions (for the intersection of stable and unstable manifolds of the equilibria) and, thus, verify that the DS considered is a Morse-Smale system. In particular, this means that the dynamics restricted to the regular attractor $\mathcal{A}$ are also preserved (up to homeomorphisms) under perturbations.

In the sequel we will apply *Theorem 2.1* or *Remark 2.1* (whenever it will be necessary) to a class of PDEs arising in mathematical physics. We especially emphasize that one of the challenging questions in the theory of attractors is, in which sense are the dynamics on the global attractor finite-dimensional? As already mentioned, the global attractor is usually not a manifold, but has a rather complicated geometric structure. So, it is natural to use the definitions of dimensions adopted for the study of fractal sets here. We restrict ourselves to the so-called fractal (or box-counting, entropy) dimension, although other dimensions (e.g., Hausdorff, Lyapunov, etc.) are also used in the attractors' theory. Here the so-called Mané theorem (which can be considered as a generalization of the classical Whitney embedding theorem for fractal sets) plays an important role in the finite-dimensional reduction theory (see [98]).

Theorem 2.2 *Let Φ be a Banach space and $\mathcal{A}$ be a compact set such that $d_f(\mathcal{A}) < N$ for some $N \in \mathbb{N}$. Then, for "almost all" $2N + 1$-dimensional planes L in Φ, the corresponding projector $\Pi_L : \Phi \to L$ restricted to the set $\mathcal{A}$ is a Hölder continuous homeomorphism.*

Thus, if the finite fractal dimensionality of the attractor is established, then, fixing a hyperplane L satisfying the assumptions of the Mané theorem and projecting the attractor $\mathcal{A}$ and the DS $S(t)$ restricted to $\mathcal{A}$ onto this hyperplane ($\bar{\mathcal{A}} := \Pi_L \mathcal{A}$ and $\bar{S}(t) := \Pi_L \circ S(t) \circ \Pi_L^{-1}$), we obtain indeed a reduced DS ($\bar{S}(t), \bar{\mathcal{A}}$) which is defined on a finite-dimensional set $\bar{\mathcal{A}} \subset L \sim \mathbb{R}^{2N+1}$. Moreover, this DS will be Hölder continuous with respect to the initial data.

Remark 2.3 Note that, good estimates on the dimension of the attractors in terms of the physical parameters are crucial for the finite-dimensional reduction described above and (consequently) there exists a highly developed machinery for obtaining such estimates. The best known upper estimates are usually obtained by the so-called volume contraction method which is based on the study of the evolution of infinitesimal k-dimensional volumes in the neighborhood of the attractor (and, if the DS considered contracts the k-dimensional volumes, then the fractal dimension of the attractor is less than k, (see [1, 98])

Remark 2.4 Lower bounds on the dimension are usually based on the observation that the global attractor always contains the unstable manifolds of the (hyperbolic) equilibria. Thus, the instability index of a properly constructed equilibrium gives a lower bound on the dimension of the attractor, (see [1, 48, 98]).

The following *Theorem 2.3* plays the decisive role in the study of the dimension of attractor, which in turn does not require differentiability of the associated semigroup in contrast to (see [1, 35, 98]). We especially emphasize for a quite large class of degenerate parabolic system arising in the modelling of life science problem (see [46]) the associate semigroup is not differentiable. We denote by $S := S(1)$.

Theorem 2.3 *Let H_1 and H be Banach spaces, H_1 be compactly embedded in H and let $K \subset\subset H$. Assume that there exists a map $S : K \to K$, such that $S(K) = K$ and the following 'smoothing' property is valid*

$$\|S(k_1) - S(k_2)\|_{H_1} \leqslant C\|k_1 - k_2\|_H \tag{2.3}$$

for every $k_1, k_2 \in K$. Then the fractal dimension of K in H is finite and can be estimated in the following way:

$$d_F(K, H) \leqslant \mathbb{H}_{1/4C}(B(1, 0, H_1), H) \tag{2.4}$$

where C is the same as in (2.3) and $B(1, 0, H_1)$ denotes the unit ball in the space H_1.

Proof Let $\{B(\varepsilon, k_i, H)\}_{i=1}^{N_\varepsilon}$, $k_i \in K$, be some ε-*covering* of the set K (here and below we denote by $B(\varepsilon, k, V)$ the ε-*ball* in the space V, centered in k). Then according to (2.3), the system $\{B(C\varepsilon, L(k_i), H_1)\}_{i=1}^{N_\varepsilon}$ of $C\varepsilon$-balls in H_1 covers the set $S(K)$ and consequently (since $S(K) = K$) the same system covers the set K.

Cover now every H_1-ball with radius $C\varepsilon$ by a finite number of $\frac{\varepsilon}{4}$-balls in H. By definition, the minimal number of such balls equals to

$$N_{\varepsilon/4}(B(C\varepsilon, S(k_i), H_1), H) = N_{\varepsilon/4}(B(C\varepsilon, 0, H_1), H) = N_{1/4C}(B(1, 0, H_1), H) \equiv \mathcal{N}.$$

Note, that the centers of $\varepsilon/4$-covering thus obtained not necessarily belongs to K but we evidently can construct the $\varepsilon/2$-covering with centers in K and with the same number of balls.

Thus, having the initial ε-covering of K in H with the number of balls N_ε we have constructed the $\varepsilon/2$-covering with the number of balls $N_{\varepsilon/2} = \mathcal{N}N_\varepsilon$.

Consequently, the ε-entropy of the set K possesses the following estimate.

In fact the assertion of the theorem is a corollary of this recurrent estimate. Indeed, since $K \subset\subset H$ then there exists ε_0 such that $K \subset B(\varepsilon_0, k_0, H)$ and consequently

$$\mathbb{H}_{\varepsilon_0}(K, N) = 0. \tag{2.5}$$

Iterating the estimate (2.5) n-times we obtain that

$$\mathbb{H}_{\varepsilon_0/2^n}(k, H) \leqslant n \log_2 \mathcal{N}.$$

Fix now an arbitrary $\varepsilon > 0$ and choose $n = n(\varepsilon)$ in such a way that

$$\frac{\varepsilon_0}{2^n} \leqslant \varepsilon \leqslant \frac{\varepsilon_0}{2^n - 1}.$$

Then

$$\mathbb{H}_{\varepsilon}(K) \leqslant \mathbb{H}_{\varepsilon/2^n}(K) \leqslant n \log_2 \mathcal{N} \leqslant \log_2 \left(1 + \frac{\varepsilon_0}{\varepsilon}\right) \log_2 \mathcal{N}. \tag{2.6}$$

Thus (2.4) is an immediate consequence of (2.6). *Theorem 2.3* is proved.

2.3   Classification of Positive Solutions of Semilinear Elliptic Equations in a Rectangle: Two Dimensional Case

As we have seen in *Chap. 1*, positive solutions of semilinear second order elliptic problems have symmetry and monotonicity properties which reflects the symmetry of the operator and of the domain, see also e.g., [65] and [18] for the case of bounded domains and [13, 19, 21, 25] and [23, 30, 65] for the case of unbounded domains.

In particular, the symmetry and monotonicity results for the case of a half space have been considered in [13, 21] and the analogous results (including the existence and uniqueness of a nontrivial positive solution) for the case of whole space have been obtained in [23, 30, 65, 75], see also the references therein.

The goal of the present section is to give a description of all bounded nonnegative solutions of the following elliptic boundary value problem in a two dimensional rectangle $\Omega_+ := \{(x, y) \in \mathbb{R}^2, \ x \geqslant 0, \ y \geqslant 0\}$:

$$\begin{cases} \Delta_{x,y} u = f(u), & (x, y) \in \Omega_+, \\ u|_{\partial\Omega_+} = 0, & u(x, y) \geqslant 0, \end{cases} \tag{2.7}$$

where we assume that $u \in C_b(\Omega)$ and the nonlinearity f is smooth enough ($f \in C^1(\mathbb{R})$) and $f(0) = 0$.

It is known (see [13]) that, under the above assumptions, every solution $u(x, y)$ of (2.7) (if it exists) should be monotonic with respect to x and y and, consequently, there exist the following limits

$$\lim_{x \to \infty} u(x, y) = \psi_u(y), \quad \lim_{y \to \infty} u(x, y) = \phi_u(x). \tag{2.8}$$

Moreover functions ψ_u and ϕ_u bounded solutions of one dimensional analogue of problem (2.7)

$$\Psi'' = f(\Psi), \quad \Psi(0) = 0, \quad \Psi(z) \geqslant 0, \quad z \geqslant 0. \tag{2.9}$$

We recall, that every solution of (2.9) stabilizes as $z \to \infty$ to some $c \geqslant 0$ such that $f(c) = 0$ and, for fixed c there exists not greater than one solution $\Psi(z) = \Psi_c(z)$ of this problem. Consequently, the functions ψ_u and ϕ_u in (2.8) should coincide: $\psi_u(z) = \phi_u(z) = \Psi_c(z)$, where the constant $c = c_u > 0$ satisfies $f(c) = 0$. Thus, we can rewrite (2.8) in the following form:

$$\lim_{(x,y) \to \infty} |u(x, y) - \Psi_c(x, y)| = 0, \quad \text{where} \quad \Psi_c(x, y) := \min\{\Psi_c(x), \Psi_c(y)\}. \tag{2.10}$$

The aim of this notes is to verify the existence and uniqueness of a solution $u(x, y)$ satisfying (2.10). We establish this fact under the following nondegeneracy assumption that

$$f'(c) \neq 0 \tag{2.11}$$

(in a fact, the existence of a solution $\Psi_c(z)$ of Eqs. (2.9) and (2.11) imply that $f'(c) > 0$, see [15]). Thus, the main result of the chapter is the following theorem.

Theorem 2.4 *Let the nonlinearity f satisfy the above assumptions, Ψ_c be a solution of (2.9) such that $f'(c) > 0$. Then, there exists a unique solution $u(x, y)$ of (2.7) which satisfies (2.10).*

The following corollary shows that, generically, Eq. (2.7) has only finite number of different positive solutions.

Corollary 2.1 *Let the above assumptions hold and let, in addition, inequality (2.11) hold, for every solution $c > 0$ of equation $f(c) = 0$. Then, problem (2.7) has the finite number of different positive bounded solutions.*

2.3.1 Sketch of the Proof of Theorem 2.4

For the proof, we need the following lemma.

Lemma 2.1 *Let the assumptions of Theorem 2.4 hold and let*

$$\Psi_c^M(x, y) := \begin{cases} c, & (x, y) \in [0, M]^2, \\ \Psi_c(x, y), & (x, y) \in \Omega_+\backslash[0, M]^2, \end{cases} \tag{2.12}$$

where M is sufficiently large positive number. Then, the spectrum of the operator $\Delta_{x,y} - f'(\Psi_c^M(x, y))$ in Ω_+ (with the Dirichlet boundary conditions) is strictly negative:

$$\sigma(\Delta_{x,y} - f'(\Psi_c^M), L^2(\Omega_+)) \leqslant -K < 0. \tag{2.13}$$

Indeed, estimate (2.13) can be easily deduced from the standard fact that

$$\sigma(\partial_z^2 - f'(\Psi_c(z)), L^2(\mathbb{R}_+)) \leqslant -K \leqslant 0 \tag{2.14}$$

(which is the corollary of the Perron-Frobenius theorem), the minimax principle and the special form of the function $\Psi_c(x, y)$. for some positive K and, consequently,

$$\int_0^\infty |\phi'(z)|^2 + f'(\Psi_c(z))|\phi(z)|^2 \, dz \geqslant K \int_0^\infty |\phi(z)|^2 \, dz, \quad \phi \in C_0^\infty(\mathbb{R}_+). \tag{2.15}$$

We claim that spectrum of the two dimensional operator $\Delta_{x,y} - f'(\Psi_c(x, y))$ in Ω_+ is also strictly negative:

$$\sigma(\Delta_{x,y} - f'(\Psi_c(x, y))) \leqslant -K \leqslant 0 \tag{2.16}$$

Indeed, in order to prove this, it is sufficient to verify that

$$\int_{\Omega_+} |(\phi_x'|^2 + |\phi_y'|^2 + f'(\Psi_c(x, y))|\phi|^2 \, dx \, dy \geqslant K \int_{\Omega_+} |\phi|^2 \, dx \, dy, \quad \phi \in C_0^\infty(\Omega_+) \tag{2.17}$$

But the last formula is an immediate corollary of (2.15) (after passing to new variables (x', y') rotated on $\pi/4$ with respect to the initial variables (x, y)). Thus, formula (2.16) is verified.

The following two corollaries of *Lemma 2.1* are of fundamental significance for what follows.

Corollary 2.2 *Let the assumptions of Theorem 2.4 hold and let $u(x, y)$ be a positive bounded solution of (2.7) which satisfies (2.10). Then:*

$$\sigma_{ess}(\Delta_{x,y} - f'(u(x, y)), L^2(\Omega)) \leqslant -K < 0. \tag{2.18}$$

Indeed, due to (2.10) and (2.12) the operator $\Delta_{x,y} - f'(u(x, y))$ is a compact perturbation of $\Delta_{x,y} - f'(\Psi_c^M)$.

Corollary 2.3 *Let the assumptions of Corollary 1.1 hold. Then, the rate of decaying in (2.10) is exponential, i.e. there exist positive constants $\varepsilon \geqslant 0$ and C depending on u such that*

$$|u(x, y) - \Psi_c(x, y)| \leqslant Ce^{-\varepsilon(x+y)}, \quad (x, y) \in \Omega_+. \tag{2.19}$$

Indeed, estimate (2.19) is more or less standard corollary of (2.13), convergence (2.10) and the maximum principle, so we left its rigorous proof to the reader.

We are now ready to verify the existence of a solution $u(x, y)$. To this end, we consider the following sequence of auxiliary problems in the domains $\Omega_N := \{(x, y) \in \Omega_+, y \leqslant N\}$:

$$\begin{cases} \Delta_{x,y} u_N = f(u_N), \quad u(x, y) \geqslant 0, \\ u(0, y) = u(x, 0) = 0, \quad u(x, N) = \Psi_c(x). \end{cases} \tag{2.20}$$

Obviously, for every $N \in \mathbb{N}$, this problem has at least one solution $u_N(x, y)$ satisfying

$$0 \leqslant u_N(x, y) \leqslant c \tag{2.21}$$

(which can be obtained using $u_- = 0$ and $u_+ = c$ as sub and super solutions respectively for problem (2.20), see e.g. [104]). Moreover, this solution is also monotonic with respect to x and y and tends exponentially as $x \to \infty$ to $\Psi_c(y)$ (analogously to Corollary 1.2). We also note that, due to the elliptic regularity theorem, estimate (2.21) implies that

$$\|u_N\|_{C_b^2(\Omega_+)} \leqslant C \tag{2.22}$$

where the constant C is independent of N.

Thus, without loss of generality, we may assume that the sequence u_N tends in $C_{loc}^2(\overline{\Omega_+})$ to a some solution $u(x, y)$ of problem (2.7) as $N \to \infty$. As we have explained in the introduction, this implies that there exists $0 \leqslant c' \leqslant c$ (may be $c' = 0$) such that $f(c') = 0$ and

$$\lim_{(x,y)\to\infty} |u(x, y) - \Psi_{c'}(x, y)| = 0. \tag{2.23}$$

We need to prove that, necessarily, $c' = c$. We prove this fact using the special integral identity. In order to derive it, we multiply Eq. (2.20) by $\partial_x u_N$. Then, we have

$$\partial_x(|\partial_x u_N|^2 - |\partial_y u_N|^2 - 2F(u_N)) = -2\partial_y(\partial_x u_N \cdot \partial_y u_N) \tag{2.24}$$

where $F(u)$ is a potential of $f(u)$. Integrating this formula over Ω_N and using the boundary conditions and the fact that $|\Psi_c'(0)|^2 = -2F(c) \geq 0$, we derive that

$$\int_0^N (|\Psi_c'(0)|^2 - |\partial_x u_N(0, y)|^2)\, dy = \int_0^N 2[F(c) - F(\Psi_c(y))] + |\Psi_c'(y)|^2\, dy$$

$$- 2\int_0^\infty \Psi_c'(x) \cdot \partial_y u_N(x, N)\, dx. \qquad (2.25)$$

Since $\Psi_c'(x) \geq 0$ and $\partial_y u_N(x, N) \geq 0$, then

$$\int_0^N (|\Psi_c'(0)|^2 - |\partial_x u_N(0, y)|^2)\, dy \leq C_{\Psi_c} \qquad (2.26)$$

where the constant C_{Ψ_c} is independent of N. Moreover, obviously, the function $\partial_x u_N(0, y)$ is strictly increasing with respect to y and $\partial_x u_N(0, N) = \Psi_c'(0)$. Consequently, (2.26) implies that

$$\int_0^N |\Psi_c'(0)^2 - \partial_x u_N(0, y)^2|\, dy \leq C_{\Psi_c}. \qquad (2.27)$$

We now note that $\partial_x u(0, y)$ is monotone increasing function (since $u(x, y)$ is monotone with respect to y and $u(0, y) = 0$) and

$$\partial_x u(0, y) < \partial_x u(0, \infty) = \Psi_{c'}'(0), \quad \forall y \in \mathbb{R}_+. \qquad (2.28)$$

Since $\Psi_{c'}'(0) < \Psi_c'(0)$ if $c' < c$, see [15] and $u_N \to u$ in $C^2_{loc}(\overline{\Omega_+})$) then estimates (2.27) and (2.28) imply that the limit function $u(x, y)$ satisfies (2.23) with $c = c'$. Thus, the existence of a solution is verified.

Let us now verify the uniqueness of the constructed solution $u(x, y)$. To this end, we need the following lemma which is of independent interest also.

Lemma 2.2 *Let $u(x, y)$ be an arbitrary solution of (2.7) which satisfies (2.10). Then the spectrum of the linearization of (2.7) on $u(t, x)$ is strictly negative, i.e.*

$$\sigma(\Delta_{x,y} - f'(u)) \leq -C_u, \qquad (2.29)$$

for some positive constant C_u, depending on the solution u.

Proof Indeed, assume that (2.29) is wrong. Then, according to (2.18), there exists a nonnegative eigenvalue $\lambda_0 \geq 0$ of this operator and the corresponding eigenvector $v \in L^2(\Omega_+)$. Moreover, it can be deduced in a standard way, using condition (2.13) and the exponential convergence (2.19) that

$$|v(x, y)| \leq C_v e^{-\varepsilon(x+y)}, \quad (x, y) \in \Omega_+, \qquad (2.30)$$

for some positive constant C_v, depending on v. We may also assume, without loss of generality, then the eigenvalue $\lambda_0 \geqslant 0$ is maximal. Then, thanks to the Perron-Frobenius theory, function $v(x, y)$ is strictly positive inside of Ω_+.

We note that the function $v_1(x, y) := \partial_x u(x, y)$ is also strictly positive and satisfies the equation

$$\Delta_{x,y} v_1 - f'(u(x, y))v_1 = 0. \tag{2.31}$$

Multiplying this equation by the eigenvector $v(x, y)$ and integrating over Ω_+, integrating by parts and using the boundary conditions, we derive that

$$\int_0^\infty v_1(0, y)\partial_x v(0, y)\, dy + \lambda_0 \int_{\Omega_+} v \cdot v_1 \, dx\, dy = 0. \tag{2.32}$$

We now recall that $v_1(x, y) := \partial_x u(x, y) \geqslant 0$, $v(x, y) \geqslant 0$ and $\partial_x v(0, y) > 0$ (due to the strict maximum principle). Consequently, (2.32) implies that

$$v_1(0, y) := \partial_x u(0, y) \equiv 0. \tag{2.33}$$

Since, $u(0, y) \equiv 0$ due to the boundary conditions, then (2.33) implies that $u(x, y) \equiv 0$ (due to the uniqueness theorem for elliptic equations). This contradiction proves estimate (2.29) and *Lemma 2.2*. □

Now we are ready to verify the uniqueness. Indeed, let $u_1(x, y)$ and $u_2(x, y)$ be two solutions of problem (2.7) which satisfy (2.10). Then, without loss of generality, we may assume that

$$u_2(x, y) \geqslant u_1(x, y). \tag{2.34}$$

Indeed, if (2.34) is not satisfied, then, using the sub and supersolution method (parabolic equation method, see e.g., [104]), we may construct the third solution $u_3(x, y)$ such that

$$c \geqslant u_3(x, y) \geqslant \max\{u_1(x, y), u_2(x, y)\} \tag{2.35}$$

which is not coincide with u_1 and u_2 and for which (2.34) is satisfied.

Let us now consider the parabolic boundary value problem in Ω_+

$$\partial_t U = \Delta_{x,y} U - f(U), \quad U|_{\partial \Omega_+} = 0, \quad U|_{t=0} = U_0 \tag{2.36}$$

with the phase space

$$W_0 := \{U_0 \in L^\infty(\Omega_+), \ u_1(x, y) \leqslant U_0(x, y) \leqslant u_2(x, y)\}. \tag{2.37}$$

Then, this problem generates a semiflow on the phase space W_0:

$$S_t : W_0 \to W_0, \quad S_t U_0 := U(t) \tag{2.38}$$

which (according to the general theory, see [1, 98] and [52] and *Sect. 2.2*) possesses a global attractor $\mathcal{A}_0 \subset W_0$. Moreover, due to (2.19) and (2.37), we have the following Lyapunov function on W_0:

$$L(U_0) := \int_{\Omega_+} |\nabla(U_0 - u_1)|^2 + 2F_{u_1}(U_0 - u_1, x, y) \, dx \, dy \tag{2.39}$$

where $F_{u_1}(z, x, y) := \int_0^z f(u_1(x, y) + z) - f(u_1(x, y)) \, dz$.

Thus, the attractor $\mathcal{A}_0$ should consist of heteroclinic orbits to the appropriate equilibria, belonging to W_0 (see [1]), but as proved in *Lemma 2.2*, all of these equilibria are exponentially stable which is possible only in the case $u_1 \equiv u_2$. Therefore, the uniqueness is also proved and *Theorem 2.4* is proved.   □

2.4   Existence of Solutions of Nonlinear Elliptic Systems

In this section and in subsequent *Sects. 2.5–2.13*, we study, mainly following [103, 108], the existence of at least one solution of the following nonlinear elliptic system in an unbounded domain $\Omega \subset \mathbb{R}^n$:

$$\begin{cases} a\Delta u - \gamma \cdot Du - f(x, u) = g(x), & x \in \Omega, \\ u|_{\partial\Omega} = u_0, \end{cases} \tag{2.40}$$

where $u = (u^1, .., u^k)$, $f = (f^1, \ldots, f^k)$ and $g = (g^1, \ldots, g^k)$, Δ is the Laplacian with respect to $x = (x^1, \ldots, x^n)$, $a \in \mathcal{L}(\mathbb{R}^k, \mathbb{R}^k)$ is the $k \times k$ matrix with constant coefficients satisfying $a + a^* > 0$, and $\gamma \cdot Du$ is a differential operator of first order with constant coefficients, that is,

$$\gamma \cdot Du = \sum_{i=1}^n \gamma^i \partial_{x_i} u, \quad \gamma^i \in \mathcal{L}(\mathbb{R}^k, \mathbb{R}^k). \tag{2.41}$$

Systems of the form (2.40) were studied in [76] under some (more restrictive) conditions both on f, g, u_0 and the geometry of $\Omega \subset \mathbb{R}^n$. In what follows, we make the following assumptions on the data in (2.40).

Condition 2.1 (Assumptions on f)

1. $f \in C(\overline{\Omega} \times \mathbb{R}, \mathbb{R})$;
2. $f(x, u) \cdot u \geqslant -C_1 + C_2 |u|^r$, $C_1, C_2 > 0$ *are some constants;*
3. $|f(x, u)| \leqslant C \left(1 + |u|^{r-1}\right)$, $r > 2$, $C > 0$ *are some constants.*

Here $\xi \cdot \eta$ denotes the inner product in $\mathbb{R}^k$.

Condition 2.2 (Assumptions on Ω) *There exists a direction $\vec{l} \in \mathbb{R}^n$ such that*

1. $T_h\Omega \subset \Omega$;
2. $\bigcup\limits_{h \geqslant 0} T_{-h}\Omega = \mathbb{R}^N$, where $T_h x = x + h\vec{l}$.

Condition 2.3 (Assumptions on u_0) $u_0 \in v_0(\partial\Omega)$.

Condition 2.4 (Assumption on g) $g = (g^1, \ldots, g^k)$ *belongs to the space $\Xi(\Omega)$:*

$$\Xi(\Omega) = [W_{loc}^{-1,2}(\Omega) \cap L_{loc}^q(\Omega)],$$

where $\frac{1}{r} + \frac{1}{q} = 1$, and r is the same as in the Assumptions on f (Condition 2.1).

For simplicity of presentation, we start with $f(x, u) \equiv f(u)$, $g \equiv 0$, $\gamma \equiv 0$ and $a \equiv Id$. As we will see from the proof the existence of solution to the problem (2.40), the general case can be studied in the similar manner. We will also give necessary hints for the proof of the general case.

Therefore, we are dealing with the existence of at least one solution of the following semilinear elliptic system:

$$\begin{cases} \Delta u = f(u), \ x \in \Omega, \\ u|_{\partial\Omega} = u_0. \end{cases} \tag{2.42}$$

We especially emphasize that no assumptions on $\partial\Omega$ are made. Since we do not assume a Lipschitz property for f, uniqueness cannot be guaranteed in general.

Theorem 2.5 *Let f, Ω, u_0 satisfy the assumptions mentioned above. Then the semilinear elliptic boundary value problem (2.42) possesses at least one solution in $\theta(\Omega)$.*

Proof We will prove *Theorem 2.5* in two stages:

Step 1 First we prove that (2.42) is solvable in any bounded domain. To prove the existence of solution in this case, we reduce (2.42) to the case with homogenous boundary condition. Indeed, let $v \in \theta(\Omega)$ such that $v|_{\partial\Omega} = u_0$ and $\|v\|_{\theta(\Omega)} \leqslant 2\|u_0\|_{v_0(\partial\Omega)}$. Existence of such $v \in \theta(\Omega)$ follows from the definition of $v_0(\partial\Omega)$ (see *Sect. 1.1* for the definitions of $\theta(\Omega)$ and $v_0(\partial\Omega)$). We set $w = u - v$. Then, w satisfies

$$\begin{cases} \Delta w - f(w + v) = \Delta v, \\ w|_{\partial\Omega} = 0. \end{cases}$$

Let $\{e_j(x)| \ j \in \mathbb{N}\}$, $e_j(x) \in C_0^\infty(\Omega)$ - orthogonal basis in $L^2(\Omega)$ which is complete in $\theta_0(\Omega)$. Let $\mathbb{P}_N$ be the orthoprojector: $\mathbb{P}_N[L^2(\Omega)] = \mathbb{V}_N := span\{e_1, \ldots, e_N\}$. We consider the following finite-dimensional problem in $\mathbb{V}_N$:

$$\mathbb{P}_N(\Delta w_N - f(w_N + v) - \Delta v) = 0. \tag{2.43}$$

Proposition 2.1 *The problem* (2.43) *possesses at least one solution* $w_N(x)$ *for each* $N \in \mathbb{N}$.

Proof We define a map $\Phi : \mathbb{V}_N \to \mathbb{V}_N$,

$$\Phi(\xi) := -\mathbb{P}_N(\Delta\xi - f(\xi + v) + \Delta v), \quad \xi \in \mathbb{V}_N.$$

It is not difficult to see that $\Phi \in C(\mathbb{V}_N, \mathbb{V}_N)$. We estimate $(\Phi(\xi), \xi)$ from below to show that $(\Phi(\xi), \xi) \geq 0$ for all $\xi \in \partial \mathbb{B}_0^R$ for sufficiently large R. Indeed,

$$- (\Delta\xi, \xi) \geq C \left(\|\xi\|^2_{W^{1,2}(\Omega)} + \|\xi\|^2_{L^2(\Omega)} \right). \tag{2.44}$$

Due to the last condition on f, we obtain

$$
\begin{aligned}
(f(\xi + v), \xi) &= (f(\xi + v), \xi + v) - (f(\xi + v), v) \\
&\geq \left(-C_1 + C_2|\xi + v|^r, 1 \right) - C \left(1 + |\xi + v|^{r-1}|v|, 1 \right) \\
&\geq \left(-C_1 + 2C_3(|\xi|^r - |v|^r), 1 \right) - \left(C + C_3|\xi|^r + C_4|v|^r, 1 \right) \\
&\geq C_0\|\xi\|^r_{L^r} - C \left(1 + \|u_0\|^r_{v_0(\partial\Omega)} \right);
\end{aligned}
\tag{2.45}
$$

$$
\begin{aligned}
|(\Delta v, \xi)| &\leq \|\Delta v\|_{\theta(\Omega)} \|\xi\|_{\Xi(\Omega)} \\
&\leq C_\varepsilon \left(1 + \|\Delta v\|^2_{\theta(\Omega)} \right) + \varepsilon\|\xi\|^2_{W^{1,2}(\Omega)} + \varepsilon\|\xi\|^2_{L^r(\Omega)}.
\end{aligned}
\tag{2.46}
$$

Taking into account (2.44)–(2.46), we obtain

$$
\begin{aligned}
(\Phi(\xi), \xi) &\geq C \left(\|\xi\|^r_{L^r(\Omega)} + \|\xi\|^2_{W^{1,2}(\Omega)} + \|\xi\|^2_{L^2(\Omega)} \right) \\
&\quad - C_1 \left(1 + \|u_0\|^r_{v_0(\partial\Omega)} \right), \quad C > 0.
\end{aligned}
\tag{2.47}
$$

From (2.47) is follows that, for sufficiently large $R > 0$, in

$$\mathbb{B}_0^R = \left\{ \xi \in \mathbb{V}_N \mid \|\xi\|_{L^2(\Omega)} < R \right\},$$

we have

$$(\Phi(\xi), \xi)\big|_{\partial\mathbb{B}_0^R} > 0. \tag{2.48}$$

Hence, from the classical theorem on Brouwer's degree (see [47, 106]) it follows that the equation $\Phi(\xi) = 0$ has a solution $\xi = w_N$ for each N in $\mathbb{V}_N$.

Corollary 2.4 *Let w_N be a solution. Then*

$$||w_N||_{\theta(\Omega)} \leqslant C \left(1 + ||u_0||_{v_0(\partial\Omega)}^r \right) \tag{2.49}$$

uniformly in N.

Indeed, it follows from (2.49). To this end, we have to use (2.47) with taking into account that $\Phi(w_N) = 0$.

Now we are in position to pass to the limit in Galerkin approximation. From (2.49) it follows that $w_N \rightharpoonup w$ in $\theta(\Omega)$. Our goal is to prove that w is a desirable solution. Accordingly to the definition of weak solution, we have to show that

$$-(\nabla w, \nabla\varphi) - (f(v + w), \varphi) = -(\Delta v, \varphi) \tag{2.50}$$

for each $\varphi \in C_0^\infty(\Omega)$. Let $\varphi \in \mathbb{V}_{N_0}$ for some $N_0 \in \mathbb{N}$. Then $\mathbb{P}_{N_0}\varphi = \varphi$. Multiplying (2.42) by φ, for sufficiently large $N \gg 1$, we have

$$-(\nabla w_N, \nabla\varphi) - (f(v + w_N), \varphi) = -(\Delta v, \varphi).$$

By definition of $\theta(\Omega)$ it follows that

$$-(\nabla w_N, \nabla\varphi) \to -(\nabla w, \nabla\varphi).$$

Next, we show that

$$(f(v + w_N), \varphi) \to (f(v + w), \varphi) \text{ as } N \to \infty.$$

According to the assumptions on f, we obtain that the map $G(\xi) := f(\xi + v)$ is a continuous map (see *Sect. 1.3*)

$$G : L^{r-1}(\Omega) \to L^1(\Omega).$$

Since $\theta(\Omega)$ is compactly embedded in $L^{r-\delta}(\Omega)$ for sufficiently small $\delta > 0$, we have that

$$w_N \to w \text{ in } L^{r-1}(\Omega),$$

and, consequently, $f(v + w_N) \to f(v + w)$ in $L^1(\Omega)$. Thus, (2.50) holds for all $\varphi = \mathbb{P}_N\varphi$ for some $N \subset \mathbb{N}$. Since $\{e_j(x)\}$ is a complete system in $\theta(\Omega)$, we obtain that (2.50) holds for any $\varphi \in C_0^\infty(\Omega)$. Thus, we proved that (2.42) has a solution belonging to $\theta(\Omega)$, in the case Ω is a bounded domain. $\quad\square$

Next, we prove uniform (in $\Omega \subset \mathbb{R}^n$) a priori estimate of the form (2.51), and, based on this a priori estimate, we will prove existence of at least one solution $u \in \theta(\Omega)$ for any unbounded domain $\mathbb{R}^n$.

Lemma 2.3 *Let Ω be any open set in $\mathbb{R}^N$, $u_0 \in v_0(\partial\Omega)$ and u is a solution of (2.42). Then*

$$\|u\|_{\theta(\Omega \cap \mathbb{B}_{x_0}^R)} \leqslant C_\varepsilon \left(1 + \|u_0\|_{v_0\left(\partial\Omega \cap \mathbb{B}_{x_0}^{R+\varepsilon}\right)}^r\right), \tag{2.51}$$

where $\varepsilon > 0$, C_ε-depend on ε, R but independent of $\Omega \subset \mathbb{R}^N$ and $x_0 \in \mathbb{R}^N$.

Proof Let us fix R, ε and $x_0 \in \mathbb{R}^N$. Due to the definition of $v_0(\partial\Omega)$ there exist $v \in \theta(\Omega)$ such that

$$\|v\|_{\theta\left(\Omega \cap \mathbb{B}_{x_0}^{R+\varepsilon}\right)} \leqslant 2\|u_0\|_{v_0\left(\partial\Omega \cap \mathbb{B}_{x_0}^{R+\varepsilon}\right)} \tag{2.52}$$

with $v|_{\partial\Omega} = u_0$. Hence, $w = u - v \in \theta_0(\Omega)$ and satisfies

$$\Delta w - f(v + w) = -\Delta v. \tag{2.53}$$

We define $\psi(x)$:

$$\psi(x) = \begin{cases} \left(R + \frac{\varepsilon}{2} - |x - x_0|\right)^{\frac{2r}{r-2}}, & |x - x_0| < R + \frac{\varepsilon}{2}, \\ 0, & |x - x_0| \geqslant R + \frac{\varepsilon}{2}. \end{cases} \tag{2.54}$$

It is easy to compute and show that

$$|\nabla\psi| \leqslant C\psi^{\frac{1}{2}+\frac{1}{r}}(x).$$

To show (2.52), we multiply (2.53) by ψw and integrate:

$$(\Delta w, \psi w) - (f(v + w), \psi w) = (-\Delta v, \psi w). \tag{2.55}$$

Since $\psi w \in \theta_0\left(\Omega \cap \mathbb{B}_{x_0}^{R+\varepsilon}\right)$, we obtain

$$\begin{aligned} -(\Delta w, \psi w) = (\nabla w, \nabla(\psi w)) &= (\psi \nabla w, \nabla w) + (\nabla w, w\nabla\psi) \\ &\geqslant 2C(\psi|\nabla w|^2, 1) - C_1\left(\psi^{\frac{1}{2}}|\nabla w|, |\nabla\psi|\psi^{-\frac{1}{2}}|w|\right) \\ &\geqslant C(\psi|\nabla w|^2, 1) - \tilde{C}\left(\psi^{\frac{2}{r}}|w|^2, 1\right). \end{aligned} \tag{2.56}$$

To obtain (2.56), we used $|\nabla \psi| \leqslant C \psi^{\frac{1}{2}+\frac{1}{r}}(x)$ and Cauchy-Schwartz inequality. Due to the assumptions on f we obtain

$$(\psi f(v+w), w) \geqslant C_1(\psi |w|^r, 1) - C_2 \|u_0\|^r_{v_0 \left(\partial \Omega \cap \mathbb{B}^{R+\varepsilon}_{x_0}\right)}. \tag{2.57}$$

Moreover,

$$|(-\Delta v, \psi w)| \leqslant \| - \Delta v\|_{\theta \left(\Omega \cap \mathbb{B}^{R+\varepsilon}_{x_0}\right)} \|\psi w\|_{\theta \left(\Omega \cap \mathbb{B}^{R+\varepsilon}_{x_0}\right)}$$

$$\leqslant \varepsilon(\psi |\nabla w|^2, 1) + \varepsilon(\psi |w|^r, 1) + C_\varepsilon \left(1 + \left(\psi^{\frac{2}{r}} |w|^2, 1\right)\right)$$

$$+ \| - \Delta v\|^2_{\theta \left(\Omega \cap \mathbb{B}^{R+\varepsilon}_{x_0}\right)} + \|u_0\|^r_{v_0 \left(\partial \Omega \cap \mathbb{B}^{R+\varepsilon}_{x_0}\right)}\right) \tag{2.58}$$

Using (2.56)–(2.58), from (2.55) we obtain ($\varepsilon << 1$).

$$(\psi |\nabla w|^2, 1) + (\psi |w|^r, 1) - C \left(\psi^{\frac{2}{r}} |w|^2, 1\right)$$

$$\leqslant C_1 \left(1 + \| - \Delta v\|^2_{\theta \left(\Omega \cap \mathbb{B}^{R+\varepsilon}_{x_0}\right)} + \|u_0\|^r_{v_0 \left(\partial \Omega \cap \mathbb{B}^{R+\varepsilon}_{x_0}\right)}\right) \tag{2.59}$$

According to the Hölder inequality

$$\left|\left(\psi^{\frac{2}{r}} |w|^2, 1\right)\right| = \left|\left(\left|\psi^{\frac{1}{r}} w\right|^2, 1\right)\right| \leqslant \varepsilon \left(\left|\psi^{\frac{1}{r}} w\right|^r, 1\right) + C_\varepsilon = \varepsilon(\psi |w|^r, 1) + C_\varepsilon. \tag{2.60}$$

Inserting (2.60) into (2.59) and taking $\varepsilon << 1$ we have:

$$(\psi |\nabla w|^2, 1) + (\psi |w|^r, 1) \leqslant C_0 \left(1 + \|u_0\|^r_{v_0 \left(\partial \Omega \cap \mathbb{B}^{R+\varepsilon}_{x_0}\right)}\right). \tag{2.61}$$

Since $\psi(x) > C > 0$ when $x \in \mathbb{B}^R_{x_0}$, (2.61) leads to

$$(|\nabla w|^2, 1) + (|w|^r, 1) \leqslant C_0 \left(1 + \|u_0\|^r_{v_0 \left(\partial \Omega \cap \mathbb{B}^{R+\varepsilon}_{x_0}\right)}\right)$$

which is the assertion of *Lemma 2.3*.   □

Step 2 Now we are ready to prove *Theorem 2.5* for any unbounded domain. Indeed, let Ω be unbounded. We consider

$$\Omega = \bigcup_{k=1}^{\infty} \Omega_k, \ \Omega_k := \Omega \cap \mathbb{B}_0^k, \ \mathbb{B}_0^k := \left\{ x \in \mathbb{R}^N \mid |x| \leqslant k \right\}.$$

Let $\varphi_k(x) \in C_0^{\infty}(\mathbb{R}^N)$ such that $0 \leqslant \varphi_k(x) \leqslant 1$, $\varphi_k(x) = 1$ for $x \in \mathbb{B}_0^{k-1}$ and $\varphi_k(x) = 0$ for $x \in \mathbb{R}^N \backslash \mathbb{B}_0^k$. Let $u_0^k := \varphi_k v|_{\partial \Omega_k}$, where $v \in \theta(\Omega)$, $v|_{\partial \Omega} = u_0$. It is not difficult to see that

$$u_0^k|_{\partial \Omega \cap \mathbb{B}_0^{k-1}} = u_0|_{\partial \Omega \cap \mathbb{B}_0^{k-1}}, \ u_0^k|_{\partial \Omega_k \backslash \partial \Omega} = 0.$$

According to Step 1, $\forall k \in \mathbb{N}$ there exists u_k solution of (2.42) for Ω_k, with $u^k|_{\partial \Omega_k} = u_0^k$ and $\|u_k\|_{\theta_0(\Omega_M)} \leqslant C_M$ uniformly in $k \geqslant M+1$. Since $\theta_0(\Omega_M)$ is reflexive, using the Cantor diagonalization procedure, one can extract a subsequence from $\{u_k\}$ (for simplicity we denote it also by $\{u_k\}$), such that

$$u_k \rightharpoonup u \text{ in } \theta(\Omega_M) \ \forall M \in \mathbb{N}, \ k \geqslant M+1.$$

Let us show that u is the desirable solution of (2.42). Indeed, let $\varphi \in C_0^{\infty}(\Omega)$ and $\operatorname{supp} \varphi \in \Omega_L$. Then, for $k > L$, we have

$$-(\nabla u_k, \nabla \varphi) = (f(u_k), \varphi).$$

Passing to the limit as $k \to \infty$ in the last equality (we justify it below), we obtain

$$-(\nabla u, \nabla \varphi) = (f(u), \varphi).$$

Hence u is a solution of (2.42). This proves *Theorem 2.5* in the case $f = f(u)$, $g \equiv 0$, $\gamma \equiv 0$ and $a \equiv Id$.

Remark 2.5 In general case, that is, in the presence of $f = f(x, u)$, $\gamma \cdot Du$, $g = g(x)$, we proceed in the Step 1 as follows:

We consider a map $\Phi : \mathbb{V}_N \to \mathbb{V}_N$:

$$\Phi(\xi) := -\mathbb{P}_N(a \Delta \xi + \gamma \cdot Du - f(x, \xi + v) - g_1(x)), \ \xi \in \mathbb{V}_N, \qquad (2.62)$$

where $g_1(x) := g(x) - a \Delta v - \gamma \cdot Dv$. We aim to prove that for sufficiently large $R > 0$, $(\Phi(\xi), \xi) \geqslant 0$ for all $\xi \in \partial B_0^R$. To this end, we have to estimate the terms: $(-a \Delta \xi, \xi)$, $(f(x, \xi + v), \xi)$, $(\gamma \cdot D\xi, \xi)$ and $(g_1(x), \xi)$. Indeed,

$$(-a \Delta \xi, \xi) = (a \nabla \xi, \nabla \xi) = \frac{1}{2}((a + a_*) \nabla \xi, \nabla \xi) \geqslant C \left(\|\xi\|_{W^{1,2}(\Omega)}^2 + \|\xi\|_{L^2(\Omega)}^2 \right). \qquad (2.63)$$

Here

$$(a\nabla\xi, \nabla\xi) = \sum_{i=1}^{k} \left(a\partial_{x_i} u \cdot \partial_{x_i} w \right).$$

The term $(f(x, \xi + v), \xi)$ can be estimated as follows

$$
\begin{aligned}
(f(x, \xi + v), \xi) &= (f(x, \xi + v), \xi + v) - (f(x, \xi + v), v) \\
&\geqslant (-C_1 + C_2|\xi + v|^r, 1) - (C(1 + |\xi + v|^{r-1}|v|), 1) \\
&\geqslant (-C_1 + 2C_3(|\xi|^r - |v|^r), 1) \\
&\quad - (C + C_3|\xi|^r + C_4|v|^r, 1) \\
&\geqslant C_0\|\xi\|_{L^r}^r - C\left(1 + \|u_0\|_{v_0(\partial\Omega)}^r\right).
\end{aligned}
\tag{2.64}
$$

As for $(\gamma \cdot D\xi, \xi)$, it can be estimated in the following way:

$$|(\gamma \cdot D\xi, \xi)| \leqslant \mu\|\xi\|_{W^{1,2}(\Omega)}^2 + \mu\|\xi\|_{L^r(\Omega)}^r + C_\mu, \ C_\mu = C_\mu(|\Omega|), \tag{2.65}$$

where μ is an arbitrary positive real number. To obtain the above inequality, we use the Hölder inequality with exponents 2, r and $\frac{2r}{r-2}$. We estimate

$$|(g_1(x), \xi)| \leqslant \|\xi, \Omega\|_+ |g_1, \Omega|_+ \leqslant C_\mu\left(1 + |g_1, \Omega|_+^2\right)$$

$$+ \mu\left(\|\xi\|_{W^{1,2}(\Omega)}^2 + \|\xi\|_{L^r(\Omega)}^r\right) \tag{2.66}$$

(for the definitions of $\|\cdot\|_+$ and $|\cdot|_+$ see *Sect. 1.1*) Inserting estimates (2.63)–(2.66) into the expression for $(\Phi(\xi), \xi)$:

$$(\Phi(\xi), \xi) = -(a\Delta\xi, \xi) - (\gamma \cdot D\xi, \xi) + (f(x, \xi + v), \xi) + (g_1(x), \xi),$$

we obtain that

$$(\Phi(\xi), \xi) \geqslant C\left(\|\xi\|_{W^{1,2}(\Omega)}^2 + \|\xi\|_{L^r(\Omega)}^r\right) - C_1\left(1 + \|u_0\|_{v_0(\partial\Omega)}^r + |g_1, \Omega|_+^2\right), \tag{2.67}$$

where $C > 0$ and Ω is a bounded domain. Hence, from (2.66) it follows that, for sufficiently large $R > 0$,

$$(\Phi(\xi), \xi) \geqslant 0 \text{ in } \mathbb{B}_0^R,$$

where $\mathbb{B}_0^R = \{\xi \in \mathbb{V}_N | \ \|\xi\|_{L^2(\Omega)} \leqslant R\}$.

Remark 2.6 Therefore, for bounded Ω, the presence of $f = f(x, u)$, $a \in \mathcal{L}(\mathbb{R}^k, \mathbb{R}^k)$, $\gamma \cdot Du$ and $g = g(x)$ do not bring additional difficulties. The case when Ω is unbounded, repeats word by word the proof of *Theorem 2.5* in the special case. Hence, the analog of the estimate (2.51) in this case has the following form:

$$\|u\|_{\theta\left(\Omega\cap\mathbb{B}_{x_0}^R\right)} \leqslant C_\varepsilon \left(1 + \|g\|_{\Sigma\left(\Omega\cap\mathbb{B}_{x_0}^{R+\varepsilon}\right)} + \|u_0\|_{\nu_0\left(\partial\Omega\cap\mathbb{B}_{x_0}^{R+\varepsilon}\right)}^r\right), \tag{2.68}$$

where ϵ is an arbitrary positive number, and the constant C_ε depends upon ϵ and R, as well as the constants appearing in the assumptions on f, but is independent of $\Omega \subset \mathbb{R}^n$ and $x_0 \in \mathbb{R}^n$.

2.5 Regularity of Solutions

In this section, we prove several theorems on the additional regularity of solutions for (2.40) depending on the additional regularity of $g(x)$. First, assume that

$$g \in [L^q(\Omega)]^k, \quad \text{where } \frac{1}{q} + \frac{1}{r} = 1 \text{ and } r \text{ is the same as in the assumptions on } f. \tag{2.69}$$

Theorem 2.6 *Let $u(x)$ be a solution of (2.40) and the condition (2.69) be satisfied and $B_{x_0}^R \subset\subset \Omega$. Then $u \in \left[H^{2,q}\left(B_{x_0}^R\right)\right]^k$ and*

$$\left\|u, B_{x_0}^R\right\|_{2,q} \leqslant C \left(1 + \left\|g, B_{x_0}^{R+\varepsilon}\right\|_{0,q}^{2r}\right), \tag{2.70}$$

where $B_{x_0}^{R+\varepsilon} \subset\subset \Omega$.

Proof Let $\varphi(x) \in C_0^\infty(\Omega)$, such that $\varphi(x) = 1$ if $|x - x_0| \leqslant R$ and $\varphi(x) = 0$ if $|x - x_0| \geqslant R + \varepsilon$. Multiplying (2.40) by φ and performing some elementary calculations, we obtain

$$\begin{cases} \Delta(\varphi u) = 2\nabla\varphi \cdot \nabla u + u\Delta\varphi + a^{-1}\left(-\varphi\gamma Du + \varphi f(x, u) + \varphi g\right) =: g_2(x), \\ \varphi u|_{B_{x_0}^{R+\varepsilon}} = 0. \end{cases}$$

Here $\nabla\varphi \cdot \nabla u = \sum_{i=1}^k \partial_i\varphi\partial_i u$. Analogously to (2.51),

$$\left\|g_2, B_{x_0}^{R+\varepsilon}\right\|_{0,q} \leqslant C \left(1 + \left\|u, B_{x_0}^{R+\varepsilon}\right\|_+^r + \left\|g, B_{x_0}^{R+\varepsilon}\right\|_{0,q}\right) \leqslant C_1 \left(1 + \left\|g, B_{x_0}^{R+\varepsilon}\right\|_{0,q}^{2r}\right). \tag{2.71}$$

From the L^q regularity of the Laplacian it follows that

$$\left\| u, B_{x_0}^R \right\|_{2,q} \leqslant C \left\| \varphi u, B_{x_0}^{R+\varepsilon} \right\|_{2,q} \leqslant C_1 \left\| g_2, B_{x_0}^{R+\varepsilon} \right\|_{0,q}. \tag{2.72}$$

Then, the estimate (2.70) is a consequence of (2.71) and (2.72). *Theorem 2.6* is thus proved. $\square$

Let us assume the matrix is self-adjoint, so that $a = a^* > 0$, and $g \in [L^p(\Omega)]^k$ for some $p > \frac{N}{2}$. Moreover, assume also that there exists a $v \in \theta(\Omega)$ such that $v|_{\partial\Omega} = u_0$, $v|_{\Omega \cap B_{x_0}^{R+\varepsilon}} \in H^{2,p}\left(\Omega \cap B_{x_0}^{R+\varepsilon}\right)$. Let us mention again that by $H^{l,p}(\Omega)$ we denote the space $W^{l,p}(\mathbb{R}^n)|_\Omega$. Then, we have a stronger result than the one stated in *Theorem 2.6*:

Theorem 2.7 *Let* $u \in \theta(\Omega)$ *be a solution of* (2.40) *and all above mentioned assumptions fulfilled. Then,* $u \in \left[L^\infty(\Omega \cap B_{x_0}^R)\right]^k$ *and it holds*

$$\left\| u, \Omega \cap B_{x_0}^R \right\|_{0,\infty} \leqslant Q \left(\left\| g, \Omega \cap B_{x_0}^{R+\varepsilon} \right\|_{0,p} \right) + Q \left(\left\| v, \Omega \cap B_{x_0}^{R+\varepsilon} \right\|_{H^{2,p}} \right), \tag{2.73}$$

where $Q : \mathbb{R}_+ \to \mathbb{R}_+$ *is some monotonic function depending on* f *and the constants* R *and* ε.

Proof For simplicity, we consider the case $v \equiv 0$. The general case is standard (see, for example, the proof of *Lemma 2.3*). Hence, in what follows, we assume that $u \in \theta_0(\Omega)$. To prove the assertion of *Theorem 2.7*, we need several lemmas, which we prove below.

Lemma 2.4 *There exists a* $\varphi \in C^2(\mathbb{R}^n)$, $\varphi \geqslant 0$, *such that*

$$\begin{cases} \varphi(x) = 1 \text{ if } |x - x_0| < R; \ \varphi(x) = 0 \text{ if } |x - x_0| \geqslant R + \varepsilon, \\ |\nabla\varphi(x)| \leqslant C\varphi^{\frac{1}{2}+\frac{1}{r}}(x), \ x \in \mathbb{R}^n, \\ |\Delta\varphi(x)| \leqslant C\varphi^{\frac{2}{r}}(x), \ x \in \mathbb{R}^n. \end{cases} \tag{2.74}$$

Proof Let $h_*(x) \in C_0^\infty(\Omega)$ such that $0 \leqslant h_*(x) \leqslant 1$ and $h_*(x) = 1$ if $x \in B_{x_0}^R$ and $h_*(x) = 0$ if $|x - x_0| > R + \frac{\varepsilon}{2}$. Let

$$\psi(t) = \begin{cases} t^\alpha & t \geqslant 0, \\ 0 & t < 0, \end{cases}$$

where α is a sufficiently large number. Then, it is not difficult to see that the function

$$\varphi(x) = h_*(x) + (1 - h_*(x))\psi(R + \varepsilon|x - x_0|)$$

satisfies all conditions of *Lemma 2.4*. $\square$

Lemma 2.5 *Let* $u \in \theta_0(\Omega)$ *be a solution of* (2.40) *and* $w = \varphi a u \cdot u$. *Then, the function*

$$\hat{w} = \begin{cases} w & x \in \Omega, \\ 0 & x \notin \Omega \end{cases}$$

belongs to $H_0^{1,l}(\mathbb{R}^n)$, $l = \frac{2r}{r+2}$ *and satisfies*

$$\begin{cases} \Delta \hat{w} = \hat{h}_u(x), \\ \hat{w}\big|_{B_{x_0}^{R+\varepsilon}} = 0, \end{cases} \tag{2.75}$$

where by $\hat{v}$ *we denote the extension of a given function* v *by the zero when* $x \notin \Omega$, *and*

$$h_u(x) := 2\varphi a \nabla u \cdot \nabla u + 2\nabla \varphi \cdot \nabla(au \cdot u) + \Delta \varphi au \cdot u - 2\gamma \, Du \cdot u$$
$$+ 2\varphi f(x, u) \cdot u + 2\varphi g \cdot u \tag{2.76}$$

Proof Since $u \in \theta_0(\Omega)$, hence $\hat{u} \in \theta(\mathbb{R}^n)$. Using the Hölder inequality with the exponents $\frac{r}{l}$ and $\frac{2}{l}$, we have

$$\left\| \nabla \hat{w}, B_{x_0}^{R+\varepsilon} \right\|_{0,l}^l \leqslant C \int_{B_{x_0}^{R+\varepsilon}} |\hat{u}|^l |\nabla \hat{u}|^l \, dx \leqslant C_1 \left(\left\| \hat{u}, B_{x_0}^{R+\varepsilon} \right\|_{0,r} \left\| \nabla \hat{u}, B_{x_0}^{R+\varepsilon} \right\|_{0,2} \right)^l .$$

Consequently, $\hat{w} \in H_0^{1,k}\left(B_{x_0}^{R+\varepsilon}\right)$ for $k = \min\left\{\frac{r}{2}, l\right\} = l$. In the sequel, we denote, as usual, by $< \cdot, \cdot >$ and $< \cdot, \cdot >_\Omega$ the scalar product in $L^2(\mathbb{R}^n)$ and $L^2(\Omega)$, respectively. Let $\Phi \in C_0^\infty(\mathbb{R}^n)$. Then,

$$< \Delta \hat{w}, \varphi > = < \hat{w}, \Delta \Phi > = - < \nabla w, \nabla \Phi >_\Omega$$
$$= - < \nabla \phi au \cdot u, \nabla \Phi >_\Omega - < \phi \nabla(au \cdot u), \nabla \Phi >_\Omega$$
$$= < \Delta \phi au \cdot u, \Phi >_\Omega + < \nabla \phi \cdot \nabla(au \cdot u), \Phi >_\Omega - < \phi \nabla(au \cdot u), \nabla \Phi >_\Omega \tag{2.77}$$

To obtain (2.77), we several times used the Green's formula

$$< \nabla U_1, U_2 >_\Omega + < U_1, \nabla U_2 >_\Omega = 0,$$

in which $U_1 \in W_0^{1,l}(\Omega \cap B_{x_0}^{R+\varepsilon})$, $U_2 \in W_0^{1,l*}(\Omega \cap B_{x_0}^{R+\varepsilon})$ and $\frac{1}{l} + \frac{1}{l_*} = 1$ (see [78]) Let us transform $< \phi \nabla(au \cdot u), \nabla \Phi >_\Omega$ in the following way:

$$< \phi \nabla(au \cdot u), \nabla \Phi >_\Omega = 2 < a \nabla u, \nabla(\varphi u \Phi) >_\Omega - 2 < \varphi a \nabla u \cdot \nabla u, \Phi >_\Omega$$
$$- < \nabla \varphi \cdot \nabla(au \cdot u), \Phi >_\Omega . \tag{2.78}$$

Since u is a solution of (2.40) and $\varphi u \Phi \in \theta_0(\Omega \cap B_{x_0}^{R+\varepsilon})$, we have

$$- < a\nabla u, \nabla(\varphi u \Phi) + < \varphi \gamma \, Du \cdot u, \Phi >_\Omega - < \varphi f(x, u) \cdot u, \Phi >_\Omega$$
$$= < \varphi g \cdot u, \Phi >_\Omega . \tag{2.79}$$

Inserting instead of $< \phi \nabla(au \cdot u), \nabla\Phi >_\Omega$ in (2.77) its expressions obtained via (2.78) and (2.79), we obtain

$$< \Delta \widehat{w}, \Phi > = < h_u, \Phi >_\Omega = < \widehat{h}_u, \Phi > .$$

This proves *Lemma 2.5*.   □

Lemma 2.6 *The functions* $\widehat{h}_u \in H^{-1,l}\left(B_{x_0}^{R+\varepsilon}\right) \cap L^1\left(B_{x_0}^{R+\varepsilon}\right)$ *and satisfies*

$$\widehat{h}_u(x) \geqslant -C_1\left(1 + |\widehat{g}(x)|\widehat{w}^{\frac{1}{2}}\right) \equiv h(x) \text{ for almost all } x \in \mathbb{R}^n. \tag{2.80}$$

Proof $\widehat{h}_u \in L^1\left(B_{x_0}^{R+\varepsilon}\right)$ follows from (2.76) and $\widehat{h}_u \in H^{-1,l}\left(B_{x_0}^{R+\varepsilon}\right)$ follows from (2.75) and the assertion of *Lemma 2.5*. Next, we prove the estimate (2.80). Similar to estimates (2.56)–(2.61), using (2.74), we obtain

$$h_u(x) \geqslant C(\varphi|\nabla u|^2 + \varphi|u|^r) - C_1\left(\varphi|\nabla u||u| + |\nabla\varphi||\nabla u||u| + |\Delta\varphi||u|^2\right.$$
$$+ 1 + \varphi|g||u|)$$
$$\geqslant C_0(\varphi|\nabla u|^2 + \varphi|u|^r) - C_2\left(1 + |g|(\varphi au \cdot u)^{\frac{1}{2}}\right)$$
$$\geqslant -C_2\left(1 + |g|w^{\frac{1}{2}}\right) .$$

This proves *Lemma 2.6*.   □

Lemma 2.7 *Let* $h_i \in H^{-1,l}\left(B_{x_0}^{R+\varepsilon}\right)$, $i = 1, 2, l > 1$ *and* $w_i \in H_0^{1,l}\left(B_{x_0}^{R+\varepsilon}\right)$ *are the solutions of*

$$\begin{cases} \Delta w_i = h_i, \\ w_i|_{\partial B_{x_0}^{R+\varepsilon}} = 0, \end{cases}$$

respectively. Let it also hold

$$< h_1, \Phi > \geqslant < h_2, \Phi > \text{ for each } \Phi \in C_0^\infty\left(B_{x_0}^{R+\varepsilon}\right). \tag{2.81}$$

Then, for almost all $x \in B_{x_0}^{R+\varepsilon}$

$$w_1(x) \leqslant w_2(x).$$

Proof For $0 < \delta << 1$, we define the 'average' operator S_δ:

$$S_\delta : D'\left(B_{x_0}^{R+\varepsilon}\right) \to C^\infty\left(B_{x_0}^{R+\varepsilon}\right),$$

$$(S_\delta h)(x) := \int_{\mathbb{R}^n} \varphi_\delta(|x - y|) h(T_\delta y)\, dy \equiv (\det T_\delta)^{-1} \left\langle h(z), \varphi_\delta\left(\left|x - T_\delta^{-1} z\right|\right)\right\rangle,$$

(2.82)

where $T_\delta x \equiv x_0 + \frac{R+\varepsilon}{R+\varepsilon+2\delta}(x - x_0)$ and $\varphi_\delta(|z|) := \frac{1}{\delta^n}\varphi\left(\frac{|z|}{\delta}\right)$, where $\varphi \in C^\infty(\mathbb{R})$, $\operatorname{supp}\varphi \subset [-1, 1]$ and $\int_{\mathbb{R}^n} \varphi(|z|)\, dz = 1$. It is not difficult to show that

$$(S_\delta^* \Phi)(z) = (\det T_\delta)^{-1} \int_{\mathbb{R}^n} \varphi_\delta\left(\left|x - T_\delta^{-1} z\right|\right)\Phi(x)\, dx, \quad < S_\delta h, \Phi > = < h, S_\delta^* \Phi >$$

and

$$S_\delta^* : C^\infty\left(B_{x_0}^{R+\varepsilon}\right) \to C^\infty\left(B_{x_0}^{R+\varepsilon}\right),$$

and for each $\Phi \in H_0^{1,p}\left(B_{x_0}^{R+\varepsilon}\right)$ and $1 \leqslant p \leqslant \infty$ $S_\delta^* \Phi \to \Phi$ as $\delta \to 0$ in $H^{1,p}\left(B_{x_0}^{R+\varepsilon}\right)$. To prove the assertion of *Lemma 2.7*, we consider $v = w_1 - w_2$, where w_1 and w_2 are due to the *Lemma 2.7*. Then,

$$\Delta v = h_1 - h_2, \quad v|_{\partial B_{x_0}^{R+\varepsilon}} = 0.$$

Let $h_\delta = S_\delta(h_1 - h_2)$. Then, $h_\delta \to h$ as $\delta \to 0$ in $H^{-1,l}\left(B_{x_0}^{R+\varepsilon}\right)$. Indeed, for all $\Phi \in H^{1,l*}\left(B_{x_0}^{R+\varepsilon}\right)$

$$|< S_\delta h, \Phi > - < h, \Phi >| = | < h, S_\delta^* \Phi - \Phi > | \leqslant ||h||_{-1,l}||S_\delta^* \Phi - \Phi||_{1,l*} \xrightarrow[\delta \to 0]{} 0.$$

Moreover, from (2.81) and (2.82), it follows that $h_\delta \in C^\infty\left(B_{x_0}^{R+\varepsilon}\right)$, $h_\delta \geqslant 0$ and

$$\begin{cases} \Delta v_\delta = h_\delta, \\ v_\delta|_{\partial B_{x_0}^{R+\varepsilon}} = 0. \end{cases}$$

Due to the maximum principle (see *Chap. 1*), $v_\delta(x) \leqslant 0$ for all $\delta > 0$ and $x \in B_{x_0}^{R+\varepsilon}$. Since Δ_x is an isomorphism between $H_0^{1,l}\left(B_{x_0}^{R+\varepsilon}\right)$ and $H^{-1,l*}\left(B_{x_0}^{R+\varepsilon}\right)$, we have $v_\delta \rightharpoonup v$ in $H_0^{1,l}\left(B_{x_0}^{R+\varepsilon}\right)$. Hence, $v \leqslant 0$ a.e. in $B_{x_0}^{R+\varepsilon}$. This proves *Lemma 2.7*. □

Lemma 2.8 *Assume that the function $\widehat{w}$ defined in Lemma 2.5 belongs to $L^m\left(B_{x_0}^{R+\varepsilon}\right)$ for some $m \geqslant 1$. Then,*

$$\widehat{w} \in L^{k(m)}\left(B_{x_0}^{R+\varepsilon}\right),$$

where $k(m)$ is defined via

$$k(m) = \begin{cases} \frac{2pnm}{pn+2(n-2p)m} & \text{if } pn + 2(n-2p)m > 0 \\ \infty & \text{if } pn + 2(n-2p)m < 0 \end{cases} \tag{2.83}$$

and the following estimate holds:

$$\left\| \widehat{w}, B_{x_0}^{R+\varepsilon} \right\|_{0,k(m)} \leqslant C \left\{ 1 + \left\| \widehat{g}, B_{x_0}^{R+\varepsilon} \right\|_{0,p} \left\| \widehat{w}, B_{x_0}^{R+\varepsilon} \right\|_{0,m}^{\frac{1}{2}} \right\} \tag{2.84}$$

Proof First, consider the following axillary problem:

$$\begin{cases} \Delta \widehat{w}_1 = \widehat{h}(x), \\ \widehat{w}_1|_{\partial B_{x_0}^{R+\varepsilon}} = 0, \end{cases}$$

where $h(x)$ is defined by (2.80). Then, with *Lemmas 2.6 and 2.7*, we obtain that

$$\widehat{w}(x) \leqslant \widehat{w}_1(x) \text{ for all most all } x \in \Omega. \tag{2.85}$$

From the Hölder inequality, it follows that

$$\left\| h, B_{x_0}^{R+\varepsilon} \right\|_{0,s} \leqslant C \left(1 + \left\| g, B_{x_0}^{R+\varepsilon} \right\|_{0,p} \left\| \widehat{w}, B_{x_0}^{R+\varepsilon} \right\|_{0,m}^{\frac{1}{2}} \right), \quad s = \frac{2pm}{p+2m}.$$

Then, the L^p regularity of solutions for the Laplace equation leads to $\widehat{w}_1 \in W^{2,s}\left(B_{x_0}^{R+\varepsilon}\right)$ and, as a consequence of the embedding $W^{2,s} \subset\subset L^q$ for $\frac{1}{q} = \frac{1}{2} - \frac{s}{n}$, $q = k(m)$. Thus, from (2.85) and $\widehat{w} \geqslant 0$, it follows that

$$\left\| \widehat{w}, B_{x_0}^{R+\varepsilon} \right\|_{0,q} \leqslant \left\| \widehat{w}_1, B_{x_0}^{R+\varepsilon} \right\|_{0,q} \leqslant C_1 \left\| \widehat{w}_1, B_{x_0}^{R+\varepsilon} \right\|_{2,s}$$

$$\leqslant C_2 \left\| \widehat{h}_1, B_{x_0}^{R+\varepsilon} \right\|_{0,s} \leqslant C_3 \left(1 + \left\| \widehat{g}, B_{x_0}^{R+\varepsilon} \right\|_{0,p} \left\| \widehat{w}, B_{x_0}^{R+\varepsilon} \right\|_{0,m}^{\frac{1}{2}} \right).$$

This proves *Lemma 2.8*. □

Now we are ready to finish the proof of *Theorem 2.7*. Analogously to (2.51), we obtain that

$$\left\| \hat{w}, B_{x_0}^{R+\varepsilon} \right\|_{0,\frac{r}{2}} \leqslant C \left(1 + \left\| \hat{g}, B_{x_0}^{R+\varepsilon} \right\|_{0,p}^4 \right).$$

Thus, all assumptions of *Lemma 2.8* are fulfilled with $m = \frac{r}{2} > 1$. Let us consider a sequence $m_0 = \frac{r}{2}$, $m_{l+1} = k(m_l)$. Next, we show that $m_l = \infty$ for sufficiently large l. Indeed, from (2.83) and the condition $n - 2p < 0$, it follows that $m_{l+1} \geqslant 2m_l$ or $m_l \geqslant 2^{l-1} r$. Substituting this inequality in (2.83), we obtain that m_∞ for $l > L = \log_2 \frac{pn}{r(2p-n)}$. In order to obtain the estimate (2.73), we have to iterate the estimate (2.84) $[L] + 1$ times. The *Theorem 2.7* is thus complete. $\qquad\square$

Theorem 2.8 *Let $u \in \theta(\Omega)$ be a solution of (2.40) and all assumptions of Theorem 2.7 fulfilled. Then $u \in \left[H^{2,p} \left(B_{x_0}^R \right) \right]^k$, and it holds*

$$\left\| u, B_{x_0}^R \right\|_{2,p} \leqslant Q \left(\left\| g, B_{x_0}^{R+\varepsilon} \right\|_{0,p} \right), \tag{2.86}$$

where $Q : \mathbb{R}_+ \to \mathbb{R}_+$ is some monotonic function depending on f and the constants R and ε.

Proof Due to the estimate (2.73) and the conditions on f, we have

$$\left\| f(x,u), B_{x_0}^R \right\|_{0,\infty} \leqslant Q_1 \left(\left\| g, B_{x_0}^{R+\varepsilon} \right\|_{0,p} \right). \tag{2.87}$$

To finish the proof, we need the following *Lemma 2.9*.

Lemma 2.9 *Let $u \in H^{1,m}(B_{x_0}^R)$ for some $m > 1$. Then, $u \in H^{1,q(m)}(B_{x_0}^R)$, where*

$$q(m) = \begin{cases} \min \left\{ \frac{mn}{n-m}, p \right\} & \text{if } n > m, \\ p & \text{if } n \leqslant m, \end{cases} \tag{2.88}$$

and the following estimate holds

$$\left\| u, B_{x_0}^R \right\|_{1,q} \leqslant C \left(\left\| g, B_{x_0}^{R+\varepsilon} \right\|_{0,p} + \left\| u, B_{x_0}^{R+\varepsilon} \right\|_{1,m} + \left\| f(x,u), B_{x_0}^{R+\varepsilon} \right\|_{0,p} \right).$$

Proof Let $\varphi(x) \in C_0^\infty(\mathbb{R}^n)$ such that $\varphi \equiv 0$ if $x \notin B_{x_0}^{R+\varepsilon}$ and $\varphi(x) = 1$ if $x \in B_{x_0}^R$. We rewrite (2.40) in the following form

$$\begin{cases} \Delta(\varphi u) = h(x), \\ u|_{\partial B_{x_0}^{R+\varepsilon}} = 0, \end{cases} \tag{2.89}$$

where

$$h(x) := 2 \sum_{i=1}^{k} \partial_i \varphi \partial_i u + \Delta \varphi u + a^{-1} \varphi \left\{ f(x, u) + g - \gamma Du \right\}.$$

From L^m-regularity of solutions of (2.89) and the embedding theorem, we have

$$\left\| u, B_{x_0}^{R} \right\|_{1,q} \leqslant C_0 \left\| u, B_{x_0}^{R} \right\|_{2,m} \leqslant C \left\| \varphi u, B_{x_0}^{R+\varepsilon} \right\|_{2,m} \leqslant C_1 \left\| h, B_{x_0}^{R+\varepsilon} \right\|_{0,m}$$

$$\leqslant C_2 \left(\left\| g, B_{x_0}^{R+\varepsilon} \right\|_{0,p} + \left\| u, B_{x_0}^{R+\varepsilon} \right\|_{1,m} + \left\| f(x, u), B_{x_0}^{R+\varepsilon} \right\|_{0,p} \right) \qquad (2.90)$$

This proves *Lemma 2.9.*   □

Now we are ready to finish the proof of *Theorem 2.8.* Let $m_0 = 2$ and $m_{l+1} = q(m_l)$. Then, it is not difficult to see that $m_l = p$ for sufficiently large l. Indeed, it follows from (2.88) that

$$m_{l+1} \geqslant \min \left\{ p, m_l \frac{n}{n - m_l} \right\} \geqslant \min \left\{ p, m_l \frac{n}{n - 2} \right\}.$$

Hence, $m_l = p$ if $l > L = \frac{\ln \frac{p}{2}}{\ln \frac{n}{n-2}}$. Then, analogously to (2.68),

$$\left\| u, B_{x_0}^{R} \right\|_{1,2} \leqslant C \left(\left\| g, B_{x_0}^{R+\varepsilon} \right\|_{0,p} + 1 \right).$$

Iterating the estimate (2.90) $[L] + 1$ times (with $m_0 = 2$ and $\frac{\varepsilon}{[L]+1}$ instead of ε), we obtain

$$\left\| u, B_{x_0}^{R} \right\|_{1,p} \leqslant C \left(\left\| g, B_{x_0}^{R+\varepsilon} \right\|_{0,p}^{2} + \left\| f(x, u), B_{x_0}^{R+\varepsilon} \right\|_{0,p} \right).$$

Applying analogously to (2.89) the theorem of L^q-regularity, we obtain a formula similar to (2.90)

$$\left\| u, B_{x_0}^{R} \right\|_{2,p} \leqslant C \left(\left\| g, B_{x_0}^{R+\varepsilon} \right\|_{0,p}^{2} + \left\| f(x, u), B_{x_0}^{R+\varepsilon} \right\|_{0,p} \right). \qquad (2.91)$$

Hence, the assertion of *Theorem 2.8* follows from (2.91) and (2.87). This proves *Theorem 2.8.*   □

Remark 2.7 If the boundary $\partial\Omega \cap B_{x_0}^{R+\varepsilon}$ is a smooth manifold and the boundary value u_0 satisfies (2.87), then, in the same manner as above, one can obtain estimates analogously to (2.86) for the point $x_0 \in \partial\Omega$.

2.6 Boundedness of Solutions as $|x| \to \infty$

In this section, we use the estimates obtained in the previous section to show the boundedness of solutions for the problem (2.40) in various functional spaces. To this end, we need several definitions.

Definition 2.4 We denote by $\sum_b(\Omega)$ the Banach space of functions $g \in \sum(\Omega)$, for which

$$|g; b|_+ = \sup_{x_0 \in \mathbb{R}^n} |g, B^1_{x_0} \cap \Omega|_+ < \infty.$$

Analogously, we define $v_b(\partial\Omega)$ with

$$|u_0; b|_0 = \sup_{x_0 \in \mathbb{R}^n} |u_0, B^1_{x_0} \cap \partial\Omega|_0 < \infty$$

and $\theta_b(\Omega)$ with

$$|u; b|_+ := \|u, \Omega; b\|_+ = \sup_{x_0 \in \mathbb{R}^n} \left\| u, B^1_{x_0} \cap \Omega \right\|_+ < \infty.$$

Definition 2.5 Let $1 \leqslant p < \infty$, $l = 0, 1, 2$. We denote by $W^{l,p}_b(\Omega)$ the Banach space of functions $g \in W^{l,p}_{loc}(\Omega)$ such that

$$\|g; b\|_{l,p} = \sup_{x_0 \in \mathbb{R}^n} \left\| g, B^1_{x_0} \cap \Omega \right\|_{l,p} < \infty.$$

We denote by $L^p_b(\Omega) := W^{0,p}_b(\Omega)$.

Theorem 2.9 *Let $g(x) \in \sum_b(\Omega)$ and $\varepsilon > 0$. Then:*

1. Any solution u of (2.40) belongs to $\theta_b(\Omega_\varepsilon)$, where

$$\Omega_\varepsilon = \{x \in \Omega | \ \text{dist}(x, \partial\Omega) > \varepsilon\},$$

and the following estimate holds:

$$\|u, \Omega_\varepsilon; b\| \leqslant C_\varepsilon \left(1 + |g, \Omega; b|^2_+\right).$$

2. Let $u_0 \in v_b(\partial\Omega)$. Then, any solution u of (2.40) belongs to $\theta_b(\Omega)$, and the following estimate holds:

$$\|u, \Omega; b\|_+ \leqslant C \left(1 + |g; b|^2_+ + |u_0; b|^r_0\right).$$

A proof of this Theorem is an immediate consequence of (2.68).

Theorem 2.10 *Let* $a = a^*$ *and* $g \in L_b^p(\Omega)$ *for* $p > \frac{n}{2}$. *Then, any solution of (2.40) belongs to* $W_b^{2,p}(\Omega_\varepsilon)$ *for each* $\varepsilon > 0$, *and the following estimate holds:*

$$\|u, \Omega_\varepsilon; b\|_{2,p} \leqslant Q_\varepsilon \left(\|g, \Omega; b\|_{0,p}\right),$$

where $Q_\varepsilon : \mathbb{R}_+ \to \mathbb{R}_+$ *is some monotonic function.*

A proof of this Theorem is an immediate consequence of (2.86).

Example 2.5 Let $\Omega = \mathbb{R}^n$ and all assumptions of *Theorem 2.9* hold. Then, any solution of (2.40) belongs to $W_b^{2,p}(\Omega)$ and, consequently, to $C_b(\Omega)$. In particular, this means that there does not exist a solution of (2.40) for which

$$\limsup_{x \to \infty} |u(x)| = \infty.$$

Example 2.6 Let $\Omega = \mathbb{R}_+ \times \mathbb{R}^{n-1}$, $x = (t, x')$ and all assumptions of *Theorem 2.9* hold. Then, due to *Theorem 2.10* and the Sobolev embedding theorem, any solution of (2.40) admits

$$\sup_{x' \in \mathbb{R}^{n-1}} |u(t, x')| = C\left(\frac{1}{t}, \|g; b\|_{0,p}\right) < \infty \text{ for all } t > 0.$$

Note that we do not impose any restriction on $u_0(x')$ as $x' \to \infty$.

Assume in addition that $u_0 \in W_{loc}^{2-\frac{1}{p},p}(\mathbb{R}^{n-1})$ and

$$\left\|u_0, \mathbb{R}^{n-1}, b\right\|_{2-\frac{1}{p},p} < \infty.$$

Then (see [101]), there exists a $v \in W_b^{2,p}(\Omega)$ such that $v|_{\partial\Omega} = u_0$. Hence, according to *Theorem 2.7* and *Remark 2.7*, any solution of (2.40) belongs to $W_b^{2,p}(\Omega)$ and, consequently, belongs to $C_b(\Omega)$.

Definition 2.6 By a local solution of (2.40) we call a function u defined in $B_{x_0}^R$ and belonging to $\theta(B_{x_0}^R)$, which, in turn, satisfies (2.40).

We especially emphasize that the proofs of *Theorems 2.7* and *2.8* by no means use that u is defined outside of $B_{x_0}^{R+\varepsilon}$ and, therefore, the assertions of *Theorems 2.7* and *2.8* remain valid for local solutions of (2.40) as well. Thus, we have the following result:

Theorem 2.11 *Let all assumptions of previous theorem hold. Then, for all* $\varepsilon > 0$ *such that* $B_{x_0}^{R+\varepsilon} \subset\subset \Omega$, *there exists a* $K = K(\varepsilon, R, \|g; b\|_{0,p})$ *such that no local solution* u *of (2.40) defined in* $B_{x_0}^R$ *exists with*

$$\sup_{x \in B_{x_0}^R} |u(x)| > K \qquad (2.92)$$

that can be extended to a local solution $\hat{u}$ in $B_{x_0}^{R+\varepsilon}$.

Proof Assume the contrary. Let $\hat{u} \in \theta \left(B_{x_0}^{R+\varepsilon} \right)$ and $\hat{u}|_{B_{x_0}^R} = u$. Then, due to Theorem 2.8 and the Sobolev embedding theorem ($W^{2,p}(B_{x_0}^R) \subset C(B_{x_0}^R)$ if $p > \frac{n}{2}$).

$$\sup_{x \in B_{x_0}^R} |u(x)| \leqslant Q \left(\left\| g, B_{x_0}^{R+\varepsilon} \right\|_{0,p} \right). \qquad (2.93)$$

Estimate (2.93) contradicts the condition (2.92) if $K > Q \left(\left\| g, B_{x_0}^{R+\varepsilon} \right\|_{0,p} \right)$. Theorem 2.11 is proved. □

2.7 Basic Definitions: Trajectory Attractor

In order to construct a trajectory attractor for (2.40), we consider together with (2.40) also

$$\begin{cases} a\Delta u + \gamma D u - f(u) = \sigma(x), \\ \sigma(\cdot) \in \Sigma. \end{cases} \qquad (2.94)$$

Here, $\Sigma \subset D'(\Omega)$ is an invariant set with respect to T_s:

$$T_s \Sigma \subset \Sigma, \ s \geqslant 0.$$

Moreover, we assume that Σ is a compact subset of some functional space $\Sigma^+ \subset D'(\Omega)$. In this sequel, we usually consider Σ endowed with its weak topology that is $\Sigma^w(\Omega)$.

Remark 2.8 In application, we take $\Sigma = \mathcal{H}^+(g)$, that is, the hull of the right-hand side of (2.40). A set Σ we call the space of symbol (2.94).

Definition 2.7 A set of solutions of (2.40) which belongs to $\theta(\Omega)$ with the right-hand side $\sigma \in \Sigma$ and $u_0 \in v_0(\partial\Omega)$, we denote by $\mathcal{K}_\sigma^+$.

It is not difficult to see that the set $\{\mathcal{K}_\sigma^+, \sigma \in \Sigma\}$ enjoys the so-called translation compatibility, that is,

$$T_s \mathcal{K}_\sigma^+ \subset \mathcal{K}_{T_s\sigma}^+ \subset \theta(\Omega).$$

Thus, the semigroup $\{T_s, s \geqslant 0\}$ acts on the trajectory phase space $\mathcal{K}_\Sigma^+$ of (2.94):

$$\mathcal{K}_\Sigma^+ = \bigcup_{\sigma \in \Sigma} \mathcal{K}_\sigma^+ \subset \theta(\Omega), \ T_s \mathcal{K}_\Sigma^+ \subset \mathcal{K}_\Sigma^+.$$

We endowed the (nonlinear) set $\mathcal{K}_\Sigma^+$ by the induced weak topology:

$$\mathcal{K}_\Sigma^+ \subset \theta_+ \equiv \theta^w(\Omega).$$

Definition 2.8 A set $\mathbb{B} \subset \mathcal{K}_\Sigma^+$ is called an attractive set for the semigroup $\{T_s, s \geqslant 0\}$ in $\mathcal{K}_\Sigma^+$, if for every neighbourhood $\mathcal{O}(\mathbb{B})$ of $\mathbb{B}$ in $\mathcal{K}_\Sigma^+$ there exists a $T = T(\mathcal{O}(\mathbb{B}))$ such that

$$T_s(\mathcal{K}_\Sigma^+) \subset \mathcal{O}(\mathbb{B}) \text{ for all } s \geqslant T. \tag{2.95}$$

Remark 2.9 Note that the definition of attracting set as given in *Definition 2.8*, is not the traditional one (compare with the definitions given in *Sect. 2.2*). Usually, the condition (2.95) is only required to be fulfilled for bounded subsets (in an appropriate topology) of $\mathcal{K}_\Sigma^+$. However, as we will later, due to the estimates for the solutions of (2.40), in some concrete situation, one can use the *Definition 2.8*, that is, the attractivity property (2.95) is satisfied for all bounded subsets in $\mathcal{K}_\Sigma^+$ with the same constant $T = T(\mathcal{O}(\mathbb{B}))$.

Definition 2.9 A set $\mathbb{A}_\Sigma \subset \mathcal{K}_\Sigma^+$ is called a trajectory attractor for (2.94) if

1. $\mathbb{A}_\Sigma$ is compact in θ_+;
2. $\mathbb{A}_\Sigma$ is strictly invariant: $T_s \mathbb{A}_\Sigma = \mathbb{A}_\Sigma$ for all $s \geqslant 0$;
3. $\mathbb{A}_\Sigma$ is an attracting set for the semigroup $\{T_s, s \geqslant 0\}$ in $\mathcal{K}_\Sigma^+$.

To formulate the main Theorem on the existence of the trajectory attractor for (2.94), we need several Definitions.

Definition 2.10 We denote by $\omega(\Sigma)$ the ω-limit set of the semigroup $\{T_s, s \geqslant 0\}$ in Σ $(T_s \Sigma \subset \Sigma)$ if

$$\omega(\Sigma) = \bigcap_{s \geqslant 0} \left[\bigcup_{h \geqslant s} T_h \Sigma \right]_{\Sigma_+},$$

where by $[\ldots]_{\Sigma_+}$ we denote the closure in the topology Σ_+.

Remark 2.10 Since Σ is compact, $\omega(\Sigma) \neq \varnothing$ (see [1]).

Definition 2.11 A family of sets $\{\mathcal{K}_\sigma^+, \sigma \in \Sigma\}$ is closed in the topology (θ_+, Σ) (sequentially closed), if its graph $\bigcup_{\sigma \in \Sigma} \mathcal{K}_\sigma^+ \times \sigma$ is sequentially closed in $\theta^+ \times \Sigma$.

Remark 2.11 Note that, in the case of Σ compact, this property is equivalent to sequential compactness of $\mathcal{K}_\Sigma^+$ in θ_+ [35].

Theorem 2.12 ([35]) *Assume that the semigroup* $\{T_s, s \geqslant 0\}$ *and* $\{\mathcal{K}_\sigma^+, \sigma \in \Sigma\}$ *defined above satisfy the following conditions:*

1. The sets $\mathcal{K}_\sigma^+, \sigma \in \Sigma$ *are closed;*
2. The semigroup $\{T_s, s \geqslant 0\}$ *possesses in* $\mathcal{K}_\Sigma^+$ *a compact attracting set.*

Then (2.94) possesses a trajectory attractor $\mathbb{A}_\Sigma$:

$$\mathbb{A}_\Sigma = \mathbb{A}_{\omega(\Sigma)} = \bigcap_{s \geqslant 0} \left[\bigcup_{h \geqslant 0} T_h \mathcal{K}_\Sigma^+ \right]_{\theta_+},$$

where $\mathbb{A}_{\omega(\Sigma)}$ *is a trajectory attractor of (2.94) with the space of symbols* $\omega(\Sigma)$, *and, as usual,* $[\ldots]$ *is the closure in* $\theta^w(\Omega)$.

The conditions 1) and 2) we check later on. Next, we formulate a theorem which describes the structure of the trajectory attractor for the semigroup $\{T_s, s \geqslant 0\}$. To this end, we need several Definitions and auxiliary lemmas.

Definition 2.12

1. We denote by $\Xi_{b^+(l)}(\Omega)$ a Fréchet space $(\Xi_{b^+(l)}(\Omega) \subset \Xi)$ with the following seminorms

$$|g; x_0, b^+(l)|_+ = \sup_{s \geqslant 0} |g, \Omega \cap T_s \mathbb{B}_{x_0}^1|_+ = C(x_0) < \infty; \quad x_0 \in \mathbb{R}^n.$$

2. By $\Xi_{b(l)}(\Omega)$ we denote a Fréchet space $(\Xi_{b(l)}(\Omega) \subset \Xi)$ with the following seminorms

$$|g; x_0, b(l)|_+ = \sup_{s \in \mathbb{R}} |g, \Omega \cap T_s \mathbb{B}_{x_0}^1|_+ = C(x_0) < \infty; \quad x_0 \in \mathbb{R}^n.$$

Analogously are defined the spaces $\theta_{b^+(l)}(\Omega)$ and $\theta_{b(l)}(\Omega)$ and the corresponding seminorms $\|u; x_0, b^+(\vec{l})\|_+$ and $\|u; x_0, b(\vec{l})\|_+$.

Lemma 2.10 *A set* $\Sigma \subset \Xi(\Omega)$ *satisfying* $T_s\Sigma \subset \Sigma$ *is relatively compact in* $\Xi^+(\Omega)$ *if and only if it is bounded in* $\Xi_{b^+(l)}(\Omega)$. *Analogous statement is valid for the space* θ.

Proof According to *Corollary 1.3*, a set Σ is relatively compact in Ξ^+ if and only if Σ is bounded in $\Xi(\Omega)$. We will show that from the boundedness in $\Xi(\Omega)$ and $\Xi(\Omega)$ and $T_s\Sigma \subset \Sigma$ it follows that Σ is bounded in the space $\Xi_{b^+(l)}(\Omega)$. Indeed, let $\mathbb{B}_{x_0}^1 \subset \mathbb{R}^n$. Then, it is not difficult to see that (compare with *Lemma 2.14*) there exists an $s_0 = s_0(x_0)$, such that $T_{s_0}\mathbb{B}_{x_0}^1 \subset \Omega$. Then, due to *Definition 2.12* and $T_s\Sigma \subset \Sigma$, we have

$$\left| \Sigma; x_0, b^+(\vec{l}) \right|_+ = \sup_{s \geqslant 0} \left| \Sigma, \Omega \cap T_s \mathbb{B}^1_{x_0} \right|_+$$

$$\leqslant \sup_{s < s_0} \left| \Sigma, \Omega \cap T_s \mathbb{B}^1_{x_0} \right| + \sup_{s \geqslant 0} \left| \Sigma, T_s T_{s_0} \mathbb{B}^1_{x_0} \right|_+$$

$$\leqslant \left| \Sigma, \Omega \cap \mathbb{B}^{s_0}_{x_0} \right|_+ + \sup_{s \geqslant 0} \left| T_s \Sigma, T_{s_0} \mathbb{B}^1_{x_0} \right|_+$$

$$\leqslant C(x_0) + \left| \Sigma, T_{s_0} \mathbb{B}^1_{x_0} \right|_+ < \infty.$$

This proves *Lemma 2.10*.   □

Definition 2.13 The function $\widehat{\xi} \in C(\mathbb{R}, \Sigma)$ is called a complete symbol for a semigroup $\{T_s, s \geqslant 0\}$, $T_s : \Sigma \to \Sigma$, if

$$T_s \widehat{\xi}(t) = \widehat{\xi}(t + s) \text{ for } s \geqslant 0, \ t \in \mathbb{R}.$$

By $Z(\Sigma)$ we denote the set of all complete symbols of a semigroup $\{T_s, s \geqslant 0\}$ acting in Σ.

Lemma 2.11 ([35]) *It is valid*

$$\omega(\Sigma) = \left\{ \widehat{\xi}(0) \middle| \widehat{\xi} \in Z(\Sigma) \right\},$$

that is, for any symbol $\sigma \in \omega(\Sigma)$ there exists at least one complete symbol $\xi(t) \in Z(\Sigma)$ such that $\widehat{\xi}(0) = \sigma$.

Lemma 2.12 *For any complete symbol $\widehat{\xi} \in Z(\Sigma)$ there exists a unique function $\xi \in \Xi_{b+(\vec{l})}(\mathbb{R}^n)$ such that*

$$\widehat{\xi}(s)(x) = \xi(x + s\vec{l})|_\Omega, \ s \in \mathbb{R}, \ x \in \Omega. \tag{2.96}$$

Remark 2.12 Since $\xi \in \Xi_{b(\vec{l})}(\mathbb{R}^n)$ is, in general, not a regular generalised function, a more precise formulation of (2.96) is the following:

$$< \widehat{\xi}(s), \varphi > = < \xi, T_s \varphi >, \ \varphi \in D(\Omega), \ s \in \mathbb{R}. \tag{2.97}$$

Proof Let $\widehat{\xi} \in Z(\Sigma)$ and $\varphi \in D(\mathbb{R}^n)$. Then it follows from the assumption on Ω and $s \geqslant 0$ (see *Lemma 2.14*) such that T_s supp $\varphi \subset \Omega$ and, consequently, $T_s \varphi \in D(\Omega)$. Consider the generalised function $\xi \in D'(\mathbb{R}^n)$ which is defined by

$$< \xi, \varphi > := < \widehat{\xi}(-s), T_s \varphi > . \tag{2.98}$$

Let us check that (2.98) is well-defined. Let $s_2 > s_1 \geqslant 0$ and $T_{s_1} \varphi, T_{s_2} \varphi \in D(\Omega)$. Then,

$$< \widehat{\xi}(-s_2), T_{s_2}\varphi >= < \widehat{\xi}(-s_2), T_{s_2-s_1} T_{s_1}\varphi >$$
$$= < T_{s_2-s_1} \widehat{\xi}(-s_2), T_{s_1}\varphi >$$
$$= < \widehat{\xi}(-s_1), T_{s_1}\varphi > .$$

Next, we check (2.97). Let $s \geqslant 0$. Then, for each $\varphi \in D(\Omega)$ it holds:

$$< \widehat{\xi}(s), \varphi >=< T_s\widehat{\xi}(0), \varphi >=< \widehat{\xi}(0), T_s\varphi >=< \xi, T_s\varphi >,$$

$$< \widehat{\xi}(-s), \varphi >=< \widehat{\xi}(-s), T_s T_{-s}\varphi >=< \xi, T_{-s}\varphi > .$$

Let us prove that $\xi \in \Xi_{b(\vec{l})}(\mathbb{R}^n)$. Indeed, from *Lemma 2.14* it follows that there exists an $s_0 = s_0(x_0) \in \mathbb{R}_+$ such that $T_{s_0}\mathbb{B}^1_{x_0} \subset \Omega$, where $\mathbb{B}^1_{x_0} \subset \mathbb{R}^n$. From *Definition 2.12*, formula (2.98) and the compactness of Σ, it follows that

$$|\xi, \mathbb{R}^n; x_0, b(\vec{l})|_+ = \sup_{s \in \mathbb{R}} |\eta, T_s\mathbb{B}^1_{x_0}|_+ = \sup_{s \in \mathbb{R}} |\xi, T_{s-s_0} T_{s_0}\mathbb{B}^1_{x_0}|_+$$

$$= \sup_{s \in \mathbb{R}} |\widehat{\xi}(s - s_0), T_s\mathbb{B}^1_{x_0}|_+ \leqslant |\Sigma, T_{s_0}\mathbb{B}^1_{x_0}| < \infty. \qquad (2.99)$$

This proves *Lemma 2.12*.   □

In the sequel, we will identify the complete symbol $\widehat{\xi}$ and the corresponding to $\widehat{\xi}$ function $\xi \in \Xi_{b(\vec{l})}(\mathbb{R}^n)$.

Corollary 2.5 *A set $\mathcal{Z}(\Sigma)$ is bounded in $\Xi_{b(\vec{l})}(\mathbb{R}^n)$, strictly invariant with respect to the group $\{T_s, s \in \mathbb{R}\}$, that is,*

$$T_s\mathcal{Z}(\Sigma) = \mathcal{Z}(\Sigma)$$

and compact in the space $\Xi^w(\mathbb{R}^n)$.

Indeed, boundedness in $\Xi_{b(\vec{l})}(\mathbb{R}^n)$ follows from (2.99), and the relative compactness in $\Xi^w(\mathbb{R}^n)$ is a consequence of *Corollary 1.3*. The strict invariance and closedness of $\mathcal{Z}(\Sigma)$ follows from the definition of the complete symbol, *Lemma 2.11* and the analogous properties of $\omega(\Sigma)$.

Definition 2.14 The function $\widehat{u}(s)$, $\widehat{u} \in C(\mathbb{R}, \mathcal{K}_{\Sigma}^+)$ is called a complete trajectory of a semigroup $\{T_s, s \geqslant 0\}$ corresponding to a complete symbol $\widehat{\xi}$, if the following conditions are satisfied:

$$\begin{cases} T_s\widehat{u}(t) = \widehat{u}(t + s), \ s \geqslant 0, \ t \in \mathbb{R}, \\ \widehat{u}(s) \in \mathcal{K}_{\widehat{\xi}(s)}^+, \ s \in \mathbb{R}. \end{cases}$$

We denote the set of all complete trajectories corresponding to a complete symbol $\hat{\xi}$ by $\mathcal{K}_{\xi}$.

Lemma 2.13 *For any complete trajectory $\hat{u} \in \mathcal{K}_{\xi}$ there exists a unique $u \in \theta_{b(\vec{l})}(\mathbb{R}^n)$ which satisfies*

$$a\Delta u + \gamma Du - f(u) = \xi(x), \ x \in \mathbb{R}^n$$

and

$$\hat{u}(t)(x) = u(x + t\vec{l})|_{\Omega}.$$

The proof of *Lemma 2.13* is analogous to the proof of *Lemma 2.12*.

Theorem 2.13 ([35]) *The trajectory attractor $\mathbb{A}_{\Sigma}$ has the following structure:*

$$\mathbb{A}_{\Sigma} = \mathbb{A}_{\omega(\Sigma)} = \bigcup_{\xi \in \mathcal{Z}(\Sigma)} \{\hat{u}(0) | \ u \in \mathcal{K}_{\xi}\}.$$

2.8   Trajectory Attractor of Nonlinear Elliptic System

In this section we will apply *Theorem 2.12* to a family of equations

$$\begin{cases} a\Delta u + \gamma Du - f(u) = \sigma(x), \\ \sigma(\cdot) \in \Sigma \end{cases} \tag{2.100}$$

which will prove existence of a trajectory attractor for system (2.100).

Theorem 2.14 *A family $\{\mathcal{K}_{\sigma}^{+}, \ \sigma \in \Sigma\}$ corresponding to (2.100) is a closed set.*

Proof Let $u_n \in \mathcal{K}_{\sigma_n}^{+}$, $u_n \to u$ in Θ^{+} and $\sigma_n \to \sigma$ in Ξ^{+}. We have to show that $u \in \mathcal{K}_{\sigma}^{+}$. In other words, we have to show that u is a solution of (2.100) with the right-hand side $\sigma(x)$. According to definition of solution of (2.100) we have

$$-\langle a\nabla u_n, \nabla\phi\rangle + \langle \gamma Du_n, \phi\rangle - \langle f(u_n), \phi\rangle = \langle \sigma_n, \phi\rangle \tag{2.101}$$

for any $\phi \in \mathcal{D}(\Omega)$. Passing to the limit in (2.101) and using arguments similar to which we used in the proof of *Theorem 2.5*, we have

$$-\langle a\nabla u, \nabla\phi\rangle + \langle \gamma Du_n, \phi\rangle - \langle f(u_n), \phi\rangle = \langle \sigma, \phi\rangle \tag{2.102}$$

for any $\phi \in \mathcal{D}(\Omega)$. *Theorem 2.14* is proved.   □

Corollary 2.6 *The set $\mathcal{K}_{\Sigma}^{+}$ is a sequentially closed subspace of Θ^{+}.*

Theorem 2.15 *A semigroup $\{T_s, \ s \geqslant 0\}$ corresponding to (2.100) possesses in $\mathcal{K}_{\Sigma}^{+}$ a compact, attracting (absorbing) set $\mathbb{B}_{abs}$.*

Proof The proof of this theorem is based on the following lemma.

Lemma 2.14 *For any $\mathbb{B}_{x_0}^{R} \subset \mathbb{R}^n$ there holds*

$$\lim_{s \to +\infty} d(T_s \mathbb{B}_{x_0}^{R}, \partial \Omega) = \infty. \tag{2.103}$$

In particular, there exists $s_0 > 0$ such that

$$T_s \mathbb{B}_{x_0}^{R} \subset \Omega \text{ for any } s \geqslant s_0. \tag{2.104}$$

Proof of Lemma 2.14 Let us start with proving (2.104). Indeed, since $\mathbb{B}_{x_0}^{R} \subset \mathbb{R}^n = \bigcup_{s>0} T_{-s}\Omega$, then for any $x \in \overline{\mathbb{B}_{x_0}^{R}}$ there exists neighborhood U_x and $s_x \geqslant 0$, such that $U_x \subset T_{-s_x}\Omega$. Let $\{U_{x_i}, \ i = 1, \ldots, N\}$ be a subcovering of $\{U_x, \ x \in \mathbb{B}_{x_0}^{R}\}$. Hence, according to *Condition 2.2*, $\mathbb{B}_{x_0}^{R} \subset T_{-s}\Omega$, where $s = s(\mathbb{B}_{x_0}^{R}) = \max\{s_{x_i}, \ i = 1, \ldots, N\}$ and consequently

$$T_s \mathbb{B}_{x_0}^{R} \subset \Omega; \ s = s(\mathbb{B}_{x_0}^{R}). \tag{2.105}$$

Since $T_{s_1}\Omega \subset T_{s_2}\Omega$ for $s_2 \geqslant s_1$, the function

$$s \longmapsto d(T_s \mathbb{B}_{x_0}^{R}, \Omega)$$

is monotone (nondecreasing), therefore the limit (2.103) exists. The next goal is to prove that this limit is ∞. Assume, on the contrary, that there is $L < \infty$ such that

$$d(T_s \mathbb{B}_{x_0}^{R}, \Omega) < L, \ \forall s \geqslant 0.$$

Then $T_s \mathbb{B}_{x_0}^{R+L} \cap \partial \Omega \neq \emptyset, \ s \in \mathbb{R}_+$, which contradicts condition (2.105), with R replaced by $R + L$. This proves *Lemma 2.14*.

Proof of Theorem 2.15 According to *Lemma 2.14*, for any ball $\mathbb{B}_{x_0}^{R} \subset \mathbb{R}^n$ there exists $S = S(\mathbb{B}_{x_0}^{R})$, such that

$$T_s \mathbb{B}_{x_0}^{2R} \subset \Omega \text{ for any } s \geqslant S.$$

Since Σ is compact in Θ^+, it follows from *Lemma 2.10* that

$$|\Sigma; x_0, R, b^+(l)|_+ \equiv \sup_{s \geqslant 0} |\Sigma, \Omega \cap T_s \mathbb{B}_{x_0}^{R}|_+ = C(x_0, R) < \infty.$$

Hence, according to estimate (2.51) with $\varepsilon = R$, and $s \geqslant S$ we have

$$\|T_s \mathcal{K}_\Sigma^+, \Omega \cap \mathbb{B}_{x_0}^R\|_+ = \|\mathcal{K}_\Sigma^+, T_s(\Omega \cap \mathbb{B}_{x_0}^R)\|_+$$

$$\leqslant \|\mathcal{K}_\Sigma^+, T_s \mathbb{B}_{x_0}^R\|_+$$

$$\leqslant C_R(1 + |\Sigma, T_s \mathbb{B}_{x_0}^{2R}|_+^2)$$

$$\leqslant C_R(1 + |\Sigma; x_0, 2R, b^+|_+^2) \equiv M(x_0, R). \qquad (2.106)$$

Note that $M(x_0, R) < \infty$. Let $\boldsymbol{\beta}_0 \subset \Theta_{b^+(\vec{l})}(\Omega)$ be the set defined by

$$\|\boldsymbol{\beta}_0, \Omega \cap \mathbb{B}_{x_0}^R\|_+ \leqslant M(x_0, R); \quad x_0 \in \mathbb{R}^n, \ R \subset \mathbb{R}_+. \qquad (2.107)$$

By *Lemma 2.10*, the set $\boldsymbol{\beta}_0$ is compact in Θ^+. Obviously, $\boldsymbol{\beta}_0$ is a closed convex subset of $\Theta(\Omega)$ and consequently (see *Lemma 1.4*) is closed in the topology of the space Θ^+. Thus, according to *Corollary 1.3*, the set $\boldsymbol{\beta}_0$ (endowed with the induced topology of Θ^+) is a compact metric space. We set

$$\mathbb{B}_{abs} := \boldsymbol{\beta}_0 \cap \mathcal{K}_\Sigma^+.$$

Our next goal is to prove that $\mathbb{B}_{abs}$ is a desirable compact attracting set. Indeed, since $\boldsymbol{\beta}_0$ is a compact metric space and the set $\mathcal{K}_\Sigma^+$ is sequentially closed, it follows that $\mathbb{B}_{abs}$ is a compact set in $\mathcal{K}_\Sigma^+$. Next we will prove the attractivity property of $\mathbb{B}_{abs}$. To this end, let us consider Q as a neighborhood of $\mathbb{B}_{abs}$ in Θ^+.

From compactness of $\mathbb{B}_{abs}$ it follows that Q contains a neighborhood Q_N of the form

$$\mathbb{B}_{abs} \subset Q_N = \bigcup_{i=1}^N \{u_i + E_i\}; \ u_i \in \mathbb{B}_{abs}, \qquad (2.108)$$

where the set E_i belong to the basis of a neighborhood of zero in the space Θ^+. Due to the definition of weak topology in $\Theta(\Omega)$ and *Theorem 1.1*, any E_i can be represented

$$E_i = E_i(R_i, \varepsilon_i; L_1^i, \ldots, L_{n_i}^i)$$

$$= \{u \in \Theta^+ : |\langle L_j^i, u|_{\Omega \cap \mathbb{B}_{x_0}^{R_i}} \rangle| < \varepsilon_i; \ j = 1, \ldots, n_i, \ L_j^i \in (\Theta(\Omega, \mathbb{B}_0^{R_i}))^* \}.$$

$$(2.109)$$

Let $R = \max\{R_i, \ i = 1, \ldots, N\}$. Then

$$Q(R) := \{u \in \Theta^+; \ u_{\Omega \cap \mathbb{B}_0^R} \subset \mathbb{B}_{abs}|_{\Omega \cap \mathbb{B}_0^R}\} \subset Q_N \subset Q.$$

Hence, it suffices to show that

$$T_s \mathcal{K}_\Sigma^+ |_{\Omega \cap \mathbb{B}_0^R} \subset \mathbb{B}_{abs} |_{\Omega \cap \mathbb{B}_0^R}$$

for sufficiently large $s \geqslant 0$. The last statement is an immediate consequence of (2.106) and the definition of $\mathbb{B}_{abs}$. *Theorem 2.15* is proved.

□

Thus, all conditions of *Theorem 2.12* are fulfilled and we have

Theorem 2.16 *A family of equations* (2.100) *possesses in* Θ^+ *a trajectory attractor* A_Σ, *which consists of restrictions to* Ω *of all solutions* $u \in \Theta_{b(\bar{l})}(\mathbb{R}^n)$ *of the following family of equations*

$$\begin{cases} a\Delta u(x) + \gamma Du(x) - f(u(x)) = \xi(x), \ x \in \mathbb{R}^n \\ \xi \in Z(\Sigma). \end{cases} \tag{2.110}$$

Remark 2.13 Due to *Lemma 2.12*, $Z(\Sigma) \subset \Xi_{b(\bar{l})}(\mathbb{R}^n)$. Therefore according to estimate (2.51) with $\Omega = \mathbb{R}^n$, it follows that, any solution $u \in \Theta(\mathbb{R}^n)$ of (2.110) belongs to the space $\Theta_{b(\bar{l})}(\mathbb{R}^n)$.

Next we will describe the character of convergence to the trajectory attractor A_Σ. In the sequel we denote A_Σ by A^{tr}.

Corollary 2.7 *It follows from Theorem 2.15, that for any ball* $\mathbb{B}_{x_0}^R \subset \Omega$ *the set* $A^{tr}|_{\mathbb{B}_{x_0}^R}$ *is an attracting set for a family of sets* $\{T_s \mathcal{K}_\Sigma^+ |_{\mathbb{B}_{x_0}^R}; \ s \geqslant \mathbb{R}_+\}$ *as* $s \to \infty$ *in the topology of* $\Theta^w(\mathbb{B}_{x_0}^R)$. *It means that, for any neighborhoods of* $A^{tr}|_{\mathbb{B}_{x_0}^R}$, *say* $\mathcal{O}(A^{tr}|_{\mathbb{B}_{x_0}^R})$ *in the topology of* $\Theta^w(\mathbb{B}_{x_0}^R)$, *there exists* $S = S(\mathcal{O})$, *such that*

$$T_s \mathcal{K}_\Sigma^+ |_{\mathbb{B}_{x_0}^R} \subset \mathcal{O}(A^{tr}|_{\mathbb{B}_{x_0}^R}) \ for \ s \geqslant S.$$

Since the embeddings

$$\begin{cases} \Theta(\mathbb{B}_{x_0}^R) \Subset H^{1-\varepsilon,2}(\mathbb{B}_{x_0}^R) \\ \Theta(\mathbb{B}_{x_0}^R) \Subset L^{2-\varepsilon}(\mathbb{B}_{x_0}^R) \end{cases} \tag{2.111}$$

are compact for any $\varepsilon > 0$, it follows from *Theorem 2.16* that, for any ball $\mathbb{B}_{x_0}^R \subset \Omega$ the following condition holds.

$$\begin{cases} \lim\limits_{s \to +\infty} dist_{H^{1-\varepsilon,2}(\mathbb{B}_{x_0}^R)} \{T_s \mathcal{K}_\Sigma^+ |_{\mathbb{B}_{x_0}^R}, A^{tr}|_{\mathbb{B}_{x_0}^R}\} = 0 \\ \lim\limits_{s \to +\infty} dist_{L^{2-\varepsilon}(\mathbb{B}_{x_0}^R)} \{T_s \mathcal{K}_\Sigma^+ |_{\mathbb{B}_{x_0}^R}, A^{tr}|_{\mathbb{B}_{x_0}^R}\} = 0, \end{cases} \tag{2.112}$$

where

$$dist_{...}\{X, Y\} = \sup_{x \in X} \inf_{y \in Y} \|x - y\|_{...}.$$

In what follows we present some useful consequences from *Theorem 2.16* and also give some useful application of this theorem. Moreover, we will also present the dependence of $\mathbb{A}^{tr}$ on the direction $\vec{l}$ and domain Ω. We start with the following lemma.

Lemma 2.15 *Let* Σ *be a compact subset of* Ξ^+, *which is invariant with respect to a semigroup* $\{T_s, \ s \geqslant 0\}$ *and let* $g \in \Sigma$. *Then a hull of* g, *that is*

$$\mathcal{H}^+(g) = [T_s g, \ s \geqslant 0]_{\Xi^+} \Subset \Xi^+ \tag{2.113}$$

is a compact set in Ξ^+.

A proof of this lemma is an immediate consequence of the inclusion $\mathcal{H}^+(g) \subset \Sigma$.

Definition 2.15 A function $g \in \Xi^+$ is called translation compact (in the direction $\vec{l}$) in Ξ^+, if the condition (2.113) is satisfied.

Analogously one can define translation compactness in the other spaces, those are invariant with respect to the semigroup $\{T_s, \ s \geqslant 0\}$. Obviously, a set $\mathcal{H}^+(g)$ is a translation invariant, that is

$$T_s \mathcal{H}^+(g) \subset \mathcal{H}^+(g), \ s \geqslant 0. \tag{2.114}$$

Consequently, the set $\mathcal{H}^+(g)$ can be chosen as a candidate of space of symbols Σ for the family of equations (2.100), if g is a translation compact function.

Remark 2.14 It is clear that, in the case of translation compactness of g, the set $\mathcal{H}^+(g)$ is a minimal set containing g, which can be chosen as a space of symbols for the following family of equations (2.100).

Remark 2.15 Since the hull $\mathcal{H}^+(g)$ of a translation compact function g is a compact metric space in Ξ^+, it follows that $\omega(g) \equiv \omega(\mathcal{H}^+(g))$ and $\mathcal{H}^+(g)$ can be represented in the following way [35]:

$$\begin{cases} \mathcal{H}^+(g) = \{T_s g; \ s \geqslant 0\} \cup \omega(g) \\ \omega(g) =: \{\xi \in \Xi^+ | \exists s_n \geqslant 0, \ s_n \to \infty \text{ as } n \to \infty, \ \xi = \lim_{n \to \infty} T_{s_n} g\}, \end{cases} \tag{2.115}$$

where by lim is denoted a limit in the topology of space Ξ^+.

Definition 2.16 By $\mathbb{A}_g^{tr}$ we denote a trajectory attractor of the equation

$$\begin{cases} a\Delta u + \gamma Du - f(u) = g(x), \\ u|_{\partial\Omega} = u_0 \end{cases} \tag{2.116}$$

with the right-hand side g (we assume that g is a translation compact in Ξ^+) and by definition we put $\mathbb{A}_g^{tr}$ to equal to the trajectory attractor $\mathbb{A}_\Sigma$ for the family of Eq. (2.100).

Let us recall that, due to the *Lemma 2.10*, a function $g \in \Xi(\Omega)$ is a translation compact in Ξ^+ if and only if, when $g \in \Xi_{b(\vec{l})}(\Omega)$. Thus we have

Theorem 2.17 *Let $g \in \Xi_{b(\vec{l})}(\Omega)$. Then the equation*

$$\begin{cases} a\,\Delta u + \gamma\,Du - f(u) = g(x), \\ u|_{\partial\Omega} = u_0 \end{cases}$$

possesses a trajectory attractor $\mathbb{A}_g^{tr}$ in the space Θ^+.

Lemma 2.16 *Let g be a translation compact in Ξ^+. Then any solution u of the Eq. (2.116) is a translation compact in Θ^+.*

Indeed, from estimate (2.51) analogously (2.106), it follows that $\mathcal{H}^+(u)$ is a bounded subset of $\Theta_{b+(\vec{l})}(\Omega)$, and consequently, according to *Lemma 2.10* is a compact in Θ^+. For convenience of the reader, below we present some examples of translation compact function g in the Eq. (2.116), for which due to *Theorem 2.17* the Eq. (2.116) possesses a trajectory attractor.

Example 2.7 Let $g = g(x)$ be a bounded function in $\mathbb{R}^n$, that is, $g \in L^\infty(\mathbb{R}^n)$. Then for any direction $\vec{l}$ and for any domain Ω satisfying the condition on Ω (see Sect. 2.1), a function g is a translation compact in $\Xi^+ = \Xi^w(\Omega)$.

Example 2.8 Let $h : \mathbb{R} \to \mathbb{R}$ be a function belonging to $L^\infty(\mathbb{R})$, and let $V_l = \mathbb{R}^{n-1}$ be the orthogonal complement of the line

$$\{s\vec{l}, \; s \in \mathbb{R}\}$$

in $\mathbb{R}^n$. Assume that a function $g_0 \in L^q_{loc}(V_l)$, where the exponent q is the same as in the definition of the space Ξ. Then it is not difficult to check that the function

$$g(x) = h(x \cdot \vec{l}) \cdot g_0\left(x - \frac{x \cdot \vec{l}}{\vec{l} \cdot \vec{l}}\vec{l}\right) \tag{2.117}$$

belongs to $\Theta_{b(\vec{l})}(\mathbb{R}^n)$, and consequently, for any domain satisfying the condition on Ω, the restriction of $g|_\Omega$ is a translation compact in Ξ^+.

In particular, let $x = (t, x')$, and let $\vec{l}$ coincide with the first coordinate vector, that is, $\vec{l} = (1, 0, \ldots, 0)$. Then

$$g(t, x') = \sin(t^2)e^{|x'|^2} \tag{2.118}$$

is a translation compact in the direction $\vec{l}$ in Ξ^+.

Remark 2.16 According to *Lemma 2.10*, a translation compact function in direction $\vec{l}$ is bounded in this direction. As was shown in the previous example, in the other directions the function need not be bounded.

2.9 Dependence of the Trajectory Attractor on the Underlying Domain

In this section we will study the dependence of the constructed attractor for nonlinear elliptic systems (2.100) on the right-hand side as well as on the domain $\Omega \subset \mathbb{R}^n$. Firstly we consider dependence of the trajectory attractor on the domain $\Omega \subset \mathbb{R}^n$.

Definition 2.17 Let Ω_1 and Ω_2 be domains satisfying the condition on domain Ω with the same direction $\vec{l}$ and let $g_i \in \Xi(\Omega_i)$, $i = 1, 2$ and $g_i(x)$ be a translation compact function in the spaces $\Xi^w(\Omega)$. Consider the trajectory attractors $\mathbb{A}_{g_i}$ of Eq. (2.100) in the domain Ω_i with the right-hand side $g_i(x)$, $i = 1, 2$. We say that $\mathbb{A}_{g_1} = \mathbb{A}_{g_2}$ if the set of corresponding complete trajectories coincide, that is

$$\hat{\mathbb{A}}_{g_1} = \bigcup_{\xi \in Z(\mathcal{H}^+(g_1))} \mathcal{K}_\xi = \hat{\mathbb{A}}_{g_2} \subset \Theta_{b(\vec{l})}(\mathbb{R}^n). \tag{2.119}$$

(In the sequel, we identify the trajectory attractor $\mathbb{A}_\Sigma$ with the corresponding sets of complete trajectories $\hat{\mathbb{A}}_\Sigma$.)

Theorem 2.18 *Let the assumptions of the previous Definition 2.17 hold for two domains Ω_i, $i = 1, 2$ and*

$$g_1|_{\Omega_1 \cap \Omega_2} = g_2|_{\Omega_1 \cap \Omega_2}. \tag{2.120}$$

Then the trajectory attractors $\mathbb{A}_{g_1}$ and $\mathbb{A}_{g_2}$ coincide, that is,

$$\mathbb{A}_{g_1} = \mathbb{A}_{g_2}. \tag{2.121}$$

Proof The proof is based on showing that

$$\mathcal{Z}(g_1) = \mathcal{Z}\left(\mathcal{H}^+_{\vec{l}(g_1)}\right) = \mathcal{Z}(g_2).$$

Let $\Omega = \Omega_1 \cap \Omega_2$. Then, it follows from (2.120) that

$$\mathcal{H}^+_{\vec{l}(g_1)}|_{\Omega_1 \cap \Omega_2} = \mathcal{H}^+_{\vec{l}(g_1)}|_{\Omega_1 \cap \Omega_2}.$$

Hence,

$$\tilde{\omega}(g_1)|_{\Omega_1 \cap \Omega_2} = \tilde{\omega}(g_2)|_{\Omega_1 \cap \Omega_2}. \tag{2.122}$$

Now, let $\xi \in \mathcal{Z}(g_1)$. Then, from definition of $\mathcal{Z}(g_1)$ it follows that there exists $\sigma_n \in \tilde{\omega}(g_1)$, such that $T_{-n}\xi|_{\Omega_1} = \sigma_n$, $n \in \mathbb{N}$. From (2.122) it follows that there exists $\tilde{\sigma}_n \in \tilde{\omega}(g_2)$, such that $\sigma_n|_{\Omega_1 \cap \Omega_2} = \tilde{\sigma}_n|_{\Omega_1 \cap \Omega_2}$. Hence, due to *Lemma 2.11*, there exists $\xi_n \in \mathcal{Z}(g_2)$ such that $\xi_n|_{\Omega_1 \cap \Omega_2} = \tilde{\sigma}_n|_{\Omega_1 \cap \Omega_2}$. Thus, we have

$$T_{-n}\xi|_{\Omega_1 \cap \Omega_2} = \xi_n|_{\Omega_1 \cap \Omega_2}, \ \xi_n \in \mathcal{Z}(g_2),$$

or, equivalently, for $\xi_n \in \mathcal{Z}(g_2)$, we have

$$\xi|_{T_{-n}(\Omega_1 \cap \Omega_2)} = T_n\xi_n|_{T_{-n}(\Omega_1 \cap \Omega_2)}.$$

According to *Lemma 2.14*, $\bigcup_{s \geq 0} T_{-s}(\Omega_1 \cap \Omega_2) = \mathbb{R}^N$, it follows that $T_n\xi_n \rightarrow \xi$ in $\Xi(\mathbb{R}^N)$, which in turn, due to *Corollary 2.5*, implies $\xi \in \mathcal{Z}(g_2)$. Hence $\mathcal{Z}(g_1) \subset \mathcal{Z}(g_2)$. The inclusion $\mathcal{Z}(g_2) \subset \mathcal{Z}(g_1)$ is obtained in the same way. □

Next assume that, a domain Ω satisfies the condition on Ω for two different directions $\vec{l}_1$ and $\vec{l}_2$. Let g_1 and g_2 be translation compact functions in the direction $\vec{l}_1$ and $\vec{l}_2$ respectively. Then according to *Theorem 2.17*, there exists trajectory attractors $\mathbb{A}_{g_1}^{l_1}$ in the directions $\vec{l}_1$ for the Eq. (2.100) with the right-hand side $g_1(x)$ and $\mathbb{A}_{g_2}^{l_2}$ in the directions $\vec{l}_2$ for the Eq. (2.100) with the right-hand side $g_2(x)$.

Theorem 2.19 *The trajectory attractors $\mathbb{A}_{g_1}^{l_1}$ and $\mathbb{A}_{g_2}^{l_2}$ coincide if and only if*

$$\omega_{l_1}(g_1) = \omega_{l_2}(g_2) \tag{2.123}$$

where $\omega_{l_i}(g_i) \equiv \omega(\mathcal{H}_{l_i}^+(g_i))$, $i = 1, 2$.

Proof Assume that $\mathbb{A}_{g_1}^{l_1} = \mathbb{A}_{g_2}^{l_2}$. Then formula (2.123) follows from *Theorem 2.16* and *Lemma 2.11*.

Now assume that, the formula (2.123) holds. Then, according to *Theorem 2.16* and *Remark 2.12*, it suffices to prove that, $Z_{l_1}(g_1) = Z_{l_2}(g_2)$. Let $\xi \in Z_{l_1}(g_1)$. Then $\forall n \in \mathbb{N}$

$$T_n^{l_1}\xi|_\Omega = \sigma_n \in \omega_{l_1}(g_1) = \omega_{l_2}(g_2). \tag{2.124}$$

Furthermore, using *Lemma 2.11* and the arguments similar to the previous theorem, we obtain

$$T_n^{l_1}\xi \longrightarrow \xi \text{ as } n \rightarrow \infty \text{ in } \Xi(\mathbb{R}^n); \ \xi_n \in Z_{l_2}(g_2). \tag{2.125}$$

Since ω_{l_i} are strictly invariant with respect to semigroup $\{T_s^{l_i}, \ s \geqslant 0\}$, then from equality (2.123) and definition of complete symbol it follows that, the set $Z_{l_2}(g_2)$ is invariant with respect to semigroup $\{T_s^{l_1}, \ s \geqslant 0\}$. Thus, $T_{-n}^{l_1}\xi_n \in Z_{l_2}(g_2)$, which in turn, according to *Corollary 2.5*, implies $\xi \in Z_{l_2}(g_2)$. *Theorem 2.19* is proved.   □

Corollary 2.8 *Let* $g = g(x)$ *be such that:*

$$\mathcal{H}_{l_1}^+(g) = \mathcal{H}_{l_2}^+(g) \equiv \mathcal{H}^+(g). \tag{2.126}$$

Then $\mathbb{A}_g^{l_1} = \mathbb{A}_g^{l_2}$.

Proof Note that, according to (2.126) and (2.114) and the definition of ω-limit set we have

$$T_p^{l_2}\omega_{l_1}(g) = T_p^{l_2}\bigcap_{s\geqslant 0}\left[\bigcup_{h\geqslant s}T_h^{l_1}\mathcal{H}^+(g)\right] \subset \bigcap_{s\geqslant 0}\left[\bigcup_{h\geqslant s}T_p^{l_2}T_h^{l_1}\mathcal{H}^+(g)\right]$$

$$= \bigcap_{s\geqslant 0}\left[\bigcup_{h\geqslant s}T_h^{l_1}T_p^{l_2}\mathcal{H}^+(g)\right]_{\Xi^+} \subset \bigcap_{s\geqslant 0}\left[\bigcup_{h\geqslant s}T_h^{l_1}\mathcal{H}^+(g)\right] = \omega_{l_1}(g)$$

and analogously $T_s^{l_1}\omega_{l_2}(g) \subset \omega_{l_1}(g)$ for $s \geqslant 0$.

Next we aim to show $\omega_{l_1}(g) = \omega_{l_2}(g)$. Let $\xi \in \omega_{l_1}(g)$. Then due to the representation (2.115) and Condition (2.126), at least one of the following three condition hold:

$$\begin{cases} 1. \ \exists s_0 \geqslant 0, \ \text{such that } T_{s_0}^{l_1}g \in \omega_{l_2}(g) \\ 2. \ T_s^{l_1}g = g \text{ for all } s \geqslant 0 \\ 3. \ \exists s_0 > 0, \ p_0 > 0, \ \text{such that } T_{s_0}^{l_1}g = T_{s_0}^{l_2}g. \end{cases} \tag{2.127}$$

In the first case, $T_s^{l_1}g \in \omega_{l_2}(g)$ for $s \geqslant s_0$ and consequently we have $\xi \in \omega_{l_2}(g)$.
In the second case, $\omega_{l_1}(g) = \omega_{l_2}(g) = \{g\}$.
In the third case, due to (2.115) we represent ξ as

$$\xi = \lim_{n\to\infty} T_{s_n}^{l_1}g.$$

Without loss of generality one can assume that, $s_n \geqslant ns_0$. Then

$$T_{s_n}^{l_1}g = T_{np_0}^{l_2}(T_{s_n-ns_0}^{l_1}g). \tag{2.128}$$

Since $T_{s_n-ns_0}^{l_1}g \in \mathcal{H}_{l_2}^+(g)$ and $np_0 \to \infty$ as $n \to \infty$ we obtain that, $\xi \in \omega_{l_2}(g)$. Hence $\omega_{l_1}(g) \subset \omega_{l_2}(g)$. The reverse inclusion can be proved analogously. Consequently, according to *Theorem 2.19*, we have $\mathbb{A}_g^{l_1} = \mathbb{A}_g^{l_2}$. *Corollary 2.8* is proved.   □

We now present some examples that illustrate the assumptions on g mentioned above.

Example 2.9 Let $g \in \Xi(\mathbb{R}^N)$ such that

$$\|g\| = \sup_{x_0 \in \mathbb{R}^N} \left\| g, \Omega \cap B_{x_0}^1 \right\|_{\Xi(\mathbb{R}^N)} < \infty$$

and there exists $h \in D'(\mathbb{R})$ such that $g(x) = h(|x|)$. Then g is translation-compact in any direction $\vec{l} \in \mathbb{R}^N$ and $H_{\vec{l}}^+(\Omega)$ does not depend on $\vec{l}$. As a consequence of *Corollary 2.8*, we obtain that $\mathcal{A}_g^{\vec{l}}$ also does not depend on $\vec{l} \in \mathbb{R}^N$.

Another interesting example is the case when $g \in \Xi_b(\mathbb{R}^N)$ is periodic with respect to $\mathbb{Z}^N$, i.e.

$$g(x_1 + k_1 \vec{l}_1, \ldots, x_N + k_N \vec{l}_N) \equiv g(x_1, \ldots, x_N), \ k_i \in \mathbb{Z}.$$

Here $\left\{ \vec{l}_i, \ i = 1, \ldots, N \right\}$ is a basis in $\mathbb{R}^N$ which generates $\mathbb{Z}^N$. Fundamental domains of $\mathbb{Z}^N$ we denote by $\mathbb{Z}_0$, i.e.

$$\mathbb{Z}_0 = \left\{ x \in \mathbb{R}^N | x = \sum_{i=1}^N x_i \vec{l}_i, \ 0 \leqslant x_i < 1 \right\}.$$

For arbitrary given direction $\vec{l} \in \mathbb{R}^N$, we denote by $O(\vec{l})$, $O(\vec{l}) \in \mathbb{T}^N$, a closure of the line $\left\{ s\vec{l}, \ s \in \mathbb{R} \right\}$ under the projection $\Pi : \mathbb{R}^N \to \mathbb{T}^N = \mathbb{R}^N / \mathbb{Z}^N$ and by $\mathbb{Z}(\vec{l})$ the set $\mathbb{Z}(\vec{l}) = \Pi^{-1} O(\vec{l})|_{\mathbb{Z}_0}$. Then, it is not difficult to see that

$$H_{\vec{l}}^+(g) = \left\{ g(x + \vec{s}), \ \vec{s} \in \mathbb{Z}(\vec{l}) \right\}.$$

In this case, $H_{\vec{l}}^+(g)$ is strictly invariant under $T_s^{\vec{l}}$, hence $\omega_{\vec{l}}(g) = H_{\vec{l}}^+(g)$. Consequently, due to *Theorem 2.19*, the attractors $\mathcal{A}_g^{\vec{l}_1}$ and $\mathcal{A}_g^{\vec{l}_2}$ coincide if and only if $O(\vec{l}_1) = O(\vec{l}_2)$.

Now let us present a sufficient condition for

$$O(\vec{l}_1) = O(\vec{l}_2).$$

Definition 2.18 A direction $\vec{l}$ is called irrational with respect to $\mathbb{Z}^N$ if all coordinates of $\vec{l}$ in the basis $\left\{ \vec{l}_i, \ i = 1, \ldots, N \right\}$ are irrational numbers.

It is known that for irrational with respect to $\mathbb{Z}^N$ directions $\vec{l}$ it holds $O(\vec{l}) = \mathbb{T}^N$. Hence, for such two directions $\vec{l}_1$ and $\vec{l}_2$, we have $\mathcal{A}_g^{\vec{l}_1} = \mathcal{A}_g^{\vec{l}_2}$.

Remark 2.17 For a direction $\vec{l}$ which is not irrational with respect to $\mathbb{Z}^N$, it is not known whether these attractors coincide or not.

In what follows, we assume that the domain Ω satisfies *Condition 2.2* and the direction $\vec{l} \in \mathbb{R}^n$ is fixed. In addition we assume that the right-hand side g is represented by

$$g(x) = g_0(x) + g_1(x), \tag{2.129}$$

where g_1 is a translation compact function in the direction $\vec{l}$ in Ξ^+, and the function $g_0 \in \Xi^+$ satisfies

$$T_s g_0 \longrightarrow 0 \text{ as } s \longrightarrow \infty \text{ in } \Xi^+. \tag{2.130}$$

Then it is not difficult to show that g is also translation compact in the direction $\vec{l}$ and

$$\omega(g) = \omega(g_1). \tag{2.131}$$

As a consequence we obtain

Theorem 2.20 *Let the assumptions* (2.129) *and* (2.130) *hold. Then*

$$\mathbb{A}_{g_0+g_1} = \mathbb{A}_{g_1}. \tag{2.132}$$

Theorem 2.20 is an immediate consequence of (2.131) and *Definition 2.12*.

Example 2.10 Let $x = (t, x')$ and let $\vec{l}$ coincide with the first coordinate in $\mathbb{R}^n$, that is, $\vec{l} = (1, 0, \ldots, 0)$. Consider

$$g_0(x) = \sin(t^2)e^{|x'|^2}.$$

Since $T_s \sin(t^2) \rightharpoonup 0$ in $C(\mathbb{R}_+)$ as $s \to \infty$, it is not difficult to check that $g_0(x)$ satisfies condition (2.130). Let $g_1 \in \Xi_{b+(\vec{l})}(\Omega)$. Then according to *Theorem 2.20*, we have

$$\mathbb{A}_{g_0+g_1} = \mathbb{A}_{g_1}. \tag{2.133}$$

2.10　Regularity of Attractor

In this section we study regularity properties of the trajectory attractor constructed above. It will be done under additional assumption, both for the matrix a and for the right-hand side g in Eq. (2.40). Moreover, we will show attraction to the trajectory attractor in the stronger topology of $W_{loc}^{2,p}(\Omega)$, $p > \frac{n}{2}$.

We state the following three conditions on the involved data:

Condition 2.5

(1) $a = a^*$;

(2) $g \in L_{loc}^p(\Omega)$ with $p > \dfrac{n}{2}$;

(3) Function g is a translation compact in the direction $\vec{l}$ either in $L_{loc}^p(\Omega)$ or $L_{loc,w}^p(\Omega)$, that is in L_{loc}^p endowed with the weak topology.

Next we formulate criteria in $L_{loc}^p(\Omega)$ and $L_{loc,w}^p(\Omega)$.

Lemma 2.17 *The following statements are equivalent:*

1) A function $g \in L_{loc}^p(\Omega)$ is a translation compact in $L_{loc}^p(\Omega)$ ($L_{loc,w}^p(\Omega)$)

2) For any ball $\mathbb{B}_{x_0}^R \subset \Omega$, the restriction of g to the semicylinder

$$\Pi_{x_0,R}^+ = \Omega \cap \{T_s \mathbb{B}_{x_0}^R; \ s \geqslant 0\}$$

is a translation compact in $L_{loc}^p(\Pi_{x_0,R}^+)$ ($L_{loc,w}^p(\Pi_{x_0,R}^+)$).

Proof Implication 1) $\Rightarrow$ 2) is obvious. Next we show 2) $\Rightarrow$ 1). Indeed, due to the Eberlein theorem (*Lemma 1.3*; see also [86]), both in the case $L_{loc}^p(\Omega)$ and $L_{loc,w}^p(\Omega)$, it suffices to show that, $\mathcal{H}^+(g)$ is a sequentially compact set in $L_{loc}^p(\Omega)$ ($L_{loc,w}^p(\Omega)$). The sequential compactness of $\mathcal{H}^+(g)$ is an immediate consequence of condition (2) and the Cantor diagonal procedure. *Lemma 2.17* is proved. $\qquad\square$

Definition 2.19 Let $\mathbb{W}_1 = W_{loc}^{2,p}(\Omega)$ and $\mathbb{W}_2 = W_{loc,w}^{2,p}(\Omega)$. Analogously we define a Fréchet space $W_{b^+(l)}^{m,p}(\Omega)$ as a subspace of $W_{loc}^{m,p}(\Omega)$. We recall that, the seminorms in $W_{loc}^{m,p}(\Omega)$ are defined by:

$$\|u; x_0, R, b^+(l)\|_{m,p} = \sup_{s \geqslant 0} \|u, \Omega \cap T_s \mathbb{B}_{x_0}^R\|_{m,p} < \infty; \ x_0 \in \mathbb{R}^n, \ R \subset \mathbb{R}_+.$$

A space $W_{b(l)}^{m,p}(\Omega) \subset W_{loc}^{m,p}(\Omega)$ is defined by the seminorms

$$\|u; x_0, R, b(l)\|_{m,p} = \sup_{s \in \mathbb{R}} \|u, \Omega \cap T_s \mathbb{B}_{x_0}^R\|_{m,p} < \infty; \ x_0 \in \mathbb{R}^n, \ R \subset \mathbb{R}_+.$$

We denote by $\mathbb{L}_{b^+(l)}(\Omega)$ and $\mathbb{L}_{b(l)}$ the spaces $W_{b^+(l)}^{0,p}(\Omega)$ and $W_{b(l)}^{0,p}(\Omega)$ respectively and by $\mathbb{W}_{b^+(l)}$ and $\mathbb{W}_{b(l)}$ the spaces $W_{b^+(l)}^{2,p}(\Omega)$ and $W_{b(l)}^{2,p}(\Omega)$ respectively. Below we formulate criteria of translation compactness on the spaces $L_{loc}^p(\Omega)$ and $L_{loc,w}^p(\Omega)$. We denote by $\mathbb{L}_1 := L_{loc}^p(\Omega)$ and $\mathbb{L}_2 := L_{loc,w}^p(\Omega)$.

Lemma 2.18

1) A function g is translation compact in $L_{loc,w}^p(\Omega)$ if and only if when $g \in \mathbb{L}_{b^+(l)}$

2) *Let a function $g \in W_{b^+(l)}^{\alpha,p}(\Omega_\varepsilon)$ for some $\alpha > 0$ and $\varepsilon \geq 0$. Then the function g is a translation compact in $\mathbb{L}_1$.*

Proof Part 1) can be proved analogously to *Lemma 2.10.* Next we prove part 2. Since the hull $\mathcal{H}^+(g)$ is metrizable in $\mathbb{L}_1$, in order to finish proof of the Lemma it suffices to prove that, from any sequence $\{s_n, \ n \in \mathbb{N}\}$, one can select a subsequence $\{s_k = s_{n_k}; \ k \in \mathbb{N}\}$ such that

$$T_{s_k} g \longrightarrow \xi \text{ in the space } \mathbb{L}_1.$$

Without loss of generality we can assume that $s_n \to \infty$ as $n \to \infty$ (otherwise, there exists a subsequence $s_k \to s_0$, and from continuity of the semigroup $\{T_s, \ s \geq 0\}$ it follows that $T_{s_k} g \to T_{s_0} g$ in $\mathbb{L}_1$). Let $\mathbb{B}_{x_0}^R$ be any ball in $\mathbb{R}^n$. According to the definition of topology in $\mathbb{L}_1^+$ and by Cantor diagonal procedure, it suffices to show relative compactness of the sequence $T_{s_n} g|_{\Omega \cap \mathbb{B}_{x_0}^R}$ in $L^p(\Omega \cap \mathbb{B}_{x_0}^R)$. From *Lemma 2.14*, it follows that, there exists $S = S(\mathbb{B}_{x_0}^R)$, such that $T_s \mathbb{B}_{x_0}^R \subset \Omega_\varepsilon$ for $s \geq S$. Thus, due to the assumption of *Lemma 2.18*, for $s_n \geq S$ we have

$$\|T_{s_n} g, \Omega \cap \mathbb{B}_{x_0}^R\|_{H^{\alpha,p}} \leq C \|g, T_{s_n} \mathbb{B}_{x_0}^R\|_{2,p} \leq C \|g, \Omega_\varepsilon; \ T_S x_0, R, b^+(l)\|_{\alpha,p}$$

$$\leq C(x_0, R).$$

Hence, the sequence $\{T_{s_n} g|_{\Omega \cap \mathbb{B}_{x_0}^R}, \ n \in \mathbb{N}, \ s_n \geq S\}$ is bounded in $H^{\alpha,p}(\Omega \cap \mathbb{B}_{x_0}^R)$. Since $H^{\alpha,p}(\Omega \cap \mathbb{B}_{x_0}^R)$ is compactly embedded in $L^p(\Omega \cap \mathbb{B}_{x_0}^R)$ for each $\alpha > 0$, this implies that the sequence $\{T_{s_n} g|_{\Omega \cap \mathbb{B}_{x_0}^R}, \ n \in \mathbb{N}, \ s_n \geq S\}$ is relatively compact in $L^p(\Omega \cap \mathbb{B}_{x_0}^R)$. This proves *Lemma 2.18*.

Theorem 2.21 *Let Conditions (1), (2), and (3) from Condition 2.5 be satisfied, that is, $a = a^*$, $g \in L_{loc}^p(\Omega)$ for $p > n/2$ and g is a translation compact in $\mathbb{L}_1 = L_{loc,w}^p(\Omega)$. Then the trajectory attractor $\hat{\mathbb{A}}_g$ of the Eq. (2.116) is a compact set in $\mathbb{W}_i(\mathbb{R}^n)$, $i = 1, 2$, and consequently, $\mathbb{A}_g$ is compact in $\mathbb{W}_i(\Omega)$, $i = 1, 2$.*

Proof Let us consider the hull $\hat{\mathcal{H}}^+(g)$ of the function g in the space $\mathbb{L}_i$. Since g is a translation compact in $\mathbb{L}_i$, hence (analogously to the case of translation compactness in Ξ^+) the hull $\hat{\mathcal{H}}^+(g)$ and ω-limit set of g, that is $\hat{\omega}(g)$ are compact in $\mathbb{L}_i(\Omega)$.

Moreover, a set of complete symbols $\hat{Z}(g)$ is a bounded subset of the space $\mathbb{L}_{b(l)}(\mathbb{R}^n)$ and is compact in $\mathbb{L}_i(\mathbb{R}^n)$, However, it is not difficult to show that

$$\hat{\mathcal{H}}^+(g) = \mathcal{H}^+(g); \ \hat{\omega}(g) = \omega(g); \ \hat{Z}(g) = Z(g). \tag{2.134}$$

Here $\mathcal{H}^+(g) = \Sigma$, $\omega(g)$, $Z(g)$- are hull, ω-limit set and set of complete symbols of g respectively; g is a translation compact function in Ξ^+. Let $\xi \in \omega(g)$. Then, according to (2.115)

$$\xi = \Xi^+ - \lim_{n \to \infty} T_{s_n} g \tag{2.135}$$

for some sequence $s_n \to \infty$ as $n \to \infty$. Since g is a translation compact in $\mathbb{L}_i$, $i = 1, 2$, then a set of limit points $\{T_{s_n} g,\ n \in \mathbb{N}\}$ (as well as any of its subsequences) is nonempty. At the same time, from (2.135) it follows that this limit set consists of a single point $\{\xi\}$, that is,

$$\xi = \mathbb{L}_i^+ - \lim_{n \to \infty} T_{s_n} g \in \widehat{\omega}(g). \tag{2.136}$$

Hence we have $\omega(g) \subset \widehat{\omega}(g)$. The inverse inclusion is obvious. Thus $\widehat{\omega}(g) = \omega(g)$. All other equalities of (2.134) can be proved analogously.

Thus, due to the estimate (2.86) we obtain that any trajectory $u(x)$, $x \in \mathbb{R}^n$ of the family of equations (2.100) belongs to $\mathbb{W}_{b(l)}(\mathbb{R}^n)$, and the set $\widehat{A}_g$ of all complete trajectories is bounded in this space, which in turn implies that (see Corollary 1.3) $\widehat{A}_g$ is relatively compact in $\mathbb{W}_2(\mathbb{R}^n)$. The closedness of $\widehat{A}_g$ in $\mathbb{W}_2(\mathbb{R}^n)$ can be proved analogously to *Theorem 2.12*. Thus, in the case of the mean topology ($i = 2$), *Theorem 2.21* is proved.

Let us prove assertion of *Theorem 2.21* in the case of the strong topology ($i = 1$), that is, prove compactness of $\widehat{A}_g$ in $\mathbb{W}_1(\mathbb{R}^n)$. To this end, we consider arbitrary $u_n \in \widehat{A}_g$ with corresponding complete symbols $\xi_n \in Z(g)$. Then as was proved above, without loss of generality we can assume that $u_n \rightharpoonup u$ (weakly) in $W^{2,p}_{loc,w}(\mathbb{R}^n)$ and $\xi_n \rightharpoonup \xi$ (weakly) in $L^p_{loc,w}(\mathbb{R}^n)$, and $u \in \mathcal{K}_\xi$. To finish the proof of *Theorem 2.21* in this case, it suffices to prove that, $u_n \rightharpoonup u$ (strongly) in $W^{2,p}_{loc}(\mathbb{R}^n)$. In other words, for any ball $\mathbb{B}^R_{x_0}$ we have

$$u_n\big|_{\mathbb{B}^R_{x_0}} \longrightarrow u\big|_{\mathbb{B}^R_{x_0}} \text{ in } W^{2,p}(\mathbb{B}^R_{x_0}). \tag{2.137}$$

Note that, firstly $\xi_n \to \xi$ in $\mathbb{L}^+_1(\mathbb{R}^n)$. Indeed, since g is a translation compact in the strong topology, it follows that $Z(g)$ is compact in $\mathbb{L}_1(\mathbb{R}^n)$. Using arguments analogous to those in the proof of equality (2.134), we obtain that $\xi_n \to \xi$.

Let us prove (2.137). Let $\varphi(x) \in C_0^\infty(\mathbb{R}^n)$, such that, $\varphi(x) = 1$ if $x \in \mathbb{B}^R_{x_0}$ and $\varphi(x) = 0$ if $x \notin \mathbb{B}^{R+\varepsilon}_{x_0}$ and rewrite Eq. (2.100) in the following way:

$$\begin{cases} \Delta_x(\varphi u_n) = 2\nabla_x \varphi \cdot \nabla_x u_n + \varphi a^{-1}[-\gamma D u_n + f(u_n) + \xi_n] \equiv \widetilde{\xi}_n(x), \\ \varphi u_n\big|_{\mathbb{B}^{R+\varepsilon}_{x_0}} = 0. \end{cases} \tag{2.138}$$

Note that, $\widetilde{\xi}_n \to \widetilde{\xi}$ in $L^p(\mathbb{B}^{R+\varepsilon}_{x_0})$. Indeed, as was proved above, $\varphi \xi_n \to \varphi \xi$ in L^p. Since $W^{2,p}(\mathbb{B}^{R+\varepsilon}_{x_0})$ is compactly embedded in $W^{1,p}(\mathbb{B}^{R+\varepsilon}_{x_0})$ and $u_n \rightharpoonup u$ in $W^{2,p}$, we obtain $\nabla_x u_n \to \nabla_x u$ in $L^p(\mathbb{B}^{R+\varepsilon}_{x_0})$. Since $p > n/2$ it follows that $W^{2,p}(\mathbb{B}^{R+\varepsilon}_{x_0})$

is compactly embedded in $C(\mathbb{B}_{x_0}^{R+\varepsilon})$ and, taking into account the continuity of f, we obtain that $f_n \to f$ in $L^p(\mathbb{B}_{x_0}^{R+\varepsilon})$. Thus $\tilde{\xi}_n(x) \to \tilde{\xi}(x)$ in $L^p(\mathbb{B}_{x_0}^{R+\varepsilon})$.

From L^p-regularity of the Laplacian we obtain that,

$$\varphi u_n \longrightarrow \varphi u \text{ in } W^{2,p}(\mathbb{B}_{x_0}^{R+\varepsilon}),$$

which in turn implies the validity of (2.137). This proves *Theorem 2.21*.   □

Our next task is to study convergence of trajectories of the equations (2.100) to the trajectory attractor $\mathbb{A}_g$ in the space $\mathbb{W}_i(\Omega)$, $i = 1, 2$. Since we do not impose any conditions (smoothness) on $\partial\Omega$, we cannot state that (in general), $T_s\mathcal{K}_\Sigma^+$ is a subset of $\mathbb{W}_i$ for $s \geqslant 0$. Consequently, the analog of *Definition 2.8* for attraction property to the attractor in the topology of space $\mathbb{W}_i$, $i = 1, 2$ must be specified. Note that, *due to Lemma 2.14* and *Theorem 2.8*, for any ball $\mathbb{B}_{x_0}^R \Subset \Omega$ we have

$$T_s\mathcal{K}_\Sigma^+|_{\mathbb{B}_{x_0}^R} \subset \mathbb{W}_i(\mathbb{B}_{x_0}^R) \text{ for } s \geqslant 0$$

and consequently the following definition is correct:

Definition 2.20 We say that the attractor $\mathbb{A}_g$ attracts the family of sets $\{T_s\mathcal{K}_\Sigma^+,\ s \geqslant 0\}$ in the topology of $\mathbb{W}_i$, if, for any $\mathbb{B}_{x_0}^R \Subset \Omega$, the restriction $\mathbb{A}_g|_{\mathbb{B}_{x_0}^R}$ attracts the family of sets $\{T_s\mathcal{K}_\Sigma^+|_{\mathbb{B}_{x_0}^R},\ s \geqslant 0\}$ in $\mathbb{W}_i(\mathbb{B}_{x_0}^R)$. In other words, in any neighborhoods of $\mathbb{A}_g$, that is, for $\mathcal{O}(\mathbb{A}_g|_{\mathbb{B}_{x_0}^R})$ in $\mathbb{W}_i(\mathbb{B}_{x_0}^R)$, there exists $S = S(\mathcal{O})$, such that for any $s \geqslant S$ the following equations holds.

$$T_s\mathcal{K}_\Sigma^+|_{\mathbb{B}_{x_0}^R} \subset \mathcal{O}(\mathbb{A}_g|_{\mathbb{B}_{x_0}^R}), \text{ for } s \geqslant S.$$

Analogously, we will say that

$$T_{s_n}u_n \longrightarrow u \text{ as } s_n \longrightarrow \infty,\ u_n \in \mathcal{K}_\Sigma^+$$

in $\mathbb{W}_i$, if for any $\mathbb{B}_{x_0}^R \Subset \Omega$

$$T_{s_n}u_n|_{\mathbb{B}_{x_0}^R} \longrightarrow u|_{\mathbb{B}_{x_0}^R} \text{ in } \mathbb{W}_i(\mathbb{B}_{x_0}^R). \tag{2.139}$$

Theorem 2.22 *Let Conditions* (1), (2), *and* (3) *from Condition 2.5 hold. Then the trajectory attractor* $\mathbb{A}_g$ *of equation* ($\Sigma = \mathcal{H}^+(g)$)

$$\begin{cases} a\Delta u + \gamma Du - f(u) = g(x), \\ u|_{\partial\Omega} = u_0 \end{cases}$$

attracts a family of semitrajectories $\{T_s\mathcal{K}_\Sigma^+,\ s \geqslant 0\}$ *in the topology of* $\mathbb{W}_i$.

Proof A proof of *Theorem 2.22* is based on the following *Lemma 2.19* (see below).

Lemma 2.19 *Let* $u_n \in \mathcal{K}_\Sigma^+$, $n \in \mathbb{N}$ *and* $s_n \to \infty$. *Then from the sequence* $\{T_{s_n} u_n,\ s \in \mathbb{N}\}$ *one can select a subsequence which converges in the topology* $\mathbb{W}_i$ *to some* $u \in \mathbb{A}_g$.

Proof From *Theorem 2.8* it follows that, for any ball $\mathbb{B}_{x_0}^R \Subset \Omega$, the sequence $T_{s_n} u_n|_{\mathbb{B}_{x_0}^R}$ is bounded in the space $W^{2,p}(\mathbb{B}_{x_0}^R)$ and consequently relative compact in $\mathbb{W}_2(\mathbb{B}_{x_0}^R)$. Using Cantor's diagonal procedure one can extract a subsequence from $T_{s_n} u_n$ (which we continue to denote as $T_{s_n} u_n$) converging in $\mathbb{W}_2(\Omega)$ to some $u \in \mathbb{W}_2(\Omega)$. By definition of trajectory attractor, it follows that $u \in \mathbb{A}_g$. Thus, in the case of the weak topology $(i = 2)$, *Lemma 2.19* is proved. Next consider the case of the strong topology $(i = 1)$. In this case, in the same manner as in the proof of *Theorem 2.22*, one can show that, from $T_{s_n} u_n \rightharpoonup u$ in $\mathbb{W}_2$ it follows that $T_{s_n} u_n \to u$ in $\mathbb{W}_1$. *Lemma 2.19* is proved.

Proof of Theorem 2.21 Assume the contrary. Then there exists a ball $\mathbb{B}_{x_0}^R \Subset \Omega$, such that, $\mathbb{A}_g|_{\mathbb{B}_{x_0}^R}$ does not attract a family $T_s \mathcal{K}_\Sigma^+|_{\mathbb{B}_{x_0}^R}$ in the space $\mathbb{W}_i(\mathbb{B}_{x_0}^R)$. This in turn (see [67]) implies that, there exists a sequence $\{T_{s_n} u_n|_{\mathbb{B}_{x_0}^R},\ n \in \mathbb{N}\}$, $u_n \in \mathcal{K}_\Sigma^+$ and $s_n \to \infty$, such that, a set of its limit points in the topology of $\mathbb{W}_i$ do not intersect $\mathbb{A}_g|_{\mathbb{B}_{x_0}^R}$. However, this contradicts the assertion of *Lemma 2.19*. This proves *Theorem 2.21*.

Corollary 2.9 *Let Conditions* (1), (2), *and* (3) *from Condition 2.5 hold. Then for any* $\mathbb{B}_{x_0}^R \Subset \Omega$ *holds*

$$\lim_{s \to +\infty} dist_{C(\mathbb{B}_{x_0}^R)} \{T_s \mathcal{K}_\Sigma^+|_{\mathbb{B}_{x_0}^R},\ \mathbb{A}_g|_{\mathbb{B}_{x_0}^R}\} = 0. \tag{2.140}$$

2.11   Trajectory Attractor of an Elliptic Equation with a Nonlinearity That Depends on x

In this section, we briefly describe the construction of the trajectory attractor of (2.40) for the case when the nonlinearity $f = f(x, u)$ depends on $x \in \Omega$. Due to the estimates (2.51) and (2.86), in this general case $(f = f(x, u))$, all estimates obtained for the case $f = f(u)$ can be easily extended. The only difference here consists in the fact that we have to change the definition of symbols (see *Sect. 2.7*) in an appropriate manner. Indeed, in order to define the corresponding trajectory dynamical systems for (2.40), we have to take the symbol σ as a pair $\sigma = (f(\cdot, \cdot), g(\cdot))$. Otherwise, a family of $\{\mathcal{K}_\sigma^+|\sigma \in \Sigma\}$ would not be translation compatible, and, consequently, we would not be in position to define a semigroup $\{T_s, s \geq 0\}$ on $\mathcal{K}_\Sigma$. In order to define a semigroup appropriate for this case and to prove the existence of the trajectory attractor, we need several definitions.

Definition 2.21 Let $\mathcal{C} = \mathbb{R}^k \times \Omega$ and denote

$$\mathcal{M} = C^{loc}(\bar{\Omega}, C^{loc}(\mathbb{R}^k, \mathbb{R}^k)) = C^{loc}(\mathcal{C}, \mathbb{R}^k). \tag{2.141}$$

A system of seminorms in $\mathcal{M}$ is defined by

$$\|f\|_{K, \mathbb{B}_{x_0}^R} = \sup \left\{ |f(x, u)| \ u \in K, \ x \in \bar{\Omega} \cap \mathbb{B}_{x_0}^R \right\},$$

for arbitrary $K \subset \mathbb{R}^k$ compact, $x_0 \in \mathbb{R}^k$, $R \in \mathbb{R}_+$.

It is obvious that a function $f(x, u)$ which satisfies the Assumptions on f (see *Condition 2.1*), belongs to $\mathcal{M}$. Consider the hull of $\mathcal{M}$:

$$\mathcal{H}(f) = \mathcal{H}_{\vec{l}}^+(f) = \left[f\left(T_s^{\vec{l}}x, u\right), s \geqslant 0 \right]_{\mathcal{M}}. \tag{2.142}$$

Definition 2.22 A function $f \in \mathcal{M}$ is called translation compact in $\mathcal{M}$ (in direction $\vec{l}$) if its hull (2.142) is compact in $\mathcal{M}$.

Criteria for translation compactness in the spaces (2.142) are formulated in [32, 33]. In what follows, we assume that f is translation compact in $\mathcal{M}$.

Remark 2.18 Let $f \in \mathcal{M}$ and satisfy

$$f\left(T_s^{\vec{l}}x, u\right) \equiv f(x, u), \ s \geqslant 0.$$

Then $\mathcal{H}_{\vec{l}}^+(f) = \{f\}$ and, consequently, f is translation compact in $\mathcal{M}$. In order to construct the trajectory attractor for the Eq. (2.40) instead of the family of equations

$$\begin{cases} a\Delta u + \gamma Du - f(u) = \sigma(x), \\ \sigma(\cdot) \in \Sigma, \end{cases}$$

we consider

$$\begin{cases} a\Delta u + \gamma Du - h(t, u) = \xi(x), \\ \sigma \equiv (h(\cdot, \cdot), \xi(\cdot)) \in \Sigma = \mathcal{H}^+(f) \times \mathcal{H}^+(g), \end{cases} \tag{2.143}$$

where the nonlinearity $\{(f, g)\}$ on the right-hand side of (2.40) respectively. Analogously to the *Sect. 2.7*, we will call the set Σ the space of symbols and its elements $\{h, \xi\}$ we call the symbols of equations (2.143).

Remark 2.19 Let f be translation compact in the space $\mathcal{M}$ and satisfy Assumptions on f (see *Condition 2.1*). Then, it is not difficult to see that any function $h \in \mathcal{H}^+(f)$ will satisfy *Condition 2.1* with the same constants C, C_1 and C_2 as were for f. Thus, all estimates (2.68) and (2.86) are fulfilled uniformly for the family of equations (2.143). Therefore, taking into the account *Remark 2.19*, all Definitions and Theorems of *Sects. 2.4–2.7* extend to the case when $f = f(x, u)$. For example, the analogs of *Theorems 2.17, 2.21, 2.22* are the following:

Theorem 2.23 *Let f be translation compact in $\mathcal{M}$ and satisfy Assumptions on f and g be translation compact in Σ^+. Then, the Eq. (2.40) possesses a trajectory attractor*

$$\mathbb{A}_{\{f,g\}} = \mathbb{A}_\Sigma, \quad \Sigma = \mathcal{H}^+(f) \times \mathcal{H}^+(g).$$

Theorem 2.24 *Let f be translation compact in $\mathcal{M}$ and all assumptions of Theorems 2.21 and 2.22 satisfied. Then, $\mathbb{A}_{\{f,g\}}$ is compact in $\mathbb{W}_i^+(\Omega)$ and attracts $\mathcal{H}_\Sigma^+$ in the topology $\mathbb{W}_i^+(\Omega)$ (in the Definition 2.20).*

2.12 Examples of Trajectory Attractors

In this section we present several examples of equations of the form (2.40) having trajectory attractors. First we start with those examples for which $\mathbb{A}_g$ is "trivial".

Theorem 2.25 *Assume that $g \equiv 0$, $\gamma = \gamma^*$ and the nonlinearity f satisfies Assumptions on f (see Condition 2.1) and in addition can be represented in the form*

$$f(u) = \alpha u + F(u), \quad F(u) \cdot u \geq 0, \quad \forall u \in \mathbb{R}^n, \tag{2.144}$$

where $\alpha > 0$. Then for any domain Ω satisfying Condition 2.2, the trajectory attractor is $\mathbb{A}_0 = \{0\}$.

Proof Note that, from (2.144), it follows that $f(0) = 0$, and thus $u \equiv 0$ is a solution of the Eq. (2.40). Our goal is to prove that $u \equiv 0$ is the only solution of the Eq. (2.40), belonging to $\Theta(\mathbb{R}^n)$. Indeed, let $u \in \Theta(\mathbb{R}^n)$ be a solution of (2.40). Since $g \equiv 0$, *Theorem 2.9* follows that, $u \in \Theta_b(\mathbb{R}^n)$. Multiplying Eq. (2.40) in $\mathbb{R}^k$ (scalar product) by the function $ue^{-\varepsilon|x|}$, $0 < \varepsilon \ll 1$ and integrating with respect to $x \in \mathbb{R}^n$ we obtain

$$\langle a\nabla u, \nabla(ue^{-\varepsilon|x|}) \rangle + \alpha \langle e^{-\varepsilon|x|}u, u \rangle - \langle \gamma Du, ue^{-\varepsilon|x|} \rangle \leq 0. \tag{2.145}$$

Since $a + a^* > 0$, we have

$$\langle a\nabla_x u, \nabla_x(ue^{-\varepsilon|x|}) \rangle \geq C\langle |\nabla_x u|^2 e^{-\varepsilon|x|}, 1 \rangle - C_1\varepsilon\langle |\nabla_x u| \cdot |u|e^{-\varepsilon|x|}, 1 \rangle$$

$$\geq C_2\langle |\nabla_x u|^2 e^{-\varepsilon|x|}, 1 \rangle - C_2\varepsilon^2\langle |u|^2 e^{-\varepsilon|x|}, 1 \rangle. \tag{2.146}$$

Integrating the third term in (2.145) by parts and taking into account self-adjointness of $\gamma_j \in L(\mathbb{R}^k, \mathbb{R}^k)$ as well as $|\nabla_x e^{-\varepsilon|x|}| \leq \varepsilon e^{-\varepsilon|x|}$, we obtain

$$|\langle \gamma D_x u, u e^{-\varepsilon|x|}\rangle| \leqslant \frac{1}{d}\sum_{i=1}^{k}|\langle \partial_{x_i}(\gamma_i u \cdot u)e^{-\varepsilon|x|}, 1\rangle|$$

$$\leqslant C\varepsilon\langle |\nabla_x u|\cdot |u|e^{-\varepsilon|x|}, 1\rangle$$

$$\leqslant C_2\langle |\nabla_x u|^2 e^{-\varepsilon|x|}, 1\rangle - C_3\varepsilon\langle |u|^2 e^{-\varepsilon|x|}, 1\rangle.$$

Inserting these estimates into (2.145) for sufficiently small $\varepsilon \ll 1$ we have

$$\langle |u|^2 e^{-\varepsilon|x|}, 1\rangle \leqslant 0. \tag{2.147}$$

Consequently $u \equiv 0$. Thus, due to *Theorem 2.16*, we obtain that $\mathbb{A}_0 = \{0\}$. This proves *Theorem 2.25*.

Remark 2.20 Note that, Condition (2.144) does not imply monotonicity of the Eq. (2.40). For example, all Conditions of *Theorem 2.25* are fulfilled by the function

$$F(u) = u|u|^2(2 + \sin |u|)^2.$$

Remark 2.21 The assumption $\gamma = \gamma^*$ is essential for the validity of the assertion of *Theorem 2.25*. Indeed, let us consider

$$y''(t) + \gamma y'(t) - y(t) = y(|y|^2 - 1)^2, \tag{2.148}$$

where $y = (y_1, y_2)$ and

$$\gamma = 2\begin{pmatrix} 0 & -1 \\ 1 & 0 \end{pmatrix}. \tag{2.149}$$

In this case, the Eq. (2.148) admits the nontrivial solution $y(t) = (\sin t, \cos t)$.

The next examples show that, in contrast to the attractors of evolution equations, an elliptic attractor (that is, a trajectory attractor corresponding to a nonlinear elliptic equation in an unbounded domain) is not necessarily connected. We now present a couple of examples that show that, in contrast to the attractors of evolution equations, $\mathcal{A}_g^l$ is not necessarily connected.

Theorem 2.26 *Assume that $N = 1$ or $N = 2$. Moreover, $a = a^* > 0$, $g = 0$ and f satisfies $f(u) \cdot u \geqslant 0$ in addition. Then, any solution $u \in \theta(\mathbb{R}^N)$ of*

$$a\Delta u = f(u) \text{ in } \mathbb{R}^N \tag{2.150}$$

is a constant, i.e., $u \equiv u_0 \in \mathbb{R}^N$ with $f(u_0) = 0$.

Proof Let $u \in \theta(\mathbb{R}^N)(\mathbb{R}^N)$ be a solution of (2.150). As is known, such solution is bounded in $\mathbb{R}^N$. Let us consider $y(x) := au(x) \cdot u(x)$. It is obvious that

$$\Delta_x y = 2a \nabla u \cdot \nabla u + 2f(u) \cdot u \geqslant 0. \tag{2.151}$$

Thus, $y(x)$ is a subharmonic function in $\mathbb{R}^N$ and for $N = 1$ and $N = 2$ it is constant, hence $y \equiv const$. Indeed, for $N = 1$ it is obvious, since y is a convex function, in the case $N = 2$ it is also clear that $y(x) = const$. Then, it follows with (2.151) that $f(u) = 0.$ □

For the convenience of the reader, we present below a proof of an analog of the Liouville theorem for subharmonic functions in $\mathbb{R}^2$.

Lemma 2.20 *Any bounded subharmonic function $w \in L_b^1(\mathbb{R}^2)$ is a constant.*

Proof Without loss of generality we will present a proof for $w \in C^\infty(\mathbb{R}^2) \cap C_b(\mathbb{R}^2)$. The general case can be reduced to $w \in C^\infty(\mathbb{R}^2) \cap C_b(\mathbb{R}^2)$ by means of averaged operator (see (2.82)). Thus, let

$$\Delta_x w = \varphi(x) \geqslant 0. \tag{2.152}$$

If $\varphi(x) \equiv 0$, then $w(x)$ is a harmonic function and the assertion of Lemma 2.20 follows from Harnack's inequality [21]. Assume now $\varphi(x) \not\equiv 0$. Then without loss of generality one can assume that, $\varphi(0) \neq 0$. Let $\mathbb{B}_0^R \subset \mathbb{R}^2$ be an arbitrary ball in $\mathbb{R}^2$ with radius R about the origin. Let

$$w(x) = w_1(x) + w_2(x), \quad x \in \mathbb{B}_0^R, \tag{2.153}$$

where

$$\begin{cases} \Delta_x w_1(x) = 0 \\ w_1|_{\partial \mathbb{B}_0^R} = w_{\partial \mathbb{B}_0^R} \end{cases} \quad \text{and} \quad \begin{cases} \Delta_x w_2(x) = \varphi(x) \\ w_2|_{\partial \mathbb{B}_0^R} = 0. \end{cases} \tag{2.154}$$

From the maximum principle (see *Chap. 1*) it follows that $w_1(x) \leqslant C$ uniformly in R, $R \in \mathbb{R}_+$. Let us estimate $w_2(x)$. Since $\varphi(x) \not\equiv 0$, then there exists $\mathbb{B}_0^\rho$, $\rho < R$, such that, $\varphi(x) \geqslant \varepsilon$ for all $x \in \mathbb{B}_0^\rho$. From the maximum principle it follows that $w_2(x) \leqslant w_*(x)$ for $x \in \mathbb{B}_0^R$, where $w_2(x)$ is a solution of the following problem:

$$\begin{cases} \Delta_x w_*(x) = \varepsilon \chi_{\mathbb{B}_0^\rho}(x), \quad x \in \mathbb{B}_0^R \\ w_*|_{\partial \mathbb{B}_0^R} = 0. \end{cases} \tag{2.155}$$

Here $\chi_{\mathbb{B}_0^\rho}(x)$ is the characteristic function of the ball $\mathbb{B}_0^\rho$ with a radius ρ about the origin.

A solution of the Eq. (2.155) can be written explicitly, namely,

$$w_*(x) = \frac{\varepsilon}{2} \left[\rho^2 \ln \frac{\max\{\rho, |x|\}}{R} - (|x|^2 - \rho^2)_- \right], \quad x \in \mathbb{B}_R(0).$$

Thus $w(x) \leqslant C + w_*(x)$ for all $x \in \mathbb{B}_0^R$. In particular, $w(0) \leqslant C - \frac{\varepsilon}{2}\rho^2 \ln \frac{R\sqrt{e}}{\rho}$. Letting $R \to \infty$, we obtain that $w(0) = -\infty$, which is a contradiction to boundedness of $w(x)$. This proves *Lemma 2.20*.   □

Remark 2.22 Simple examples show that, in the case of dimension $n \geqslant 3$ the assertion of *Theorem 2.26* cannot be true (in general). Moreover, $\gamma = 0$ is also essential. Indeed, let us consider the following equation ($k = 1, n = 1$):

$$y''(t) + 2y'(t) = f(y), \tag{2.156}$$

where

$$f(y) = \begin{cases} y^3 & y < 0 \\ \sin^2 y(1 - \sin y \cos y) & 0 \leqslant y \leqslant \pi \\ (y - \pi)^3 & y > \pi. \end{cases} \tag{2.157}$$

One can easily check that this equation admits the family of solutions of the form

$$y(t) = \frac{\pi}{2} + \arctan(t + t_0). \tag{2.158}$$

Example 2.11 Let us consider in an arbitrary domain $\Omega \subset \mathbb{R}^2$ that satisfies the conditions on the domain Ω (see *Sect. 2.4*). Then, due to the *Theorem 2.26*, the trajectory attractor of the equation

$$\begin{cases} \Delta u = u(u - 1)^2 \cdots (u - M)^2, \\ u|_{\partial \Omega} = u_0. \end{cases}$$

consists only of the corresponding equlibria,

$$\mathcal{A}_0^{\vec{l}} = \{u \equiv 0, 1, \ldots, M\}.$$

Example 2.12 Let us consider in $\mathbb{R}^N$ the following 'autonomous' equation (i.e., its coefficients do not depend explicitly on x):

$$\Delta u - u = -2u\theta(u) + u|u|^2(1 - \theta(u)). \tag{2.159}$$

Here, $\theta(u)$ is a C^∞ cut-off function, which is equal to 1 in $|u| \leqslant 1$ and 0 in $|u| > 2$. It is obvious that (2.159) satisfies all assumptions of *Theorem 2.17* and possesses a trajectory attractor $\mathcal{A}$. We show that $\dim_F \mathcal{A} = \infty$. Indeed, for $|u| \leqslant 1$, the Eq. (2.159) takes the form

$$\Delta u + u = 0. \tag{2.160}$$

Let W_0 be a space of solutions of (2.160) belonging to $C_b(\mathbb{R}^N)$ - the space of continuous functions bounded as $|x| \to \infty$. Obviously, W_0 is a closed subspace of $C_b(\mathbb{R}^2)$, and its dimension is infinite since W_0 contains infinitely many linearly independent functions

$$w_\alpha(x) = \sin(\alpha x_1) \cdot \ldots \cdot \sin(\alpha x_N), \quad \sum_{i=1}^{N} \alpha_i^2 = 1.$$

Hence, $\mathcal{A}$ contains a unit ball of W_0 and thus $\dim_F \mathcal{A} = \infty$.

2.13   The Trajectory Dynamical Approach for the Nonlinear Elliptic Systems in Non-smooth Domains

In this section our goal is to study how nonsmoothness of the unbounded domain (see below) is inherited by the corresponding trajectory attractor for elliptic systems. Indeed, in the half cylinder $\Omega_+ = \mathbb{R}_+ \times \omega$, where ω is a bounded polyhedral domain in $\mathbb{R}^n$, we consider the following elliptic system:

$$\begin{cases} a(\partial_t^2 u + \Delta u) + \gamma \partial_t u - f(u) = g(t), \\ u|_{t=0} = u_0, \ \partial_n u|_{\partial\omega} = 0. \end{cases} \tag{2.161}$$

Here (t, x) are the variables in Ω_+, $u = u(t, x) = (u^1, \cdots, u^k)$, $g = (g^1, \cdots, g^k)$, and $f(u)$ are vector-valued functions, Δ is the Laplacian with respect to the variable $x = (x^1, \cdots, x^n)$ and γ and a are constant $k \times k$ matrices with $a = a^* > 0$.

Notice that domains ω with non-degenerate edges and corners on the boundary are admitted. To be more rigorous, the domain ω is said to be polyhedral if any of its boundary points b is either regular or there are polyhedron $P \subset \mathbb{R}^n$, a non-regular boundary point b_1 of P, open subsets U, V of $\mathbb{R}^n$ with $b \in U$, $b_1 \in V$, and a C^∞-diffeomorphism $\chi : U \to V$ such that $\chi(b) = b_1$ and $\chi(\overline{\omega} \cap U) = \overline{P} \cap V$.

On the nonlinear term $f(u)$ we impose the following conditions:

$$\begin{cases} (1) \ f \in C(\mathbb{R}^k, \mathbb{R}^k); \\ (2) \ f(u) \cdot u \geqslant -C_1 + C_2|u|^p, \ 2 < p < 2 + 4/(n-3); \\ (3) \ |f(u)| \leqslant C(1 + |u|^{p-1}). \end{cases} \tag{2.162}$$

Here and below $u \cdot v$ denotes the inner product in $\mathbb{R}^k$.

We suppose that the right-hand side g belongs to the space $[L^2_{loc}(\Omega_+)]^k$ and has a finite norm

$$|g|_b = \sup_{T \geqslant 0} \|g, \Omega_T\|_{0,2} < \infty, \tag{2.163}$$

where $\Omega_T = (T, T + 1) \times \omega$. Further, we write $H^1(U)$ instead of $H^{(1,2)}(U)$ and $\|\cdot, U\|_{t,p}$ instead of $\|\cdot\|_{H^{l,p}(U)}$.

A solution $u(t, x)$ to problem (2.161) is defined as a function that belongs to the space

$$[H^2_Q(\Omega_T)]^k = [H^2\left((T, T + 1), L^2(\omega)\right) \cap L^2\left((T, T + 1), H^2_Q(\omega)\right)]^k$$

for each $T \geqslant 0$ and satisfies Eq. (2.161) in the sense of distributions. As in *Sect. 1.8*, we denote by $H^2_Q(\omega)$ the domain of the Laplace operator $-\Delta$ in $L^2(\omega)$ with the homogeneous Neumann boundary condition. As we noted in *Sect. 1.8*, for polyhedral domains in general $H^2_Q(\omega) \neq \{u \in H^2(\omega), \partial_n u|_{\partial\omega} = 0\}$, in contrast to the case of smooth $\partial\omega$ (see [40]).

Note that the third assumption of (2.162) implies that, for $u \in [H^2_Q(\Omega_T)]^k$, the function $f(u)$ belongs to $[L^2(\Omega_T)]^k$. Hence Eq. (2.161) can be considered as an equality in the space $[L^2_{\text{loc}}(\Omega_+)]^k$.

The initial data u_0 are assumed to belong to the trace space V_0 on $\{t = 0\}$ of functions in $[H^2_{Q,\text{loc}}(\Omega_+)]^k$.

For domains ω with smooth boundary, the problem (2.161) has been investigated under different assumptions on the nonlinear part f and the right-hand side g in [8, 29, 31, 103].

The main objective of this section is to study the behaviour of the solutions to Eq. (2.161) as $t \to +\infty$ for polyhedral cross-sections ω. The following estimate is of fundamental significance in that connection:

$$\|u, \Omega_T\|_{2,Q} \leqslant C(1 + \chi(1 - T)\|u_0\|_{V_0}^{p-1} + \|g, \tilde{\Omega}_T\|_{0,2}). \tag{2.164}$$

Here $\tilde{\Omega}_T = (\max\{0, T - 1\}, T + 2) \times \omega$, $\chi(z)$ is the Heaviside function, i.e., $\chi(z) = 1$ for $z \geqslant 0$ and $\chi(z) = 0$ for $z < 0$, and the constant C is independent of u_0. This estimate makes it possible to apply the methods of the theory of attractors (see [1, 32–34, 36]) to the problem (2.161).

Furthermore, estimate (2.164) implies that every solution $u(t, x)$ to the problem (2.161) is bounded as $t \to \infty$, i.e.,

$$\|u\|_b = \sup_{T \geqslant 0} \|u, \Omega_T\|_{2,Q} < \infty. \tag{2.165}$$

The subspace of functions $u \in [\mathcal{D}'(\Omega_+)]^k$ which have finite norm (2.165) is denoted by F_0^+.

Since the conditions that we impose on the nonlinear function f (see (2.162)) guarantee in general only the existence (but not the uniqueness) of a solution to the problem (2.161), in order to describe the behaviour of solutions as $t \to +\infty$ we construct a trajectory attractor for the dynamical system generated by the semigroup $\{T_s, \ s \geqslant 0\}$ of positive shifts of the solutions to (2.161) along the t-axis (see *Sect. 2.7* and [32, 34, 47, 103]). Here, since g explicitly depends on t, it is natural to study the family of equations of the form (2.161) generated by all positive shifts of this equation with respect to t and their limits in a suitable topology (see *Sect. 2.10*). The trajectory attractor $\mathbb{A}$ attracts the set K^+ of all trajectories of the above family as $t \to +\infty$.

Recall that the attracting property is usually required only for those subsets of the phase space K^+ which are, in a certain sense, *bounded*. But in our case estimate (2.164) allows us to verify the following improved version (see *Sect. 2.7* and also [108]): for any neighbourhood $\mathcal{O}(\mathbb{A})$ of the attractor $\mathbb{A}$ in the space K^+, there exists a number $T = T(\mathcal{O})$ such that

$$T_s K^+ \subset \mathcal{O}(\mathbb{A}) \ \text{ for all } \ s \geqslant T.$$

Moreover, like an ordinary attractor, the trajectory attractor $\mathbb{A}$ is strictly invariant with respect to the semigroup $\{T_s, \ s \geqslant 0\}$ and is generated by all the trajectories of this semigroup that are defined and bounded for $t \in \mathbb{R}$.

We also study in this Section the problem of stabilization of the solutions to Eq. (2.161) as $t \to +\infty$ in the case that the nonlinear part f has a potential ($f = \nabla F$, $F : \mathbb{R}^k \to \mathbb{R}$). Especially, it is shown that in the autonomous case ($g(t, x) \equiv g(x)$) every solution to the problem (2.161) in the whole cylinder $\Omega = \mathbb{R} \times \omega$ is a heteroclinic orbit connecting two stationary solutions (see below).

Furthermore, there is a non-canonical splitting of K^+ into a regular part and the space K^+_{sing} which contains the edge asymptotics of solutions belonging to K^+. This splitting is chosen in a way such that K^+_{sing} becomes invariant with respect to the semigruop $\{T_s, \ s \geqslant 0\}$ of positive shifts along the t-axis:

$$T_s : \ K^+_{\text{sing}} \to K^+_{\text{sing}} \ s \geqslant 0.$$

We then show that the semigroup $\{T_s, \ s \geqslant 0\}$ restricted to K^+_{sing} possesses an attractor $\mathbb{A}_{\text{sing}}$ which in turn is interpreted as the singular part of the trajectory attractor $\mathbb{A}$ of Eq. (2.161).

We start to obtain a priori estimates for solutions to the problem (2.161). In the sequel these estimates will be used to prove existence of solutions and to construct the trajectory attractor.

Theorem 2.27 *Let u be a solution to* (2.161). *Then*

$$\|u, \Omega_T\|_{1,2}^2 \leqslant C(1 + \chi(1 - T)\|u_0\|_{V_0}^p + \|g, \tilde{\Omega}_T\|_{0,2}^2), \tag{2.166}$$

where C is independent of u.

Proof By definition of V_0, there exists a function $v \in [H^2_{Q,b}(\Omega_+)]^k$ (see *Sect. 1.8*) such that supp $v \subset \Omega_0$, $v|_{t=0} = u_0$, and

$$\|v, \Omega_0\|_{2,Q} \leqslant C\|u_0\|_{V_0}, \tag{2.167}$$

where the constant C is independent of u_0.

Let us rewrite Eq. (2.161) for the function $w = u - v$,

$$\begin{cases} a(\partial_t w + \Delta w) + \gamma \partial_t w - f(w + v) = g(t) - a(\partial_t v + \Delta v) - \gamma \partial_t v \equiv h(t), \\ w|_{t=0} = 0. \end{cases} \tag{2.168}$$

From the choice of v it follows that

$$\|h, \Omega_T\|_{0,2} \leqslant C(\|g, \Omega_T\|_{0,2} + \chi(1 - T)\|u_0\|_{V_0}). \tag{2.169}$$

Let $\phi(t) = \phi_T(t)$ be the following cut-off function:

$$\phi(t) = \begin{cases} (1 - |t - T - 1/2|)^{2p/(p-2)}, & \text{for } t \in (T - 1/2, T + 3/2), \\ 0, & \text{for } t \notin (T - 1/2, T + 3/2). \end{cases}$$

It is readily seen that $\phi' \in L^\infty(\mathbb{R})$. Moreover, the following estimate is valid:

$$|\phi'(t)| \leqslant C\phi(t)^{1/2+1/p}, \text{ for } t \in \mathbb{R}. \tag{2.170}$$

Multiplying Eq. (2.168) by ϕw in $\mathbb{R}^k$ and integrating over Ω_+ gives us

$$\langle a\partial_t^2 w, \phi w \rangle + \langle a\Delta w, \phi w \rangle + \langle \gamma \partial_t w, \phi w \rangle - \langle f(v + w), \phi w \rangle = \langle h, \phi w \rangle. \tag{2.171}$$

From the positivity of a and (2.170) it follows that

$$-\langle a\partial_t^2 w, \phi w \rangle \geqslant C_1 \langle \phi|\partial_t w|^2, 1 \rangle - \langle |\phi'||\partial_t w|, |w| \rangle$$

$$\geqslant C_1 \langle \phi|\partial_t w|^2, 1 \rangle - \frac{1}{2}C_1 \langle \phi|\partial_t w|^2, 1 \rangle - C \langle \phi^{2/p}|w|^2, 1 \rangle$$

$$\geqslant C_2 \langle \phi|\partial_t w|^2, 1 \rangle - C \langle \phi^{2/p}|w|^2, 1 \rangle. \tag{2.172}$$

Applying Hölder's inequality to the third term in (2.171) we obtain the estimate

$$|\langle \gamma \partial_t w, \phi w \rangle| \leqslant \mu \langle \phi|\partial_t w|^2, 1 \rangle + C_\mu \langle \phi|w|^2, 1 \rangle \leqslant \mu \langle \phi|\partial_t w|^2, 1 \rangle + C_\mu \langle \phi^{2/p}|w|^2, 1 \rangle$$

for any $\mu > 0$ with some constant $C > 0$.

In view of assumption (2.162) on the nonlinear term $f(u)$, we further have

$$
\begin{aligned}
\langle f(w+v), \phi w \rangle =& \langle f(w+v) \cdot (w+v), \phi w \rangle - \langle f(v+w), v\phi \rangle \\
\geqslant & -C + C_1 \langle \phi |w+v|^p, 1 \rangle - C \langle 1 + |w+v|^{p-1}, \phi |v| \rangle \\
\geqslant & -C_2(1 + \langle \phi |v|^p, 1 \rangle) + C_3 \langle \phi |w|^p, 1 \rangle \\
\geqslant & -C_4(1 + \chi(1-T)\|u_0\|_{V_0}^p) + C_3 \langle \phi |w|^p, 1 \rangle.
\end{aligned} \tag{2.173}
$$

Here we have employed (2.167) and the embedding $H_Q^2\left(\Omega_{T_1,T_2}\right) \subset L^q\left(\Omega_{T_1,T_2}\right)$. Using the positivity of a again, after integration by parts we find

$$
-\langle a\Delta w, \phi w \rangle \geqslant C \langle \phi |\nabla w|^2, 1 \rangle.
$$

Finally, from (2.169) and Hölder's inequality we conclude

$$
\begin{aligned}
\langle h, \phi w \rangle \leqslant & \langle \phi |h|^2, 1 \rangle + \langle \phi |w|^2, 1 \rangle \\
\leqslant & C(\langle \phi |g|^2, 1 \rangle + \chi(1-T)\|u_0\|_{V_0}^2) + C_1 \langle \phi^{2/p} |w|^2, 1 \rangle.
\end{aligned} \tag{2.174}
$$

By inserting all the estimates (2.172)–(2.174) into (2.171), a short calculation yields

$$
\begin{aligned}
& \langle \phi |\partial_t w|^2, 1 \rangle + \langle \phi |\nabla w|^2, 1 \rangle + \langle \phi |w|^p, 1 \rangle - C \langle \phi^{2/p} |w|^2, 1 \rangle \\
& \leqslant C_1(1 + \langle \phi |g|^2, 1 \rangle + \chi(1-T)\|u_0\|_{V_0}).
\end{aligned} \tag{2.175}
$$

We estimate the last term of the left-hand side in (2.175) using Hölder's inequality,

$$
\langle \phi^{2/p} |w|^2, 1 \rangle = \langle |\phi^{1/p} w|^2, 1 \rangle \leqslant C(\langle \phi |w|^p, 1 \rangle^{2/p} \leqslant \mu \langle \phi |w|^p, 1 \rangle + C_\mu,
$$

which holds for any $\mu > 0$. Choosing $\mu > 0$ sufficiently small, this estimate inserted into (2.175) yields

$$
\langle \phi |\partial_t w|^2, 1 \rangle + \langle \phi |\nabla w|^2, 1 \rangle + \langle \phi |w|^p, 1 \rangle \leqslant C_2(1 + \langle \phi |g|^2, 1 \rangle + \chi(1-T)\|u_0\|_{V_0}). \tag{2.176}
$$

Recall that $\phi(t) > C_0 > 0$ for $t \in (T, T+1)$. Hence from (2.176) we infer that

$$
\|w, \Omega_T\|_{1,2}^2 \leqslant C(1 + \chi(1-T)\|u_0\|_{V_0}^p + \|g, \tilde{\Omega}_T\|_{0,2}^2).
$$

Theorem 2.27 is proved. $\qquad\qquad\qquad\qquad\qquad\qquad\qquad\qquad\qquad\qquad\square$

Remark 2.23 In a similar manner it follows from (2.176) that

$$
\|u, \Omega_T\|_{0,p}^p \leqslant C(1 + \chi(1-T)\|u_0\|_{V_0}^p + \|g, \tilde{\Omega}_T\|_{0,2}^2). \tag{2.177}
$$

Theorem 2.28 *Let u be a solution to (2.161). Then for each $T \geqslant 0$ we have*

$$\|u, \Omega_T\|_{0,2(p-1)}^{2(p-1)} \leqslant C(1 + \chi(1-T)\|u_0\|_{V_0}^{2(p-1)} + \|g, \tilde{\Omega}_T\|_{0,2}^2 + \|u, \tilde{\Omega}_T\|_{0,p}^p).$$

(2.178)

The exponent p is defined in (2.162).

Proof We fix some $T \geqslant 0$ and take another cut-off function $\varphi(t) \in C_0^\infty(\mathbb{R})$ such that $\varphi(t) = 1$ for $t \in (T, T+1]$, $\varphi(t) = 0$ for $t \notin (T-1, T+2)$, and $0 \leqslant \varphi(t) \leqslant 1$.

Multiplying Eq. (2.168) by $\varphi w |w|_a^{p-2}$, where $|w|_a \equiv (aw \cdot w)^{1/2}$, and afterwards integrating over Ω_+, we obtain the following equality:

$$\langle a(\partial_t^2 w + \Delta w), \phi w |w|_a^{p-2}\rangle = -\langle \varphi \gamma \partial_t w, w|w|_a^{p-2}\rangle + \langle \varphi f(w+v) \cdot w, |w|_a^{p-2}\rangle$$

$$+ \langle \varphi h, w|w|_a^{p-2}\rangle. \qquad (2.179)$$

By definition of $[H_Q^2(\tilde{\Omega}_T)]^k$, $\partial_t^2 w + \Delta w \in [L^2(\tilde{\Omega}_T)]^k$. It is not difficult to see that the functions $w|w|_a^{p-2}$ and $f(w+v)$ also belong to $[L^2(\tilde{\Omega}_T)]^k$. Hence, all the integrals in (2.179) are correctly defined. Moreover, by virtue of *Theorem 1.35*, we have $w|w|_a^{p-2} \in H^1(\tilde{\Omega}_T)$. Thus on the left-hand side of (2.179) we can integrate by parts and get

$$\langle a\partial_t^2 w, \phi w|w|_a^{p-2}\rangle$$

$$= -\langle a\partial_t w, \partial_t(\phi w|w|_a^{p-2})\rangle$$

$$= -\frac{1}{p}\langle \phi', \partial_t(|w|_a^p)\rangle - \langle \phi|\partial_t w|_a^2, |w|_a^{p-2}\rangle - (p-2)\langle \phi(a\partial_t w, w)^2, |w|_a^{p-4}\rangle$$

$$= \frac{1}{p}\langle \phi'', |w|_a^p\rangle - \langle \phi|\partial_t w|_a^2, |w|_a^{p-2}\rangle - \frac{4(p-2)}{p^2}\langle \phi \partial_t(|w|_a^{p/2}), \partial_t(|w|_a^{p/2})\rangle$$

$$\leqslant C_1\|w, \tilde{\Omega}_T\|_{0,p}^p - C_2\langle \phi \partial_t(|w|_a^{p/2}), \partial_t(|w|_a^{p/2})\rangle.$$

Analogously,

$$\langle a\Delta w, \phi w|w|_a^{p-2}\rangle \leqslant -C_2\langle \phi\nabla(|w|_a^{p/2}), \nabla(|w|_a^{p/2})\rangle.$$

Hence we obtain

$$-\langle a(\partial_t^2 w + \Delta w), \phi w|w|_a^{p-2}\rangle \geqslant -C_1\|w, \tilde{\Omega}_T\|_{0,p}^p + C_2(\langle \phi \partial_t(|w|_a^{p/2}), \partial_t(|w|_a^{p/2})\rangle$$

$$+ \langle \phi\nabla(|w|_a^{p/2}), \nabla(|w|_a^{p/2})\rangle). \qquad (2.180)$$

It follows from Hölder's inequality that

$$|\langle \gamma \partial_t w, \phi w |w|_a^{p-2}\rangle| \leqslant \mu\langle \phi \partial_t(|w|_a^{p/2}), \partial_t(|w|_a^{p/2})\rangle + C_\mu\langle \phi |w|^p, 1\rangle$$

and

$$|\langle h, \phi w |w|_a^{p-2}\rangle| \leqslant \mu\langle \phi |w|^{2(p-1)}, 1\rangle + C_\mu\langle \phi h, 1\rangle$$
$$\leqslant \mu\langle \phi |w|^{2(p-1)}, 1\rangle + C_\mu(\|g, \tilde{\Omega}_T\|_{0,2}^2 + \chi(1-T)\|u_0\|_{V_0}^2). \tag{2.181}$$

Here $\mu > 0$ is an arbitrary number.

Arguing as for (2.173) above we obtain

$$\langle f(w+v), \phi w |w|_a^{p-2}\rangle \geqslant -C_1(1 + \langle \phi |v|^{2(p-1)}, 1\rangle) + C_2\langle \phi |w|^{2(p-1)}, 1\rangle$$
$$\geqslant -C_3(1 + \chi(1-T)\|u_0\|_{V_0}^{2(p-1)}) + C_2\langle \phi |w|^{2(p-1)}, 1\rangle. \tag{2.182}$$

Now replacing all terms in equality (2.179) by their corresponding bounds in (2.180) to (2.182) and taking $\mu > 0$ sufficiently small, after a short calculation we get

$$\langle \phi |w|^{2(p-1)}, 1\rangle \leqslant C(1 + \chi(1-T)\|u_0\|_{V_0}^{2(p-1)} + \|g, \tilde{\Omega}_T\|_{0,2}^2 + \|w, \tilde{\Omega}_T\|_{0,p}^p).$$

This proves *Theorem 2.28*.   □

Corollary 2.10 *Let u be a solution to* (2.161). *Then for each $T \geqslant 0$ we have*

$$\|f(u), \tilde{\Omega}_T\|_{0,2} \leqslant C(1 + \chi(1-T)\|u_0\|_{V_0}^{p-1} + \|g, \tilde{\Omega}_T\|_{0,2}). \tag{2.183}$$

This follows from the estimates (2.177), (2.178).

Theorem 2.29 (The Main Estimate) *Let u be a solution to the problem* (2.161). *Then the following estimate holds.*

$$\|u, \tilde{\Omega}_T\|_{2,Q} \leqslant C(1 + \chi(1-T)\|u_0\|_{V_0}^{p-1} + \|g, \tilde{\Omega}_T\|_{0,2}). \tag{2.184}$$

Proof Rewrite Eq. (2.168) in the following form:

$$\begin{cases} \partial_t^2(\varphi w) + \Delta(\varphi w) = h_w(t), \\ \varphi w|_{t=\max\{T-1,0\}} = 0, \ \varphi w|_{t=T+2} = 0, \ \partial_n(\varphi w)|_{\partial\omega} = 0. \end{cases} \tag{2.185}$$

Here φ is a cut-off function as in the proof of *Theorem 2.28* and

$$h_w(t) = \varphi'' w + 2\varphi' \partial_t w - a^{-1}(\varphi h(t) + \varphi f(u) - \gamma \partial_t w).$$

By (2.166) and (2.183) we have the estimate

$$\|h_w, \tilde{\Omega}_T\|_{0,2} \leqslant C(1 + \chi(2 - T)\|u_0\|_{V_0}^{p-1} + \|g, \hat{\Omega}_T\|_{0,2}),$$

where $\hat{\Omega}_T = (\max\{0, T - 2\}, T + 3) \times \omega$. By the L^2-regularity theorem (see *Sect. 1.8*) we obtain

$$\begin{aligned}
\|w, \Omega_+ \cap \tilde{\Omega}_T\|_{2,Q} &\leqslant C_1 \|\varphi w, \tilde{\Omega}_T\|_{2,Q} \\
&\leqslant C \|h_w, \tilde{\Omega}_T\|_{0,2} \\
&\leqslant C_2(1 + \chi(2 - T)\|u_0\|_{V_0}^{p-1} + \|g, \hat{\Omega}_T\|_{0,2}).
\end{aligned}$$

This completes the proof of *Theorem 2.29*.   □

Remark 2.24 Let the condition (2.163) be satisfied. Then each solution u to (2.161) that is in $[H^2_{Q,\text{loc}}(\Omega_+)]^k$ belongs automatically to the space $[H^2_{Q,b}(\Omega_+)]^k$. More precisely, we have the estimate

$$\|u\|_b \equiv \sup_{T \geqslant 0} \|u, \Omega_T\|_{2,Q} \leqslant C(1 + \|u_0\|_{V_0}^{p-1} + \|g\|_b). \tag{2.186}$$

In fact, (2.186) is a consequence of (2.184).

2.13.1   Existence of Solutions

In this section we shall prove solvability for the problem (2.161). We first solve the following auxilliary problem in a finite cylinder:

$$\begin{cases}
a(\partial_t^2 u + \Delta u) + \gamma \partial_t u - f(u) = g(t), \\
u|_{t=0} = u_0, \; u|_{t=M} = u_1, \; \partial_n u|_{\partial \omega} = 0.
\end{cases} \tag{2.187}$$

Here $u_0, \; u_1 \in V_0$ and $u \in [H^2_Q(\Omega_{0,M})]^k$. Then we shall obtain the solution u to the main problem (2.161) as the limit as $M \to \infty$ of solutions u_M to the corresponding auxiliary problems (2.187).

Theorem 2.30 *Let u be the solution to the problem (2.187). Then the following estimate holds uniformly with respect to $M \to \infty$:*

$$\|u, \Omega_T\|_{2,Q} \leqslant C(1 + \chi(1-T)\|u_0\|_{V_0}^{p-1} + \chi(T-M+1)\|u_1\|_{V_0}^{p-1}$$

$$+ \|g, \tilde{\Omega}_T \cap \Omega_{0,M}\|_{0,2}). \tag{2.188}$$

The proof of (2.188) is analogous to that of (2.184) given above in the case of the semibounded cylinder:

Theorem 2.31 *For every* u_0, $u_1 \in V_0$, *the problem* (2.187) *has at least one solution.*

Proof Introduce the space

$$\mathcal{W}_M = \{w \in [H_Q^2(\Omega_{0,M})]^k : w|_{t=0} = w|_{t=M} = 0\}$$

and reformulate problem (2.187) with respect to the new function $w = u - v$, where $w \in \mathcal{W}_M, v \in [H_Q^2(\Omega_{0,M})]^k$:

$$\begin{cases} \partial_t^2 w + \Delta w = a^{-1}(-\gamma\partial_t w + f(v+w) + g(t)), \\ w|_{t=0} = 0, \ w|_{t=M} = 0, \ \partial_n w|_{\partial\omega} = 0. \end{cases} \tag{2.189}$$

Here $g_1 = -a(\partial_t^2 v + \Delta v) - \gamma\partial_t v + g$.

Let A denote the inverse to the Laplace operator with respect to the variables $(t, x) \in \Omega_{0,M}$ and the boundary conditions $w|_{t=0} = 0$, $w|_{t=M} = 0$, $\partial_n w|_{\partial\omega} = 0$. Then from the results of *Sect. 1.8* we get

$$A : [L^2(\Omega_{0,M})]^k \to \mathcal{W}_M.$$

Applying the operator A to both sides of Eq. (2.189) we obtain

$$w + F(w) = h \equiv -A(\partial_t^2 + \Delta v),$$

where

$$F(w) = -Aa^{-1}(-\gamma\partial_t w + f(v+w) + g - \gamma\partial_t v).$$

Now we use the Leray-Schauder principle in the following form (see [49, 69]):

Leray-Schauder Principle *Let D be a bounded open set in a Banach space W and let* $F : \overline{D} \to W$ *be a compact and continuous operator. Further let the point* $h \in D$ *be such that*

$$w + sF(w) \neq h, \ \text{for all } w \in \partial D, \ s \in [0, 1]. \tag{2.190}$$

Then the equation

$$w + F(w) = h$$

has at least one solution in D.

Let B_R be an open ball in $\mathcal{W}_M$ of sufficiently large radius and suppose that

$$w_s + sF(w_s) = h, \text{ for some } w_s \in \partial D, \ s \in [0, 1]. \tag{2.191}$$

Equation (2.191) can be rewritten in the form

$$\begin{cases} a(\partial_t^2 u_s + \Delta u_s) + s\gamma \partial_t u_s - sf(u_s) = sg(t), \\ u_s|_{t=0} = 0, \ u_s|_{t=M} = 0, \ \partial_n u_s|_{\partial\omega} = 0, \end{cases} \tag{2.192}$$

where $u_s = w_s + v$. Now (2.192) is of the form (2.187). Moreover, it is not difficult to see that the estimate (2.188) holds uniformly with respect to $s \in [0, 1]$. Hence

$$\|w_s\|_{\mathcal{W}_M} \leqslant K$$

for all solutions to (2.192) uniformly with respect to $s \in [0, 1]$. Therefore, condition (2.190) is fulfilled if the radius of B_R is chosen larger than K.

We prove compactness for the operator F. It is sufficient to prove compactness for the nonlinear part $Aa^{-1}f(w + v)$. To do this decompose the nonlinear part as a composition of three continuous operators $A \circ F_2 \circ F_1$, with one of them being compact: $F_1 : \mathcal{W}_M \rightarrow [L^{2(p-1)}(\Omega_{0,M})]^k$ is the embedding which is compact because $2(p-1) < q_0$ (see *Theorem 1.35*) and $F_2 w = a^{-1}f(v + w)$. The operator F_2 is continuous from $[L^{2(p-1)}(\Omega_{0,M})]^k$ to $[L^2(\Omega_{0,M})]^k$ in view of condition (2.162) and Krasnoselski's theorem (see *Theorem 1.14* and [74]). Hence the operator F is compact and according to the Leray-Schauder principle the problem (2.187) has at least one solution. □

Theorem 2.32 *The problem (2.161) has at least one solution $u \in [H^2_{Q,b}(\Omega_+)]^k$.*

Proof Consider a sequence u_M, $M = 1, 2, \ldots$, of solutions to the auxiliary problems (2.187) with $u_1|_{t=M} = 0$. It follows from Theorem 2.30 that, for every fixed N,

$$\|u_M, \Omega_{0,N}\|_{2,Q} \leqslant C(u_0, N, g)$$

holds uniformly with respect to $M \geqslant N$. Using Cantor's diagonalization procedure we extract a subsequence from u_M, again denoted by u_M, obeying the following property:

$$u_M|_{\Omega_{0,N}} \rightarrow u|_{\Omega_{0,N}} \text{ weakly in } [H^2_Q(\Omega_{0,N})]^k$$

for a certain $u \in [H^2_{Q,b}(\Omega_+)]^k$. We finally show that u is a solution to (2.161). It is sufficient to prove that, for every $\Phi \in [C^\infty_0(\Omega_+)]^k$, the following equality holds:

$$-\langle a\partial_t u, \partial_t \Phi \rangle - \langle a\nabla u, \nabla\Phi \rangle + \langle \gamma \partial_t u, \Phi \rangle - \langle f(u), \Phi \rangle = \langle g, \Phi \rangle. \qquad (2.193)$$

From the definition of u_M we conclude that

$$-\langle a\partial_t u_M, \partial_t \Phi \rangle - \langle a\nabla u_M, \nabla\Phi \rangle + \langle \gamma \partial_t u_M, \Phi \rangle - \langle f(u_M), \Phi \rangle = \langle g, \Phi \rangle \qquad (2.194)$$

when M is sufficiently large. Taking the limit $M \to \infty$ in (2.194) we obtain (2.193). In fact, the only non-trivial part in its proof is to show that

$$\langle f(u_M), \Phi \rangle = \langle f(u), \Phi \rangle$$

holds. Suppose that supp $\Phi \subset \Omega_{0,N}$. By *Theorem 1.35*, the embedding

$$[H^2_Q(\Omega_{0,N})]^k \subset [L^{2(p-1)}(\Omega_{0,N})]^k$$

is compact. Hence $u_M \to u$ in $[L^{2(p-1)}(\Omega_{0,N})]^k$ and, by condition (2.162), $f(u_M) \to f(u)$ in $[L^2(\Omega_{0,N})]^k$. Theorem 2.32 is proved. $\qquad\qquad \square$

2.13.2   Trajectory Attractor for the Nonlinear Elliptic System

Now we are going to construct the trajectory attractor for the problem (2.161). For the convenience of the reader we recall and adopt below the main concepts and definitions as well as theorems from *Sect. 2.7* to (2.161). See also [33, 34] for more details.

Definition 2.23 The right-hand side g of (2.161) is said to be translation-compact in

$$\Xi = [L^2_{\mathrm{loc}}(\mathbb{R}_+, L^2(\omega))]^k$$

if its hull

$$\mathcal{H}^+(g) = [T_s g, \ s \geqslant 0]_{\Xi^+}, \quad (T_s g)(t) = g(t+s)$$

is compact in Ξ^+. Here $[\cdot]_{\Xi^+}$ means the closure in the space Ξ^+.

The right-hand side g of (2.161) is said to be weakly translation-compact in the space Ξ^+ if its weak hull

$$\mathcal{H}_w^+(g) = [T_s g, \ s \geqslant 0]_{\Xi_w^+}$$

is compact in Ξ_w^+. Here Ξ_w^+ denotes the space Ξ_w^+ equipped with the weak topology.

Remark 2.25 If the function g is translation-compact for the strong topology, then it is weakly translation-compact and

$$\mathcal{H}^+(g) = \mathcal{H}_w^+(g)$$

(see [103]).

Remark 2.26 A functions g that is almost-periodic in t with values in $L^2(\omega)$ in the sense of Bochner-Amerio, in particular, a periodic or a quasi-periodic function, is evidently translation-compact in the space Ξ^+ (for its strong topology). Hence translation-compactness is a generalization of the concept of almost-periodicity.

Remark 2.27 It follows from the definition of the hull that

$$T_s\mathcal{H}^+(g) \subseteq \mathcal{H}^+(g), \ T_s\mathcal{H}_w^+(g) \subseteq \mathcal{H}_w^+(g) \text{ for } s \geqslant 0, \tag{2.195}$$

i.e., the semigroup $\{T_s, \ s \geqslant 0\}$ of shifts acts on $\mathcal{H}^+(g)$ and $\mathcal{H}_w^+(g)$, respectively.

Next we formulate necessary and sufficient conditions for translation-compactness and weak translation-compactness in the space Ξ^+.

Theorem 2.33 ([32])

(1) *A function g is weakly translation-compact in Ξ^+ if and only if it is bounded with respect to $t \to \infty$, i.e., $\|g\|_b < \infty$.*
(2) *A function g is translation-compact in Ξ^+ if and only if the following conditions hold:*

   (a) *for any fixed $t > 0$ the set $\{\int_s^{t+s} g(z)dz, \ s \in \mathbb{R}_+\}$ is precompact in the space $[L^2(\omega)]^k$;*
   (b) *there exists a function $\beta(s), s \geqslant 0, \beta(s) \to 0$ as $s \to +0$, such that*

$$\int_t^{t+1} \|g(z) - g(z+l)\|_{L^2(\omega)}^2 \, dz \leqslant \beta(|l|) \text{ for all } t \in \mathbb{R}_+ \text{ with } t+l \in \mathbb{R}_+.$$

$$\tag{2.196}$$

Remark 2.28 Condition (2.196) is fulfilled, e.g., if

$$\|T_s g, \ (0, 1) \times \omega\|_{\delta,2} \leqslant C, \ s \geqslant 0,$$

for a suitable $\delta > 0$.

To construct the trajectory attractor for the problem (2.161), we consider the family of problems of the form (2.161) obtained from all positive shifts of the initial problem (2.161) together with all limits in the appropriate topology:

$$
\begin{cases}
a(\partial_t^2 u + \Delta u) + \gamma \partial_t u - f(u) = \sigma(t), \ \sigma \in \Sigma, \\
u|_{t=0} = u_0, \ \partial_n u|_{\partial\omega} = 0.
\end{cases}
\tag{2.197}
$$

Here we take $\Sigma = \mathcal{H}^+(g)$, if g is translation-compact for the strong topology, and $\Sigma = \mathcal{H}_w^+(g)$ otherwise.

Definition 2.24 For each $\sigma \in \Sigma$, K_σ denotes the space of all solutions to (2.197) with an arbitrary $u_0 \in V_0$. Further define K_Σ^+ as the union of all K_σ^+:

$$
K_\Sigma^+ = \bigcup_{\sigma \in \Sigma} K_\sigma^+.
$$

It follows from (2.195) that the semigroup $\{T_s, \ s \geqslant 0\}$ of non-negative shifts along the t-axis $((T_s v)(t) \equiv v(t + s))$ acts on the space K_Σ^+, i.e.,

$$
T_s K_\Sigma^+ \subseteq K_\Sigma^+ \text{ for } s \geqslant 0.
$$

The set K_Σ^+ is endowed with the relative topology induced from the embedding $K_\Sigma^+ \subset \Theta_0^+$ if $\Sigma = \mathcal{H}^+(g)$ (in case of the strong topology) and induced from the embedding $K_\Sigma^+ \subset (\Theta_0^+)^w$ if $\Sigma = \mathcal{H}_w^+(g)$ (in case of the weak topology), respectively. (For the definition of Θ_0^+, see *Sect. 1.8*.)

Definition 2.25 The (global) attractor of the semigroup $\{T_s, \ s \geqslant 0\}$ acting on the topological space K_σ^+ is called the trajectory attractor of the family (2.197). That means that a set $\mathcal{A}_\Sigma \subset K_\Sigma^+$ is the trajectory attractor of the family (2.197) if the following conditions hold:

(1) $\mathcal{A}_\Sigma$ is compact in K_Σ^+;
(2) $\mathcal{A}_\Sigma$ is strongly invariant with respect to $\{T_s, \ s \geqslant 0\}$, i.e.,

$$
T_s \mathcal{A}_\Sigma \subseteq \mathcal{A}_\Sigma \text{ for all } s \geqslant 0;
$$

(3) $\mathcal{A}_\Sigma$ is attracting for $\{T_s, \ s \geqslant 0\}$, i.e., for any neighbourhood $\mathcal{O} = \mathcal{O}(\mathcal{A}_\Sigma)$ in the topology of K_Σ^+ there is an $s_\mathcal{O} > 0$ such that

$$
T_s K_\Sigma^+ \subset \mathcal{O} \text{ for all } s \geqslant s_\mathcal{O}.
\tag{2.198}
$$

Remark 2.29 The attracting property is usually required only for (in some sense) bounded subsets of K_Σ^+. In view of estimate (2.185), however, the set $T_1 K_\Sigma^+$ is already bounded both in F_0^+ and Θ_0^+. Hence the attracting property (2.198) is automatically implied for all subsets of K_Σ^+, with the same constant s_O (see also *Remark 2.9*).

Theorem 2.34 ([32]) *Let the following conditions be satisfied:*

(1) *There exists a compact attracting set $P \subset K_\Sigma^+$ for the semigroup $\{T_s, \ s \geqslant 0\}$;*
(2) *the set K_Σ^+ is closed in the space Θ_0^+ in case of the strong topology and sequentially closed in the space $(\Theta_0^+)^w$ in case of the weak topology, respectively.*

Then the family (2.198) possesses a trajectory attractor $\mathbb{A} = \mathcal{A}_\Sigma$ in K_Σ^+.

Definition 2.26 The trajectory attractor $\mathcal{A}^w$ of the family (2.198) with $\Sigma = \mathcal{H}_w^+(g)$ (the case of the weak topology) is called the weak trajectory attractor of the initial problem (2.161).

Analogously the trajectory attractor $\mathcal{A} = \mathcal{A}^s$ of the family (2.198) with $\Sigma = \mathcal{H}^+(g)$ (the case of the strong topology) is called the (strong) trajectory attractor of the initial problem (2.161).

Theorem 2.35

(1) *Let condition (2.163) hold. Then the problem (2.161) possesses a weak trajectory attractor $\mathcal{A}^w$.*
(2) *Let the right-hand side g of the problem (2.161) be translation-compact in Ξ^+ (endowed with the strong topology). Then the problem (2.161) possesses a strong trajectory attractor $\mathcal{A} = \mathcal{A}^s$.*

To this end, we check the conditions of *Theorem 2.34*.

Lemma 2.21 *The set K_Σ^+ is sequentially closed in the space $(\Xi_0^+)^w$.*

Proof Let $u_m \in K_{\sigma_m}^+$, $u_m \to u$ in $(\Xi_0^+)^w$. Without loss of generality we may suppose that $\sigma_m \to \sigma$ weakly in Ξ^+, since Σ is compact in Ξ_w^+. We have to prove that $u \in K_\sigma^+$. By definition, the functions $u_m(t)$ are bounded solutions to the following problems:

$$
\begin{cases}
a(\partial_t^2 u_m + \Delta u_m) + \gamma \partial_t u_m - f(u_m) = \sigma_m(t), \\
u_m|_{t=0} = u_m^0, \ u_m^0 \in V_0.
\end{cases}
\tag{2.199}
$$

As in the proof of *Theorem 2.32*, taking the limit $m \to \infty$ in (2.199) we obtain that $u \in K_\sigma^+$. $\qquad\square$

Hence the second condition of *Theorem 2.34* holds. Let us check the first condition. From the estimate (2.184) we infer that the set

$$
P = B_R \cap K_\Sigma^+,
$$

where B_R is a sufficiently large ball in the space F_0^+, is an absorbing set for the semigroup $\{T_s, \ s \geqslant 0\}$. First we consider the case of the weak topology. The set

B_R is a compact and metrizable subset of $(\Theta_0^+)^w$. In fact, the ball B_R is bounded in Θ_0^+ and Θ_0^+ is a reflexive and separable Fréchet space, hence B_R is semi-compact and metrizable for the weak topology. Due to convexity, B_R is a metrizable compact. From *Lemma 2.21* we finally conclude that the set P is compact in $(\Theta_0^+)^w$.

Now let us suppose that the right-hand side g of the problem (2.161) is translation-compact for the strong topology.

Lemma 2.22 *Let the previous conditions hold. Then the set $T_s P$ is compact in Θ_0^+ for all $s > 0$.*

Proof Without loss of generality we suppose that $s = 1$. Let $\{u_m\}$, $u_m \in B_R \cap K_{\sigma_m}^+$, be an arbitrary sequence. We can suppose that $\sigma_m \to \sigma$ in Ξ^+, since the hull Σ is compact in Ξ^+ (for the strong topology). Further we can suppose that $\{u_m\}$ is weakly convergent to some $u \in K_\sigma^+$, since P is weakly compact.

To prove the lemma it suffices to show that $u_m \to u$ in $[H_Q^2(\Omega_T)]^k$ for an arbitrary $T \geqslant 1$. The functions u_m satisfy Eq. (2.199). Multiplying (2.199) by a cut-off function $\phi \in C_0^\infty(\mathbb{R})$ such that $\phi(t) = 1$ for $t \in (T, T+1)$ and $\phi(t) = 0$ for $t \neq (T-1, T+2)$, we obtain

$$\begin{cases} \partial_t^2(\phi u_m) + \Delta(\phi u_m) = a^{-1}(\phi \sigma_m + \phi f(u_m) - \gamma \phi \partial_t u_m) + 2\partial_t \phi \partial_t u_m + (\partial_t^2 \phi)u_m \\ \equiv h(t), \\ \phi u_m|_{t=T-1} = 0, \ u_m|_{t=T+2} = 0, \ \partial_n(\phi u_m)|_{\partial \omega} = 0. \end{cases}$$

Then, arguing as in the proof of *Theorem 2.32*, we can show that $f(u_m) \to f(u)$ and $\partial_t u_m \to \partial_t u$ in Ξ^+. So $h_m \to h$ in $[L^2(\tilde{\Omega}_T)]^k$ and

$$\phi u_m \to \phi u \text{ in } H_Q^2(\tilde{\Omega}_T)$$

(see *Sect. 1.8*). Hence,

$$u_m \to u \text{ in } [H_Q^2(\Omega_T)]^k$$

The proof is finished.   □

Therefore, all conditions of *Theorem 2.34* hold: so the proof of *Theorem 2.35* is complete.

Corollary 2.11 *Let the right-hand side g of the problem (2.161) be translation-compact in Ξ^+. Then*

$$\text{dist}_{2,Q}(\Pi_{T_1,T_2} T_s K_\Sigma^+, \Pi_{T_1,T_2} \mathbb{A}) \to 0 \text{ as } s \to +\infty,$$

where

$$\text{dist}_{2,Q}(M, N) = \sup_{x \in M} \inf_{y \in N} \|x - y, w_{T_1,T_2}\|_{2,Q}.$$

Here Π_{T_1,T_2} denotes the restriction in t to the interval (T_1, T_2).

This corollary is immediate from the definition of trajectory attractor.

Corollary 2.12 *Let the right-hand side g of the problem* (2.161) *be weakly translation-compact in* Ξ^+. *Then*

$$\text{dist}_{3/2+\varepsilon,2}(\Pi_{T_1,T_2} T_s K_\Sigma^+, \Pi_{T_1,T_2} \mathbb{A}) \to 0 \text{ as } s \to +\infty,$$

and

$$\text{dist}_{0,q}(\Pi_{T_1,T_2} T_s K_\Sigma^+, \Pi_{T_1,T_2} \mathbb{A}) \to 0 \text{ as } s \to +\infty,$$

where $\varepsilon > 0$ *is sufficiently small and* $q < 2(n+1)/(n-3)$.

This corollary follows from the compactness of the embeddings $[H_Q^2(w_{T_1,T_2})]^k \subset [H^{3/2+\varepsilon,2}(w_{T_1,T_2})]^k$ and $[H_Q^2(w_{T_1,T_2})]^k \subset [L^q(w_{T_1,T_2})]^k$ proved in *Sect. 1.8*.

Next we investigate the structure of the trajectory attractor $\mathbb{A}$. To this end for the convenience of the reader we recall several concepts from *Sect. 2.7* again.

Let $\omega(\Sigma)$ be the ω-limit set (the attractor) of the semigroup $\{T_s, \ s \geqslant 0\}$ acting on the compact space Σ. It is non-empty and can be represented in the form

$$\omega(\Sigma) = \bigcap_{t>0} \left[\bigcup_{s>t} T_s \Sigma \right]_\Sigma$$

(see [1]). Here $[\cdot]_\Sigma$ denotes the closure in the space Σ.

Definition 2.27 A function $\xi(t)$, $t \in \mathbb{R}$, is called a complete symbol of (2.197) if

$$\Pi_+ \xi_s(\cdot) \in \omega(\Sigma) \text{ for all } s \in \mathbb{R}.$$

Here $\xi_s(t) = \xi(t+s)$. The operator Π_+ is the restriction to the half-axis $\mathbb{R}_+$.

The set of all complete symbols of (2.197) is denoted by $Z(\Sigma)$.

Lemma 2.23 ([32]) *For each* $\sigma \in \omega(\Sigma)$, *there exists a complete symbol* $\xi \in Z(\Sigma)$ *such that* $\Pi_+ \xi = \sigma$.

Definition 2.28 For $\xi \in Z(\Sigma)$, K_ξ is the set of all bounded solutions to Eq. (2.197) on the whole axis $t \in \mathbb{R}$, where $\sigma(t)$ is replaced by $\xi(t)$.

Theorem 2.36 ([32]) *The attractor* $\mathbb{A}$ *has the following structure:*

$$\mathbb{A} = \Pi_+ \bigcup_{\xi \in Z(\Sigma)} K_\xi$$

Corollary 2.13 ([103]) *Let the right-hand side g be translation-compact in* Ξ^+. *Then the weak trajectory attractor of the problem* (2.161) *coincides with the strong trajectory attractor:*

$$\mathbb{A}^s = \mathbb{A}^w.$$

2.13.3 Stabilization of Solutions in the Potential Case

In this section we shall study the long-term behaviour of solutions when the right-hand side $g(t)$ of (2.161) has the form

$$g(t, x) = g_+(x) + g_1(t, x), \qquad (2.200)$$

where $g_+ \in L^2(\omega)$ is independent of t and g_1 satisfies the condition

$$T_s g_1 \rightarrow 0 \text{ as } s \rightarrow +\infty \qquad (2.201)$$

in Ξ^+ and $(\Xi^+)^w$, respectively. It is not difficult to see that in the first case the function g is strongly translation-compact in Ξ^+, while in the second case it is weakly translation-compact in Ξ^+.

Theorem 2.37 *Suppose that the condition (2.201) holds. Then the problem (2.161) with right-hand side (2.200) possesses a strong and weak trajectory attractor $\mathbb{A} = \mathbb{A}_g$, respectively. It coincides with the attractor of the limit autonomous equation*

$$a(\partial_t^2 u + \Delta u) - \gamma \partial_t u - f(u) = g_+,$$

i.e.,

$$\mathbb{A} = \mathbb{A}_g. \qquad (2.202)$$

Proof The existence of the trajectory attractor follows immediately from *Theorem 2.35*. Thus we show (2.202).

From condition (2.201) we get that

$$Z(g) \equiv Z(\Sigma) = \omega(\Sigma) = g_+.$$

Here Σ is the respectively strong and weak hull of the right-hand side g in the space Ξ^+. Hence formula (2.202) holds in view of *Theorem 2.36. Theorem 2.37* is proved.

$\square$

Now we assume that the nonlinear term $f(u)$ on the left-hand side of Eq. (2.161) is gradient-like, i.e.,

$$f(u) = -\nabla F(u), \quad F \in C(\mathbb{R}^k, \mathbb{R}). \qquad (2.203)$$

For $u \in [H_{Q,b}^2(\Omega_+)]^k$, we introduce the function $\mathcal{F}_u(t)$ by

$$\mathcal{F}_u(t) = \frac{1}{2}(a\partial_t u(t), \partial_t u(t)) - \frac{1}{2}(a\nabla u(t), \nabla u(t)) + (F(u(t)), 1) - (g_+, u(t)),$$

$$(2.204)$$

where $(\cdot, \cdot)$ is the L^2-scalar product in the cross-section.

Theorem 2.38

(1) *For every* $u \in [H_{Q,b}^2(\Omega_+)]^k$, *the function* $\mathcal{F}_u$ *is well-defined and belongs to the space* $H_b^{1,1}(\mathbb{R}_+)$.
(2) *If* u *is a solution to the problem* (2.161), *then*

$$\frac{d\mathcal{F}_u(t)}{dt} = -(\gamma \, \partial_t u(t), \partial_t u(t)) + (g_1(t), \partial_t u(t)). \tag{2.205}$$

Proof Let $u \in [H_{Q,b}^2(\Omega_+)]^k$. Then, according to *Corollary 1.12*, the first, the second, and the fourth term on the right-hand side of (2.204) are well-defined. It remains to consider the third term. From (2.162) and (2.203) we infer that

$$|F(u)| \leqslant C(1 + |u|^p).$$

Then *Corollary 1.11* yields that

$$(F(u(t)), 1) \in C_b(\overline{\mathbb{R}}_+).$$

Hence $\mathcal{F}_u(t)$ is well-defined.

When calculating the derivative, using standard methods of distribution theory, we obtain that $\mathcal{F}_u \in H_b^{1,1}(\mathbb{R}_+)$,

$$\frac{d\mathcal{F}_u(t)}{dt} = (a(\partial_t^2 u + \Delta u) - f(u) - g_+, \partial_t u). \tag{2.206}$$

Hence the first part of the proof is finished.

Now suppose that u is a solution to the problem (2.161). Then (2.203) follows immediately from (2.206). The second part of the proof is also finished.   □

Theorem 2.39 *Suppose that the conditions* (2.201) *and* (2.203) *hold. Further suppose that the matrix* γ *on the left-hand side of* (2.161) *is sign-definite, i.e.,*

$$either \ \gamma + \gamma^* > 0 \ or \ \gamma + \gamma^* < 0,$$

and the function $g_1(t) = g_1(t, x)$ *satisfies at least one of the following conditions:*

$$\begin{cases} \text{(i)} & \int_0^\infty \|g_1(t)\|_{0,2} \, dt < \infty; \\ \text{(ii)} & \partial_t g_1 \in L_{\text{loc}}^1(\mathbb{R}_+, L^2(\omega)) \ and \ \int_0^\infty \|\partial_t g_1(t)\|_{0,2} \, dt < \infty; \\ \text{(iii)} & \sum_{N=0}^\infty \|G_1, \Omega_N\|_{0,2} < \infty \ for \ some \ G_1 \ such \ that \ \partial_t G_1 = g_1. \end{cases} \tag{2.207}$$

Then every solution u to the problem (2.161) *possesses the finite dissipative integral*

$$\int_0^\infty \|\partial_t u(t)\|_{0,2}^2 \, dt < \infty. \tag{2.208}$$

Proof We integrate (2.204) over $t \in [0, T]$ and obtain

$$\int_0^T (\gamma \partial_t u, \partial_t u) \, dt = \mathcal{F}_u(0) - \mathcal{F}_u(T) + \int_0^T (g_1, \partial_t u) \, dt.$$

Now it follows from the sign-definiteness of the matrix γ that

$$\int_0^T \|\partial_t u(t)\|_{0,2}^2 \, dt = C|\mathcal{F}_u(T) - \mathcal{F}_u(0)| + C\left|\int_0^T (g_1, \partial_t u) \, dt\right|. \tag{2.209}$$

Theorem 2.38 implies that function $\mathcal{F}_u(T)$ is bounded as $T \to \infty$. Hence it suffices to show the boundedness of the integral on the right-hand side of (2.209).

Suppose that condition (i) of (2.207) holds. Then

$$\left|\int_0^T (g_1, \partial_t u) \, dt\right| \leqslant \int_0^T \|g_1(t)\|_{0,2} \|\partial_t u(t)\|_{0,2} \, dt$$

$$\leqslant \sup_{t \in [0,T]} \|\partial_t u(t)\|_{0,2} \int_0^T \|g_1(t)\|_{0,2} \, dt$$

$$\leqslant \|u\|_b \int_0^\infty \|g_1(t)\|_{0,2} \, dt. \tag{2.210}$$

Thus, $|\int_0^T (g_1, \partial_t u) \, dt|$ is bounded as $T \to \infty$.

Now suppose that condition (ii) of (2.207) holds. Then we obtain by integration by parts

$$\left|\int_0^T (g_1, \partial_t u) \, dt\right| \leqslant |(g_1(T), u(T))| + |(g_1(0), u(0))| + \left|\int_0^T (\partial_t g_1(t), u(t)) \, dt\right|. \tag{2.211}$$

The integral on the right-hand side of (2.211) is estimated in the same manner as the integral in (2.210). To estimate the first two terms on the right-hand side it suffices to prove that under above assumptions $g_1 \in C_b(\overline{\mathbb{R}}_+, L^2(\omega))$. Let $[N, N + 1] \subset \overline{\mathbb{R}}_+$ be an arbitrary interval and let $[t, T]$ be in that interval. Then

$$\|g_1(T)\|_{0,2} \leqslant \|g_1(t)\|_{0,2} + \|g_1(T) - g_1(t)\|_{0,2}$$

$$\leqslant \|g_1(t)\|_{0,2} + \int_t^T \|\partial_t g_1(t)\|_{0,2} \, dt$$

$$\leqslant \|g_1(t)\|_{0,2} + \int_t^\infty \|\partial_t g_1(t)\|_{0,2} \, dt. \tag{2.212}$$

Integrating (2.212) over $t \in [N, N + 1]$, we get

$$\|g_1(T)\|_{0,2} \leqslant C\|g_1, \Omega_N\|_{0,2} + \int_t^{\infty} \|\partial_t g_1(t)\|_{0,2} \, dt \leqslant \|g_1\|_b + \|\partial_t g_1\|_{L^1(\mathbb{R}_+, L^2(\omega))}.$$

Since the constant N was arbitrarily chosen, we find $g_1 \in C_b(\mathbb{R}_+, L^2(\omega))$.

Now suppose that condition (iii) of (2.207) holds. Again integrating by parts we obtain that

$$\left| \int_0^T (g_1, \partial_t u) \, dt \right| \leqslant |(G_1(T), \partial_t u(T))| + |(G_1(0), \partial_t u(0))|$$

$$+ \left| \int_0^T (G_1(t), \partial_t^2 u(t)) \, dt \right|.$$

The first two terms on the right-hand side can be estimated as before. The third term is estimated as follows:

$$\left| \int_0^T (G_1(t), \partial_t^2 u(t)) \, dt \right| \leqslant \int_0^T \|G_1(t)\|_{0,2} \|\partial_t^2 u(t)\|_{0,2} \, dt$$

$$\leqslant \sum_{N=0}^{[T]} \|G_1, \Omega_N\|_{0,2} \|\partial_t^2 u, \Omega_N\|_{0,2}$$

$$\leqslant C\|u\|_b \sum_{N=0}^{\infty} \|G_1, \Omega_N\|_{0,2}.$$

Theorem 2.39 is proved.   □

Theorem 2.40 *Suppose that all the assumptions of the previous theorem hold. Further suppose that the limit problem in the cross-section* w

$$\begin{cases} a\Delta v_+ - f(v_+(x)) = g_+(x), \\ \partial_n v_+|_{\partial\omega} = 0, \end{cases} \qquad (2.213)$$

has only a finite number of solutions

$$v_+ \in V_+ = \{v_+^1(x), \cdots, v_+^l(x)\}.$$

Then, for every solution u *to the problem* (2.161). *there exists an equilibrium* $v_+^N(x) \in V_+$ *such that*

$$(T_s u)(t, x) \to v_+^N(x) \text{ in } \Theta^+ \text{ as } s \to +\infty. \qquad (2.214)$$

Here Θ^+ denotes the space Θ_0^+ if g is strongly translation-compact in Ξ and the space $(\Theta_0^+)^w$ if g is weakly translation-compact in Ξ, respectively.

Remark 2.30 As is known (see, for instance, [1]), there exists an open dense subset in $L^2(\omega)$ such that $\mathcal{V}_+$ is finite for every g_+ belonging to this set.

Proof Let u be a solution of the problem (2.161). Consider the ω-limit set[1] $\omega(u)$ of $u \in \Theta^+$ under the action of the semigroup $\{T_s, \ s \geqslant 0\}$. Recall that $u_+ \in \omega(u)$ if and only if there exists the sequence $\{s_j\}_{j \in \mathbb{N}}, s_j \to \infty$, such that

$$T_{s_j} u \to u_+ \text{ in } \Theta^+. \tag{2.215}$$

By *Theorem 2.37*, $\{T_s, \ s \geqslant 0\}$ possesses an attractor $\mathbb{A}$ in $K_\Sigma^+ \subset \Theta^+$, hence $\omega(u)$ is a nonempty, compact, and connected subset of Θ^+ (see [1]).

Let u_+ be in $\omega(u)$. Let $\{s_j\}_{j \in \mathbb{N}}$ be a sequence as in 2.215. Then, for every $T > 0$,

$$T_{s_j} u \to u_+ \text{ weakly in } H_Q^2(\Omega_T) \text{ as } s_j \to \infty.$$

In particular,

$$\|T_{s_j} \partial_t u - \partial_t u_+, \Omega_T\|_{0,2} \to 0 \text{ as } s_j \to \infty.$$

Now the finiteness of the dissipative integral (2.208) implies that

$$\|T_{s_j} \partial_t u, \Omega_T\|_{0,2} = \|\partial_t u, T_{s_j} \Omega_T\|_{0,2} \to 0 \text{ as } s_j \to \infty.$$

Therefore, $\|\partial_t u_+, \Omega_T\|_{0,2} = 0$ and $u_+(t, x) \equiv u_+(x)$.

From condition (2.201) and *Lemma 2.21*, however, we conclude that $u_+(x)$ is a solution to the limit problem (2.213). Thus

$$\omega(u) \subset \mathcal{V}_+. \tag{2.216}$$

Since $\omega(u)$ is connected and $\mathcal{V}_+$ is discrete, we eventually get

$$\omega(u) = \{v_+^N\} \text{ for some } N \in \{1, \cdots, l\}.$$

Finally (2.214) is a consequence of the attracting property for $\{T_s, \ s \geqslant 0\}$. *Theorem 2.40* is proved.   □

---

[1] Actually we use here the letter ω in two different senses: for the ω-limit set and for a bounded polyhedral domain in $\mathbb{R}^n$; $\Omega_+ := \mathbb{R} \times \omega, \Omega = \mathbb{R} \times \omega$. We hope that it will not lead to a misunderstanding for the readers.

Corollary 2.14 *Both in the case of strong translation-compactness and of weak translation-compactness of g, (2.216) implies that*

$$
\begin{cases}
\lim_{t \to +\infty} \| u(t, \cdot) - v_+^N(\cdot) \|_{0, p_0} = 0, \\
\lim_{t \to +\infty} \| \partial_t u(t, \cdot) \|_{\varepsilon, 2} = 0,
\end{cases}
$$

where the exponent p_0 is defined in Corollary 1.11 and $\varepsilon < 1/2$.

This is obtained similarly to the proof of *Corollary 2.12*.

Corollary 2.15 *Suppose that the function g_+ satisfies the conditions of Theorem 2.40. Then, any solution $u(t)$, $t \in \mathbb{R}$, to Eq. (2.202) in the full cylinder $\Omega = \mathbb{R} \times \omega$, which is not an equilibrium itself, is a heteroclinic orbit, i.e., there exist two different equilibria w_u^+ and w_u^- belonging to $\mathcal{V}_+$ such that*

$$
T_s u \to w_u^+ \text{ as } s \to +\infty, \ T_s u \to w_u^- \text{ as } s \to -\infty. \tag{2.217}
$$

In fact, in view of estimate (2.184), *see Remark 2.23*, any solution $u(t)$ to the problem (2.197) is bounded with respect to both $t \to \infty$ and $t \to -\infty$. Thus the convergence (2.217) follows from *Theorem 2.40*. Hence, it remains to prove that $w_u^+ \neq w_u^-$. Integrating (2.205) over $\mathbb{R}$, where $g_1 \equiv 0$, we get

$$
\mathcal{F}_u(+\infty) - \mathcal{F}_u(-\infty) = \mathcal{F}_{w^+} - \mathcal{F}_{w^-} = -\int_{\mathbb{R}} (\gamma \partial_t u, \partial_t u) \, dt \neq 0.
$$

Thus $w^+ \neq w^-$.

We now give examples for the perturbation term $g_1(t, x)$ satisfying the conditions of *Theorem 2.40*.

Example 2.13 Let

$$
g_1(t, x) = \varphi(t) g_0(x), \tag{2.218}
$$

where $g_0 \in L^2(\omega)$ and

$$
\varphi(t) = \frac{|\sin(t^2)|}{1 + t^2}.
$$

Then condition (i) of (2.207) is fulfilled. (2.201) holds for the strong topology.

Example 2.14 Let $g_1(t, x)$ be as in (2.218), where

$$
\varphi(t) = \frac{t}{1 + t^2}.
$$

Then condition (ii) of (2.207) is fulfilled. (2.201) holds for the strong topology.

Example 2.15 Let $g_1(t, x)$ be as in (2.218), where

$$\varphi(t) = \sin(t^3).$$

Then condition (iii) of (2.207) is fulfilled. (2.201) holds for the strong topology.

2.13.4   Regular and Singular Part of the Trajectory Attractor

In this final section we show that the trajectory attractor $\mathbb{A}$ of the problem (2.161) decomposes into a regular part $\mathbb{A}_{\text{reg}}$ and a singular part $\mathbb{A}_{\text{sing}}$. For brevity we suppose that the right-hand side g of the problem (2.161) is strongly translation-compact in Ξ^+. The case of weak translation-compactness is treated analogously.

Let $K^+ = K_{\Sigma}^+$ be the union of all solutions to the family (2.197) (see Definition 2.24). Let Π_2^+ be the same as in *Theorem 1.39*. The regular and the singular part of the union K^+ are introduced by

$$K_{\text{reg}}^+ = \Pi_1^+ K^+, \; K_{\text{sing}}^+ = \Pi_2^+ K^+, \text{ where } \Pi_1^+ = \text{Id} - \Pi_2^+.$$

Notice that

$$K_{\text{reg}}^+ \subset [H_{N,\text{loc}}^2(\Omega_+)]^k, \tag{2.219}$$

and the topology on K_{reg}^+ induced by the embedding $K_{\text{reg}}^+ \subset \Theta_0^+$ coincides with the topology induced by the embedding (2.219).

From *Theorem 1.39* it follows that the semigroup $\{T_s, \; s \geqslant 0\}$ of positive shifts acts both in K_{reg}^+ and in K_{sing}^+, i.e.,

$$T_s K_{\text{reg}}^+ \subseteq K_{\text{reg}}^+ \text{ and } T_s K_{\text{sing}}^+ \subseteq K_{\text{sing}}^+ \text{ for all } s \geqslant 0.$$

Definition 2.29 The attractor $\mathbb{A}_{\text{reg}}$ of the semigroup $\{T_s, \; s \geqslant 0\}$ acting in the topological space K_{reg}^+ is called the regular trajectory attractor for the problem (2.161) (see Definition 2.24).

Analogously, the attractor $\mathbb{A}_{\text{sing}}$ of the semigroup $\{T_s, \; s \geqslant 0\}$ acting in the topological space K_{sing}^+ is called the singular trajectory attractor for the problem (2.161)

Theorem 2.41 *Under the above assumptions, the problem (2.161) possesses a regular trajectory attractor $\mathbb{A}_{\text{reg}}$ as well as a singular trajectory attractor $\mathbb{A}_{\text{sing}}$. Moreover,*

$$\mathbb{A}_{\text{reg}} = \Pi_1^+ \mathbb{A}, \; \mathbb{A}_{\text{sing}} = \Pi_2^+ \mathbb{A},$$

where $\mathbb{A}$ is the trajectory attractor for the problem (2.161).

Proof We only verify that $\mathbb{A}_{\text{sing}} = \Pi_2^+ \mathbb{A}$. $\mathbb{A}_{\text{reg}} = \Pi_1^+ \mathbb{A}$ is completely analogous.

First we are concerned with the attracting property. Let $\mathcal{O} = \mathcal{O}(\Pi_2^+ \mathbb{A})$ be a neighbourhood of $\Pi_2^+ \mathbb{A}$ in K_{sing}^+. By *Theorem 1.39*, $(\Pi_2^+)^{-1} \mathcal{O}$ is a neighbourhood of $\mathbb{A}$ in K^+. Hence from the attracting property for $\mathbb{A}$ we obtain that there exists a $s_{\mathcal{O}} > 0$ such that

$$T_s K^+ \subset \Pi_2^{-1} \mathcal{O} \text{ for } s \geqslant s_{\mathcal{O}}. \tag{2.220}$$

Applying Π_2^+ to both sides of (2.220) and using assertion (b) of *Theorem 1.39*, we find

$$T_s K_{\text{sing}}^+ \subset \Pi_2^+ (\Pi_2^+)^{-1} \mathcal{O} \subseteq \mathcal{O} \text{ for } s \geqslant s_{\mathcal{O}}.$$

This is the attracting property for $\Pi_2^+ \mathbb{A}$. Secondly, by definition, $T_s \mathbb{A} = \mathbb{A}$ for all $s \geqslant 0$. Applying Π_2^+ to both sides and using (b) of *Theorem 1.39* again we get

$$T_s \Pi_2^+ \mathbb{A} = \Pi_2^+ \mathbb{A} \text{ for } s \geqslant 0.$$

This is the strict invariance of $\Pi_2^+ \mathbb{A}$ under the action of $\{T_s, \ s \geqslant 0\}$. Finally, the compactness of $\Pi_2^+ \mathbb{A}$ in K_{sing}^+ is immediate from the compactness of the attractor $\mathbb{A}$ and the continuity of Π_2^+.

Thus $\Pi_2^+ \mathbb{A}$ is the singular trajectory attractor for the problem (2.161). *Theorem 2.41* is proved. □

Corollary 2.16 *Let τ_j^+, $1 \leqslant j \leqslant \kappa$, be the trace operators supplied by Corollary 1.17. Then the semigroup $\{T_s, \ s \geqslant 0\}$ of positive shifts acts in the spaces $\tau_j^+ K^+ \subset [H_{\text{loc}}^{1-\pi/\alpha_j}(\overline{\mathbb{R}}_+)]^k$ and possesses the attractors $\mathbb{A}_j = \tau_j^+ \mathbb{A}$ there.*

This is immediate from the topological isomorphism

$$(\tau_1^+, \cdots, \tau_\kappa^+) : \Pi_2^+ H_{Q,\text{loc}}^2(\overline{\mathbb{R}}_+ \times \omega) \to \bigoplus_{j=1}^{\kappa} H_{\text{loc}}^{1-\pi/\alpha_j}(\overline{\mathbb{R}}_+)$$

derived in *Sect. 1.9*. Note that $\tau_j^+ \mathbb{A}$ has an invariant meaning, while $\Pi_2^+ \mathbb{A}$ depends on the choice of the projection Π_2^+.

Finally we are concerned with the question of stabilization of asymptotics in the case when stabilization of solutions takes place. For that we impose all assumptions of *Theorem 2.39*, in particular, that $f(u) = -\nabla F(u)$ is gradient-like (see (2.203)) and the limit equation

$$a\Delta v_+ - f(v_+) = g_+, \ \partial_n v_+|_{\partial \omega} = 0$$

has only a finite number of solutions $v_+ = v_+^N$ in $[H_Q^2(w)]^k$, $N = 1, \ldots, l$.

Let $\{d_j^N\}_{j=1}^\kappa$ be the sequence of singular coefficients to v_+^N, i.e.,

$$v_+^N = v_0^N + \sum_{j=1}^\kappa d_j^N (\chi_j)^* (\psi_j(r) r^{\pi/\alpha_j} \cos(\pi\theta/\alpha_j))$$

where $v_0^N = \Pi_1^+ v_+^N \in [H_N^2(w)]^k$ and $d_j^N \in \mathbb{C}^k$ (see Remark 1.24).

Theorem 2.42 *Let the assumptions of Theorem 2.40 be fulfilled. Then, for each solution u to the problem (2.161), there exists an equilibrium v_+^N such that*

$$T_s u \to v_+^N \ \text{in}\ \Theta_0^+ \ \text{as}\ s \to \infty. \tag{2.221}$$

Moreover,

$$T_s \Pi_1^+ u \to \Pi_1^+ v_+^N = v_0^N, \ T_s \Pi_2^+ u \to \Pi_2^+ v_+^N$$

and

$$T_s \tau_j^+ u \to d_j^N \ \text{in}\ H_{\text{loc}}^{1-\pi/\alpha_j}(\overline{\mathbb{R}_+}) \ \text{as}\ s \to \infty.$$

Proof (2.221) is a consequence of *Theorem 2.40*. The other assertions are immediate from the continuity of the operators Π_1^+, Π_2^+ and τ_j^+ in the appropriate spaces.

$\square$

2.14 The Dynamics of Fast Nonautonomous Travelling Waves and Homogenization

We study the elliptic boundary value problem

$$\begin{cases} a(\partial_t^2 + \Delta_x u) - \frac{\gamma}{\varepsilon}\partial_t u - f(u) = g(t, x), \\ u|_{\partial\Omega} = 0; \ u|_{t=0} = u_0(x) \end{cases} \tag{2.222}$$

in a semicylinder $(t, x) \in \Omega_+ := \mathbb{R}_+ \times \omega$, where ω is a smooth bounded domain in $\mathbb{R}^n$, $u = (u^1, \ldots, u^k)$ is an unknown vector function, f, g and u_0 are given vector functions, and a and γ are given constant $k \times k$ matrices such that $a + a^* > 0$ and $\gamma = \gamma^* > 0$, and ε is a small positive number. In this section we follow [107], thereby preserving the notation (for a similar homogenization setting for parabolic systems see [48]).

The problems of (2.222) type arise in the study of travelling wave solutions of nonautonomous evolution equations in a cylindrical domain $\Omega := \mathbb{R} \times \omega$. Indeed, consider the second order nonautonomous parabolic equation in $(t, x) \in \Omega$.

$$\partial_\eta v = a(\partial_t^2 + \Delta_x u) - f(v) - g\left(t - \frac{\gamma}{\varepsilon}\eta, x\right) \tag{2.223}$$

with the fast travelling wave external force $g\left(t - \frac{\gamma}{\varepsilon}\eta, x\right)$ (where $\frac{\gamma}{\varepsilon} \gg 1$ is the wave speed, and the variable η plays the role of time). Then, the problem of finding the travelling wave solution (modulated by the external travelling wave g)

$$v(\eta, t, x) := v\left(t - \frac{\gamma}{\varepsilon}\eta, x\right)$$

leads to the elliptic boundary value problem (2.222) in a cylinder Ω. Applying the dynamical approach to studying the problem in the full cylinder, we obtain the axillary problem of the type (2.222) which is of independent interest. We assume that the nonlinear term $f(u)$ in (2.222) satisfies the following assumptions:

$$\begin{cases} f(u) \cdot v \geqslant -C, \\ f'(v) \geqslant -K, \\ |f(v)| \leqslant C(1 + |v|^q) \text{ for all } v \in \mathbb{R}^k \end{cases} \tag{2.224}$$

for some appropriate constants C and K and with the growth exponent $q < q_{max} = \frac{n+2}{n-2}$. Moreover, we assume that the external force $g(t, x)$ is almost periodic with respect to t with values in $L^2(\omega)$, that is, $g \in C_b(\mathbb{R}, L^2(\omega))$. Recall that (see *Sect. 2.8* and [48]) it means that the hull

$$\mathcal{H}(g) := [T_h g, h \in \mathbb{R}]_{C_b(\mathbb{R}, L^2(\omega))}, \quad (T_h g)(t) := g(t + h)$$

is compact in $C_b(\mathbb{R}, L^2(\omega))$. Here, we denote by $[\ldots]_H$ the closure in the space H. The solution of (2.222) is a function which belongs to $W^{2,2}(\Omega_T)$ for every $T \geqslant 0$ ($\Omega_T := [T, T+1] \times \omega$) and has the finite norm

$$\|u\|_{W_b^{2,2}(\mathbb{R}_+)} := \sup_{T \geqslant 0} \|u, \Omega_T\|_{2,2} < \infty,$$

and, therefore, we restrict ourselves to consider only the solutions of the problem (2.222) that are bounded with respect to $t \to \infty$. It is natural to assume that the initial data u_0 belongs to the space $\mathbb{V}_0 := W^{\frac{3}{2},2}(\omega) \cap \{u_0|_{\partial\omega} = 0\}$, which is in fact the space of traces of the functions from $W_b^{2,2}(\Omega_+) \cap \{u|_{\partial\omega} = 0\}$ at $t = 0$ (see, e.g., [101]). The equations of type (2.222) have been studied, under various assumptions on $a, \gamma, f, g,$ and ε, in [31]. It is known that (see [31]) under the assumptions (2.224) for each fixed $\varepsilon < \varepsilon_0 \ll 1$ the problem (2.222) possesses a unique (bounded with respect to $t \to \infty$) solution $u(t, x)$, which satisfies the estimate

$$\|u, \Omega_T\|_{2,2} \leqslant Q_\varepsilon\left(\|u_0\|_{\mathbb{V}_0}\right) e^{-\alpha T} + Q_\varepsilon\left(\|g\|_{L_b^2}\right) \tag{2.225}$$

with a certain monotonic function Q_ε depending only on f, a, γ and ε (and independent of u_0 and g) and a positive α. Consequently, the problem (2.222) defines a process $U_g^\varepsilon(t, \tau)$ in the phase space $\mathbb{V}_0$ by the formula

$$U_g^\varepsilon(t, \tau)u_\tau(x) = u(t, x),$$

where $u(t, x)$, $t \geqslant \tau$ is a solution of the problem (2.222) with $u(\tau, x) = u_\tau(x)$. It is well-known (see [35, 48]) that the process can be extended to a semigroup S_t^ε acting in the extended phase space $\mathbb{V}_0 \times \mathcal{H}(g)$ in the following way:

$$S_t^\varepsilon(u_0, \xi) = \left(U_\xi^\varepsilon(t, 0)u_0, T_t\xi \right), \ t \geqslant 0, \ \xi \in \mathcal{H}(g), \ u_0 \in \mathbb{V}_0. \tag{2.226}$$

It is not difficult to prove (see [1, 48]) using the dissipative estimate (2.225) and the compactness of $\mathcal{H}(g)$ in $C_b(\mathbb{R}, L^2(\omega))$ that the semigroup (2.226) possesses a (global) attractor $\mathcal{A}^\varepsilon$ in $\mathbb{V}_0 \times \mathcal{H}(g)$. The projection of $\mathbb{A}^\varepsilon := \Pi_1\mathcal{A}^\varepsilon$ of this attractor to the first component ($\mathbb{V}_0$) is called (uniform) attractor for (2.222) (see [1, 48]). Note that the attractor $\mathbb{A}^\varepsilon$ is generated by all bounded solutions of the family of equations

$$a(\partial_t^2 u + \Delta_x u) - \frac{\gamma}{\varepsilon}\partial_t u - f(u) = \xi(t, x), \ \xi \in \mathcal{H}(g), \tag{2.227}$$

which is defined in a full cylinder $\Omega = \mathbb{R} \times \omega$:

$$\mathbb{A}^\varepsilon = \bigcup_{\xi \in \mathcal{H}(g)} K_\xi^\varepsilon|_{t=0}, \tag{2.228}$$

where K_ξ^ε is a set of all bounded in $W_b^{2,2}(\Omega)$ solutions of (2.227) with the right-hand side ξ, or, which is the same, the set of all travelling wave solutions of the evolution equation (2.223) (with g replaced by ξ). This justifies the attractor's approach to study the travelling wave solutions. The main goal of this section is to investigate the behaviour of the attractors $\mathbb{A}^\varepsilon$ as $\varepsilon \to 0$. To this end, we make the time rescaling $t \to \varepsilon t$ and write the problem in the following more convenient form:

$$\begin{cases} a(\varepsilon^2\partial_t^2 u + \Delta_x u) - \gamma\partial_t u - f(u) = g_\varepsilon(t, x), \\ u|_{t=0} = u_0, \ u|_{\partial\Omega_+} = 0, \end{cases} \tag{2.229}$$

where $g_\varepsilon(t, x) = g\left(\frac{t}{\varepsilon}, x\right)$. Obviuosly, the attractors of the Eqs. (2.222) and (2.229) coincide. The feature of (2.229) is that the right-hand side becomes rapidly t-oscilating external force $g\left(\frac{t}{\varepsilon}, x\right)$. Let

$$\hat{g}(x) := \lim_{T \to \infty} \frac{2}{T} \int_{-T}^{T} g(t, x)\,dt$$

and write the limit ($\varepsilon = 0$) equation in the following form:

$$\begin{cases} \gamma \partial_t u = a\Delta_x - f(u) - \hat{g}(x) \\ u|_{t=0} = u_0, \ u|_{\partial\Omega_+} = 0, \end{cases} \tag{2.230}$$

Equation (2.230) is an autonomous dissipative reastion-diffusion equation and, consequently (see [1, 35, 48, 98]), possesses a (global) attractor $\mathbb{A}^0$ in the phase space $L^2(\omega)$. The main result of this section is the following Theorem

Theorem 2.43 *Let the above assumptions hold. Then, the attractors $\mathbb{A}^\varepsilon$ converge to the attractors $\mathbb{A}^0$ in the space $W^{1-\delta,2}(\omega)$ for every $\delta > 0$ in the following sense:*

$$\mathrm{dist}_{W^{1-\delta,2}(\omega)}\left(\mathbb{A}^\varepsilon, \mathbb{A}^0\right) \to 0 \text{ as } \varepsilon \to 0. \tag{2.231}$$

(see [1, 48, 67, 98]).

Assume now that the limit attractor $\mathbb{A}^0$ is exponential, that is,

$$\mathrm{dist}_{L^2(\omega)}(S_t\mathbb{B}, \mathbb{A}^0) \leqslant C(\mathbb{B})e^{-\nu t} \tag{2.232}$$

for every bounded subset $\mathbb{B} \subset L^2(\omega)$. Here, S_t is a semigroup generated by the autonomous equation (2.230), $\nu > 0$, and the constant $C(\mathbb{B})$ depends on $||B||_{L^2(\omega)}$. It is known that (2.232) is valid for generic $\hat{g}(x)$ at least if the Eq. (2.230) possesses a Lyapunov function, that is, $a = a^*$ and $f(u) = \nabla_u F(u)$.

Theorem 2.44 *Let the assumptions of the previous theorem hold and let the limit attractor be exponential. Moreover, let the almost-periodic function $g(t, x) - \hat{g}(x)$ have a bounded primitive in $L^2(\omega)$, that is,*

$$\mathcal{C}(T) := \int_0^T (g(t, x) - \hat{g}(x))\, dt, \ ||\mathcal{C}(T)||_{L^2(\omega)} \leqslant C(g), \text{ for all } T \geqslant 0. \tag{2.233}$$

Then, the following error estimate holds:

$$\mathrm{dist}_{L^2(\omega)}(\mathbb{A}^\varepsilon, \mathbb{A}^0) \leqslant C_g \varepsilon^\chi, \tag{2.234}$$

where $0 < \chi < 1$, and C_g can be calculated explicitly.

Remark 2.31 Note that the assumption (2.233) is obviously valid for any periodic function g but may be wrong for more general almost periodic ones. Some sufficient conditions on g satisfying (2.233) will be given in the end of this section.

Remark 2.32 Error estimate (2.234) for the differences between the regular attractors for semigroups possessing global Lyapunov functions and depending regularly on the parameter ε is given in [1]. These results have been extended in [48] to regular attractors for a quite large class of autonomous reaction-diffusion equations with spatially oscillating coefficients $\left(\frac{x}{\varepsilon}\right)$ in the main diffusion term and their homogenizations.

To prove *Theorems 2.43* and *2.44*, we derive several estimates for the solutions of the following auxiliary problem

$$\begin{cases} a(\varepsilon^2 \partial_t^2 u + \Delta_x u) - \gamma \partial_t u - f(u) = g_\varepsilon(t, x), \\ u|_{t=0} = u_0, \ u|_{\partial\Omega_+} = 0, \end{cases} \tag{2.235}$$

Lemma 2.24 *Let the above assumptions hold. Then, for every $\varepsilon < \varepsilon_0 \ll 1$ the problem (2.235) possesses a unique solution which satisfies the following estimate:*

$$\|u, \Omega_T\|_{\mathbb{V}^\varepsilon} \leqslant Q_\varepsilon \left(\|u_0\|_{\mathbb{V}_0}\right) e^{-\alpha T} + Q_\varepsilon \left(\|g\|_{L_b^2}\right), \tag{2.236}$$

where, by definition,

$$\|u, \Omega_T\|_{\mathbb{V}^\varepsilon}^2 := \varepsilon^4 \left\|\partial_t^2 u, \Omega_T\right\|_{0,2}^2 + \|\partial_t u, \Omega_T\|_{0,2}^2 + \|u, \Omega_T\|_{2,2}^2,$$

$$\|u_0\|_{\mathbb{V}_0^\varepsilon}^2 := \varepsilon \|u_0\|_{\mathbb{V}_0}^2 + \|u_0\|_{0,2}^2.$$

Note that in (2.236) the monotonic function Q and the exponent α are independent of $\varepsilon < \varepsilon_0$.

Remark 2.33 The uniform estimates of the form (2.236) are more or less known (see [31]), therefore we omit their proof here.

Lemma 2.25 *Let the assumptions of Lemma 2.24 hold. In addition, assume that the right-hand side g satisfies (2.233). Let $u_\varepsilon(t, x)$ and $\hat{u}(t, x)$ be the solutions of the problems (2.235) and (2.230), respectively, with $u_\varepsilon(0, x) = \hat{u}(0, x) = u_0(x)$. Then:*

1.

$$\|u_\varepsilon(t, x) - \hat{u}(t, x)\|_{0,2} \leqslant C_1 \varepsilon^{\frac{1}{2}} e^{K_1 t}, \tag{2.237}$$

where the constants C_1 and K_1 are independent of ε and uniform with respect to bounded in $\mathbb{V}_0^\varepsilon$ sets of $u_0(x)$.
2. *If the nonlinear term satisfies the additional regularity*

$$|f'(u)| \leqslant C \left(1 + |u|^{\frac{4}{n-2}}\right), \tag{2.238}$$

and primitive the $C_s(T)$ is bounded not only in $L^2(\omega)$, but in $W_0^{1,2}(\omega)$ as well, then the estimate (2.237) remains true with ε^1 instead of $\varepsilon^{\frac{1}{2}}$ on the right-hand side, that is,

$$||u_\varepsilon(t, x) - \hat{u}(t, x)||_{0,2} \leqslant C\varepsilon e^{K_1 t}. \tag{2.239}$$

Proof Let $v(t, x) := u_\varepsilon(t, x) - \hat{u}(t, x)$. Then, $v(t, x)$ satisfies

$$\begin{cases} \gamma \partial_t v = a\Delta_x v - (f(u_\varepsilon) - f(\hat{u})) - h_\varepsilon(t, x), \\ v|_{t=0} = 0, \quad v|_{\partial\Omega_+} = 0, \end{cases} \tag{2.240}$$

where $h_\varepsilon(t, x) := \varepsilon^2 \partial_t u_\varepsilon(t, x) + (g_\varepsilon(t, x) - \hat{g}(x))$. Multiplying (2.240), by $v(t, x)$ and integrating over $(t, x) \in [0, T] \times \omega$ and using quasimonotony assumption $f' \geqslant -K$, as well as the positivity of matrices γ and a, we obtain

$$\alpha\left(||v(T, x)||_{0,2}^2 + \int_0^T ||v(t, x)||_{1,2}^2\, dt\right)$$

$$\leqslant K \int_0^T ||v(t, x)||_{0,2}^2\, dt + \int_0^T (h_\varepsilon(t, x), v(t, x))\, dt \tag{2.241}$$

with an appropriate positive α. Next, we estimate the last integral on the right-hand side of (2.241). To this end, we decompose it in a sum of two integrals

$$I_1(T) := \varepsilon^2 \int_0^T (\partial_t^2 u_\varepsilon, v)\, dt,$$

$$I_2(T) := \int_0^T (g_\varepsilon(t, x) - \hat{g}(x), v)\, dt.$$

Let $C_\varepsilon(T) := \int_0^T (g_\varepsilon(t, x) - \hat{g}(x))\, dt$. Then, obviously due to (2.233), we have $C_\varepsilon(T) = \varepsilon C\left(\frac{T}{\varepsilon}\right)$, and consequently,

$$||C_\varepsilon(T)||_{0,2} \leqslant C\varepsilon, \tag{2.242}$$

and the assumptions of *Lemma 2.25* (2) imply that $||C_\varepsilon(T)||_{1,2} \leqslant C\varepsilon$. Integrating I_1 by parts and using the facts that $\partial_t u_\varepsilon$ and $\partial_t v$ are uniformly bounded with respect to ε in $L_b^2(\Omega_+)$ (according to *Lemma 2.24*), we obtain

$$I_1 = -\varepsilon^2 \int_0^T (\partial_t u_\varepsilon, \partial_t v)\, dt + \varepsilon^2 (\partial_t u_\varepsilon(T, x), v(T, x)) \leqslant C_\mu T \varepsilon^2 + \mu ||v(T, x)||_{0,2}^2, \tag{2.243}$$

where $\mu > 0$ can be chosen to be arbitrary small. Moreover, to prove (2.243), we also used uniform boundedness with respect to ε of the term $\varepsilon\, ||\partial_t u_\varepsilon(T, x)||_{0,2}$. Indeed, it follows from the interpolation inequality

$$\varepsilon \, \|\partial_t u_\varepsilon(T,x)\|_{0,2} \leqslant C \left(\varepsilon^2 \left\| \partial_t^2 u_\varepsilon \right\|_{L^2([T,T+1],L^2)} \right)^{\frac{1}{2}} \left(\|\partial_t u_\varepsilon\|_{L^2([T,T+1],L^2)} \right)^{\frac{1}{2}} \leqslant C_1.$$

Thus, it remains to estimate I_2. Integrating by parts again, we have

$$I_2 = - \int_0^T (\mathcal{C}_\varepsilon(t), \partial_t v) \, dt + (\mathcal{C}_\varepsilon(T), v(T,x)). \qquad (2.244)$$

The second term on the right-hand side of (2.244) can easily be estimated by Hölder inequality and (2.242), leading to

$$(\mathcal{C}_\varepsilon(T), v(T,x)) \leqslant C_\mu \varepsilon^2 + \mu \|v(T,x)\|_{0,2}^2.$$

Estimating the first term on the right-hand side of (2.244), which we denote by I_2^1, by Hölder inequality as well as (2.242) and the uniform boundedness of $\partial_t v$ in $L_b^2(\Omega_+)$ one can easily derive $I_2^1 \leqslant c\varepsilon T$, which in contrast to the previous estimate leads to a linear growth with respect to ε and not ε^2. Inserting all previously obtained estimates into (2.241) and applying Gronwall inequality, we obtain the rough estimate (2.237), with the rate of converging $\varepsilon^{\frac{1}{2}}$, instead of ε^1. Our next task is to derive a sharper estimate for the I_2^1 under the assumption *Lemma 2.25* (2). For this purpose, we take the expression for the term $\partial_t v$ from the Eq. (2.240) and insert it to I_2^1:

$$I_2^1 = - \int_0^T (\mathcal{C}_\varepsilon(T), \gamma^{-1} a \Delta_x v(t,x)) \, dt + \int_0^T (\mathcal{C}_\varepsilon(T), \gamma^{-1}(f(u_\varepsilon) - f(\widehat{u})) \, dt$$
$$+ \int_0^T (\mathcal{C}_\varepsilon(t), \gamma^{-1} \mathcal{C}_\varepsilon') \, dt = J_1 + J_2 + J_3.$$

Let us now derive estimate the terms J_j, $j = 1, 2, 3$. To estimate J_1, we integrate by parts with respect to x and use the estimate $\|\nabla_x \mathcal{C}_\varepsilon(t)\|_{0,2} \leqslant C\varepsilon$ together with Hölder inequality, which leads to

$$J_1 = \int_0^T (\nabla_x \mathcal{C}_\varepsilon(t), \gamma^{-1} a \Delta_x v(t,x)) \, d \leqslant C_\mu \varepsilon^2 T + \mu \int_0^T \|v(t,x)\|_{1,2}^2 \, dt \qquad (2.245)$$

Here, we essentially used the fact that $\mathcal{C}_\varepsilon$ in zero on the boundary $\partial \omega$ (without this fact, we would obtain the additional boundary term which would be an obstacle for the desired estimate). To estimate J_2, we use uniform boundedness of $u_\varepsilon(t,x)$ with respect to ε in $W^{1,2}(\omega)$ (due to the *Lemma 2.24*), the growth assumption (2.238), and Sobolev embedding theorem. Indeed, since $f(u_\varepsilon) - f(\widehat{u}) = v \int_0^1 f'(su_\varepsilon + (1 - s)\widehat{u}) \, ds$, then, applying Hölder inequality with the exponents $p_1 = p_2 = \frac{2n}{n-2}$ and $p_3 = \frac{n}{2}$ and the embedding $W^{1,2} \subset L^{p_1}$, we will have

$$J_2 \leqslant C \int_0^T \left(1 + ||u_\varepsilon(t, x)||_{0, p_1} + ||\hat{u}(t, x)||_{0, p_1}\right)^{\frac{4}{n-2}} ||v(t, x)||_{0, p_1} ||\mathcal{C}_\varepsilon(t)||_{0, p_1} dt$$

$$\leqslant C_1 \int_0^T \left(1 + ||u_\varepsilon(t, x)||_{1, 2} + ||\hat{u}(t, x)||_{1, 2}\right)^{\frac{4}{n-2}} ||v(t, x)||_{1, 2} ||\mathcal{C}_\varepsilon(t)||_{1, 2} dt$$

$$\leqslant C_2 \varepsilon^2 T + \mu \int_0^T ||v(t, x)||_{1, 2}^2 dt.$$

As for the estimate for J_3:

$$J_3 = \frac{1}{2}(\mathcal{C}_\varepsilon(T), \gamma^{-1} \mathcal{C}_\varepsilon(T)) \leqslant C\varepsilon^2. \tag{2.246}$$

Inserting the estimates (2.245)–(2.246) in (2.244), we obtain that

$$I_2(T) \leqslant C\varepsilon^2(1 + T) + \mu \int_0^T ||v(t, x)||_{1, 2}^2 dt + \mu ||v(T, x)||_{0, 2}^2. \tag{2.247}$$

To prove *Lemma 2.25*, it remains to insert the estimates (2.243) and (2.247) in (2.241), taxing μ small enough and applying the Gronwall inequality, which in turn leads to (2.239). □

After these preliminaries, we are in position to start the proof of *Theorem 2.43*.

Proof It follows from the estimate (2.236) that the attractors $\mathbb{A}^\varepsilon$ are uniformly bounded in the norms of $\mathbb{V}_0^\varepsilon$ and, in particular, in the norm of $W_0^{1,2}(\omega)$:

$$||\mathbb{A}^\varepsilon||_{1, 2} \leqslant C, \ \varepsilon < \varepsilon_0. \tag{2.248}$$

Thus, in order to prove the upper semi-continuity (2.231), it is sufficient to verify that if $u_\varepsilon \in \mathbb{A}^\varepsilon$ and $u_{\varepsilon_n} \rightharpoonup u_*$ as $\varepsilon_n \to 0$ weakly in $W_0^{1,2}$, then $u_* \in \mathbb{A}^0$. It would indicate that we have the upper semicontinuity in the weak topology of $W^{1,2}$ and, consequently, due to compact embedding of $W^{1-\delta, 2}$ in $W^{1,2}$, it would also indicate the upper semicontinuity in $W^{1-\delta, 2}(\omega)$. Indeed, due to the (2.228), there exists a sequence $\xi_n \in \mathcal{H}(g)$ and complete bounded solutions $u_{\varepsilon_n}(t, x), t \in \mathbb{R}$ of the equations

$$a(\varepsilon_n^2 \partial_t^2 u_{\varepsilon_n} + \Delta_x u_{\varepsilon_n}) - \gamma \partial_t u_{\varepsilon_n} - f(u_{\varepsilon_n}) = \xi_n \left(\frac{t}{\varepsilon_n}\right). \tag{2.249}$$

Our goal now is to pass to the limit as $n \to \infty$ in (2.249). For this purpose, we recall that due to *Lemma 2.24* $u_{\varepsilon_n}(t, x)$ are uniformly bounded in $\mathbb{V}^\varepsilon$ for each $T \in \mathbb{R}$. Hence, passing to a subsequence if necessary, we may assume that for each $T \in \mathbb{R}$,

$$\partial_t u_{\varepsilon_n}(t, x) \rightharpoonup \partial_t \hat{u}(t, x) \text{ and } u_{\varepsilon_n}(t, x) \rightharpoonup \hat{u}(t, x) \text{ in } W^{2,2}(\Omega_T), \tag{2.250}$$

and the limit function $\hat{u}(t, x)$ is bounded with respect to t, that is, $\partial_t \hat{u}, \Delta_x \hat{u} \in L_b^2(\Omega)$ and $\hat{u}(0, x) = u_*(x)$. It remains to prove that the function $\hat{\omega}(t, x)$ satisfies the limit equation (2.230). Then, (2.228) implies that $u_* \in \mathbb{A}^0$. Note that the convergence (2.250) together with restriction (2.224) admits to pass to the limit in the left-hand side of (2.249) in a standard way (see [1]). In order to pass to the limit on the right-hand side of (2.249), we use the following Lemma.

Lemma 2.26 *Let* $g \in C_b(\mathbb{R}, L^2(\omega))$ *be almost periodic with a hull* $\mathcal{H}(g)$. *Let also* $\xi_n \in \mathcal{H}(g)$. *Then, for each* $T \in \mathbb{R}$

$$\xi_n \left(\frac{t}{\varepsilon_n} \right) \rightharpoonup \hat{g} \text{ as } \varepsilon_n \to 0$$

weakly in $L^2(\Omega_T)$.

Proof Indeed, since the almost periodic flow is strictly ergodic, it follows from Birkhoff-Khinchin ergodic theorem that the L^2-limit

$$\hat{g}(x) = \lim_{T \to \infty} \int_{-T}^{T} \xi(t, x) \, dt$$

is uniform with respect to $\xi \in \mathcal{H}(g)$. The assertion of *Lemma 2.26* is a simple corollary of this fact. $\square$

Proof We are now ready to prove *Theorem 2.44*. In fact, the assertion of this Theorem is a simple corollary of *Lemma 2.25*. Indeed, assume that (2.233) is valid for the initial right-hand side $g(t, x)$. Then, it is not difficult to verify that it is valid uniformly with respect to $\xi \in \mathcal{H}(g)$ and, consequently, the estimate (2.237) is also valid uniformly with respect to $\xi \in \mathcal{H}(g)$. Namely, let $u_{\varepsilon,\xi}(t, x)$ be a solution of the Eq. (2.235) with right hand side $\xi \left(\frac{t}{\varepsilon}, x \right)$, $\xi \in \mathcal{H}(g)$ and let $\hat{u}(t, x)$ be the corresponding solution $(u_{\varepsilon,\xi}(0, x) = \hat{u}(0, x) = u_0(x))$ of the limit problem (2.230). Then, it holds

$$\|u_{\varepsilon,\xi}(t, x) - \hat{u}(t, x)\|_{0,2} \leqslant C \varepsilon^{\frac{1}{2}} e^{K_1 t} \tag{2.251}$$

uniformly with respect to $\xi \in \mathcal{H}(g)$ and bounded in $\mathbb{V}_0^\varepsilon$ sets of initial data $u_0(x)$. Assume now that $\varphi \in \mathbb{A}^\varepsilon$. According to the attractor's structure theorem, there exists a complete bounded trajectory $u_\varepsilon(t, x)$, $t \in \mathbb{R}$ of the Eq. (2.235) with the right-hand side $\xi \in \mathcal{H}(g)$. Let us fix an arbitrary $T > 0$ and consider the trajectory $\hat{u}(t, x)$ of the limit equation such that $\hat{u}(0, x) = u_\varepsilon(-T, x)$. Then (since $\mathbb{A}^\varepsilon$ are uniformly bounded in $\mathbb{V}_0^\varepsilon$), (2.251) implies that

$$\|\varphi - \hat{u}(T, x)\|_{0,2} \leqslant C \varepsilon^{\frac{1}{2}} e^{K_1 T}. \tag{2.252}$$

On the other hand, since $\mathbb{A}^0$ is an exponential attractor, it follows that

$$\text{dist}_{L^2(\omega)}(\hat{u}(T, x), \mathbb{A}^0) \leqslant C_1 e^{-\nu T}, \tag{2.253}$$

where $\nu > 0$ is some positive constant. Combining (2.252) and (2.253), we deduce that

$$\text{dist}_{L^2(\omega)}(\varphi, \mathbb{A}^0) \leqslant C_1 e^{-\nu T} + C\varepsilon^{\frac{1}{2}} e^{K_1 T}. \tag{2.254}$$

Taking the optimal value for T (solving the equation $C_1 e^{-\nu T} = C\varepsilon^{\frac{1}{2}} e^{K_1 T}$) in the estimate (2.254), we have

$$\text{dist}_{L^2(\omega)}(\varphi, \mathbb{A}^0) \leqslant C_2 \varepsilon^{\chi}, \ \chi = \frac{\nu}{2(K_1 + \nu)}.$$

Since $\varphi \in \mathbb{A}^\varepsilon$ is arbitrary, we obtain

$$\text{dist}_{L^2(\omega)}(\mathbb{A}^\varepsilon, \mathbb{A}^0) \leqslant C_2 \varepsilon^{\chi}.$$

This proves *Theorem 2.44*.　　　　　　　　　　　　　　　　　　　□

Let us formulate some sufficient conditions (see *Proposition 2.2* below) for almost periodic right-hand sides g satisfying the assumption of *Theorem 2.44*. For this purpose, we recall (see [48] and references therein) that every almost periodic function belonging to $C_b(\mathbb{R}, L^2(\omega))$ possesses a Fourier expansion

$$g(t, x) = \sum_{k=-\infty}^{\infty} g_{\alpha_k}(x) e^{i\alpha_k t}, \tag{2.255}$$

where $\{\alpha_k\} \subset \mathbb{R}$ is a countable set of Fourier modes for g and the corresponding amplitudes $g_{\alpha_k}(x) \in L^2(\omega)$ satisfy

$$\sum_{k=-\infty}^{\infty} \|g_{\alpha_k}(x)\|_{0,2}^2 < \infty.$$

Moreover, $\hat{g}(x) = g_0$ (here, we define $g_\alpha \equiv 0$ if $\alpha \notin \{\alpha_k\}$).

Proposition 2.2 *Let the Fourier amplitudes g_{α_k} of an almost periodic function $g \in C_b(\mathbb{R}, L^2(\omega))$ satisfy*

$$\sum_{\alpha_k \neq 0} \frac{1}{|\alpha_k|} \|g_{\alpha_k}(x)\|_{0,2} < \infty. \tag{2.256}$$

Then, the function $C(T)$ satisfies the inequality (2.233). Analogously, if

$$\sum_{\alpha_k \neq 0} \frac{1}{|\alpha_k|} \|g_{\alpha_k}(x)\|_{1,2} < \infty, \tag{2.257}$$

then $\|C(T)\|_{1,2} \leqslant C$.

Proof Let us verify (2.233) using the Fourier expansion (2.255). Indeed, subtracting $\hat{g}(x) = g_0$ in (2.255) and integrating over t, we derive that

$$C(t) = \sum_{\alpha_k \neq 0} g_{\alpha_k}(x) \frac{1}{i\alpha_k} \left(e^{i\alpha_k t} - 1 \right). \tag{2.258}$$

Taking the L^2-norm from both sides of (2.258) and using (2.256), we obtain (2.233). The second part of the *Proposition 2.2* can be verified analogously. *Proposition 2.2* is proved.   □

Corollary 2.17 *Let the assumptions of Lemma 2.24 hold and the function g satisfy (2.256). Then, the estimate (2.237) holds. If, in addition, $C(t)$ belongs to $W_0^{1,2}(\omega)$ and the assumptions (2.238) and (2.257) hold, then the improved estimate (2.239) is valid.*

Example 2.16 Let us consider the case of quasiperiodic functions. By definition, $g \in C_b(\mathbb{R}, L^2(\omega))$ is called a quasiperiodic function if there exists a finite vector of frequencies $\beta = (\beta_1, \ldots, \beta_m) \in \mathbb{R}^m$, $m > 1$ such that $\alpha_k = (\beta, l(k)) = \sum_{i=1}^m \beta_i (l(k))_i$ for the appropriate $l(k) \in \mathbb{Z}^m$ and β_i are rationally independent. Then, (2.255) reads as follows

$$g(t, x) = \sum_{l \in \mathbb{Z}^m} g_l(x) e^{i(\beta, l)t}.$$

Moreover, it is known that for every such $g \in C_b(\mathbb{R}, L^2(\omega))$ there exists a 2π-periodic with respect to every z_i, $i = 1, .., m$ function $\Phi \in C_b(\mathbb{R}^m, L^2(\omega))$ such that

$$g(t, x) = \Phi(\beta_1 z_1, \ldots, \beta_m z_m), \quad \Phi(z, x) = \sum_{l \in \mathbb{Z}^m} g_l(x) e^{i(z, l)t}. \tag{2.259}$$

In fact, (2.259) gives another definition of a quasiperiodic function. The condition (2.256) reads in our case as follows:

$$I := \sum_{l \in \mathbb{Z}^m, l \neq 0} \frac{\|g_l(x)\|_{0,2}}{|(\beta, l)|} < \infty. \tag{2.260}$$

Recall that, due to the theory of Diophantine approximation for every $\delta > 0$ and for almost every $\beta \in \mathbb{R}^m$ (with respect to the Lebesgue measure) the following estimate is valid:

$$|(\beta, l)| \geq C_\beta |l|^{-m-\delta}, \ l \neq 0. \tag{2.261}$$

Assume that $\beta \in \mathbb{R}^m$ is chosen such a way that (2.261) holds. Then, the sum I in (2.260) can be estimated by

$$I \leq C \sum_{l \in \mathbb{Z}^m} |l|^{m+\delta} \|g_l(x)\|_{0,2} \leq C \left(\sum_{l \neq 0} |l|^{2(m+\delta-\alpha)} \right)^{\frac{1}{2}} \left(\sum_{l \neq 0} |l|^{2\alpha} \|g_l(x)\|_{0,2}^2 \right)^{\frac{1}{2}} \tag{2.262}$$

Note that the first term on the right-hand side of (2.262) is finite if $2(m + \delta - \alpha) < -m$, that is, if $\alpha > \frac{3m}{2+\delta}$, and the second term is a finite for every function g such that the function Φ from the representation (2.259) belong to $C_b^\alpha(\mathbb{R}^m, L^2(\omega))$. Thus, for every β satisfying (2.261) and every periodic $\Phi \in C_b^\alpha(\mathbb{R}^m, L^2(\omega))$ with $\alpha > \delta + \frac{3m}{2}$, the function (2.259) satisfies the assumption (2.256).

Remark 2.34 The assumption (2.233) can weakened in the following way:

$$\|\mathcal{C}(T)\|_{0,2} \leq CT^{1-\beta}, \ \beta > 0. \tag{2.263}$$

Then $\|\mathcal{C}_\varepsilon(T)\|_{0,2} \leq \varepsilon^\beta CT^{1-\beta}$, and, arguing as in the proof of *Lemma 2.25* and *Theorem 2.44*, we derive the estimate (2.234) with $\chi = \frac{\beta v}{2(k_1 + v)}$. Note that (2.263) looks not very restrictive because this estimate with $\beta = 0$: $\|\mathcal{C}(T)\|_{0,2} \leq CT$ is obviously valid for every almost periodic function g.

Chapter 3
Symmetry and Attractors: The Case $N \leqslant 3$

3.1 Introduction

As it was shown in *Chap. 1*, positive solutions of semilinear second order elliptic problems have symmetry and monotonicity properties which reflect the symmetry of the operator and of the domain, see e.g. [18, 65] for the case of bounded domains and [13, 19, 21] for the case of unbounded domains (such as $\Omega = \mathbb{R}^n$, $\Omega = \mathbb{R}_+ \times \mathbb{R}^{n-1}$, cylindrical domains, etc.). These results have been extended to the case of positive solutions of second order parabolic problems in bounded symmetric domains in [10, 11]. Moreover, the symmetrization and stabilization properties of such solutions as $t \to \infty$ were investigated using a combination of moving planes method with the classical methods of dynamical systems theory (such as ω-limit sets, attractors, etc.).

The main goal of this chapter is to apply the dynamical system approach to study the symmetrization and stabilization (as $|x| \to \infty$) properties of positive solutions of elliptic problems in asymptotically symmetric unbounded domains. To the best of our knowledge, the use of dynamical systems methods for elliptic problems was initiated in the pioneering paper of K. Kirchgässner [72], where a local center manifold for a semilinear elliptic equation on a strip was constructed.

As it was indicated in the Preface, one of the main difficulties which arises in the dynamical study of elliptic equations, is the fact that the corresponding Cauchy problem is not well posed for such equations, and, consequently, the straightforward interpretation of the elliptic equation as an evolution equation leads to semigroups of multivalued maps even in the case of cylindrical domains, see [8] (for global attractors of multivalued flows associated with subdifferentials see [71]). The use of multivalued maps can be overcome using the so-called trajectory dynamical approach [15, 61, 81] (see also an alternative approach in [12, 31, 92]). Recall that, under this approach, which we introduced in *Chap. 2*, one fixes a signed direction $\vec{l}$ in $\mathbb{R}^n$, which will play the role of time. The space K^+ of all bounded solutions of the elliptic problem in the unbounded domain Ω is then considered as a trajectory

© Springer Nature Switzerland AG 2018

M. Efendiev, *Symmetrization and Stabilization of Solutions of Nonlinear Elliptic Equations*, Fields Institute Monographs 36, https://doi.org/10.1007/978-3-319-98407-0_3

phase space for the semi-flow $T_h^{\vec{l}}$ of translations along the direction $\vec{l}$ defined via

$$(T_h^{\vec{l}} u)(x) := u(x + h\vec{l}), \ h \in \mathbb{R}_+, \ u \in K^+.$$

In order for the trajectory dynamical system $(T_h^{\vec{l}}, K^+)$ to be well defined, one evidently needs the domain Ω to be invariant with respect to positive translations along the $\vec{l}$ directions:

$$T_h^{\vec{l}} \Omega \subset \Omega, \ T_h^{\vec{l}} x := x + h\vec{l}.$$

In this chapter, we apply the trajectory dynamical systems approach, which we developed in *Chap. 2*, to a more detailed study of the asymptotic behavior of positive solutions of the following model elliptic boundary problem in an unbounded domain $x := (x_1, x_2, x_3) \in \Omega_+ := \mathbb{R}_+ \times \mathbb{R}_+ \times \mathbb{R}^n$:

$$\begin{cases} \Delta_x u - f(u) = 0; \\ u|_{x_1=0} = u_0; \ u|_{x_2=0} = 0; \end{cases} \tag{3.1}$$

It is assumed that the nonlinear term $f(u)$ satisfies the following conditions:

$$\begin{cases} 1. \ f \in C^1(\mathbb{R}, \mathbb{R}), \\ 2. \ f(v).v \geqslant -C + \alpha |v|^2, \ \alpha > 0, \\ 3. \ f(0) \leqslant 0 \end{cases} \tag{3.2}$$

(see *Remark 3.2* for some relaxed conditions).

As mentioned above, we consider non-negative solutions of problem (3.1):

$$u(x) \geqslant 0, \ x \in \Omega_+$$

and study their behavior as $x_1 \to +\infty$. Thus, in this case, the x_1-axis will play the role of time ($\vec{l} := (1, 0, 0)$). Moreover, we restrict our consideration to bounded (with respect to $x \to \infty$) solutions of (3.1). More precisely, a bounded solution of (3.1) is understood to be a function $u \in C_b^{2+\beta}(\overline{\Omega_+})$, for some fixed $0 < \beta < 1$, which satisfies (3.1) in the classical sense (in fact, due to interior estimates, this assumption is equivalent to $u \in C_b(\overline{\Omega_+})$, but we prefer to work with classical solutions). Therefore, the boundary data is assumed to be non-negative $u_0(x_2, x_3) \geqslant 0$ and to belong to the space

$$u_0 \in C_b^{2+\beta}(\Omega_0), \ (x_2, x_3) \in \Omega_0 := \mathbb{R}_+ \times \mathbb{R}^n.$$

Here and below, we use the notation

$$C_b^{2+\beta}(V) := \{u_0 : \|u_0\|_{C_b^{2+\beta}} := \sup_{\xi \in V} \|u_0\|_{C^{2+\beta}(B_\xi^1 \cap V)} < \infty\},$$

where B_ξ^r denotes the ball of radius r, centered in ξ.

3.2 A Priori Estimates and Solvability Results

We start with proving a dissipative estimate with respect to the a priori estimate for positive solutions of (3.1), which allows us to apply the trajectory dynamical systems approach.

Theorem 3.1 *Let $u_0 \in C_b^{2+\beta}(\Omega_0)$ and let the first and second compatibility conditions be valid on $\partial\Omega_0$ (i.e. $u_0(0, x_3) = 0$ and $\partial_{x_2}^2 u_0(0, x_3) = f(0)$). Then, (3.1) has at least one non-negative bounded solution, every such solution u satisfies the estimate*

$$\|u\|_{C^{2+\beta}(B_x^1 \cap \Omega_+)} \leqslant Q(\|u_0\|_{C_b^{2+\beta}})e^{-\gamma x_1} + C_f, \tag{3.3}$$

$x = (x_1, x_2, x_3) \in \Omega_+$, where $\gamma > 0$, Q is an appropriate monotonic function and C_f is independent of u_0.

Proof Let us first verify the a priori estimate (3.3). To this end, we consider, as usual, the function $w(t, x) = u^2(t, x)$, which evidently satisfies the equation

$$\Delta_x w = 2f(u).u + 2\nabla_x u.\nabla_x u \geqslant -2C + 2\alpha w, \quad w|_{x_1=0} = u_0^2, \quad w|_{x_2=0} = 0. \tag{3.4}$$

Consider also the auxiliary linear problem

$$\Delta_x w_1 = -2C + 2\alpha w_1, \quad w_1|_{x_1=0} = w|_{x_1=0} = u_0^2, \quad w_1|_{x_2=0} = 0, \tag{3.5}$$

with the same boundary conditions as w.

Lemma 3.1 *The linear equation (3.5) has a unique bounded solution $w_1(x)$, which satisfies the following estimate:*

$$\|w_1\|_{C(B_x^1)} \leqslant C_1 \|u_0\|_{C_b(\Omega_0)}^2 e^{-\alpha x_1} + C_2. \tag{3.6}$$

Proof The proof of *Lemma 3.1* is standard and is based on the maximum principle. Indeed, let us decompose $w_1(x) = v(x) + v_1(x)$, where $v_1(x)$ is a solution of the non-homogeneous equation (3.5), with zero boundary conditions and $v(x)$ is a solution of the homogeneous equation, with non-zero boundary conditions. Then, evidently,

$$\|v_1\|_{C_b(\Omega_+)} \leqslant C_2, \tag{3.7}$$

where C_2 depends only on the constant C on the right-hand side of (3.5). In order to obtain the exponential decay of $v(x)$, we introduce the functions $\psi_h(x) := 1/\cosh(\varepsilon(x_1 - h))$, where $\varepsilon > 0$ is sufficiently small and $h \geqslant 0$. Then, it is not difficult to verify that

$$|\psi_h'(x)| \leqslant \varepsilon \psi_h(x), \quad |\psi_h''(x)| \leqslant 3\varepsilon^2 \psi_h(x), \quad \psi_h(0) \leqslant e^{-\varepsilon h}. \tag{3.8}$$

Let $v_h(x) := \psi_h(x)v(x)$. Then this function satisfies the equation

$$\Delta_x v_h + L_h^1(x)v_h + L_h^2(x)\partial_{x_1} v_h - 2\alpha v_h = 0, \quad v_h|_{x_1=0} = \psi_h(0)u_0^2, \tag{3.9}$$

and (3.8) implies that $|L_h^1(x)| \leqslant C\varepsilon^2$ and $|L_h^2(x)| \leqslant C\varepsilon$ (where C is independent of h). Consequently, if $\varepsilon > 0$ is small enough, the classical maximum principle works for Eq. (3.9) and, therefore,

$$\|v_h\|_{C_b(\Omega_+)} \leqslant \|v_h\|_{C_b(\Omega_0)}. \tag{3.10}$$

Estimate (3.6) is an immediate corollary of (3.7), (3.10) and the third estimate of (3.8). *Lemma 3.1* is proved.

$\square$

Having estimate (3.6), applying the comparison principle to the solutions w and w_1 of (3.4) and (3.5), respectively, and using the evident fact that $w = u^2$ is non-negative, we derive that

$$\|u\|_{C(B_x^1)}^2 \leqslant \|w\|_{C(B_x^1)} \leqslant \|w_1\|_{C(B_x^1} \leqslant C_1\|u_0\|_{C_b^{2+\beta}(\Omega_0)} e^{-\alpha x_1} + C_2. \tag{3.11}$$

Recall that, due to classical interior estimates for the Laplace equation (see e.g. [3, 76]), we have the following estimate for every small positive $\delta > 0$:

$$\|u\|_{C^{2-\delta}(\Omega_+ \cap B_x^1)}$$
$$\leqslant C(\|f(u)\|_{L^\infty(\Omega \cap B_x^1)} + \|u\|_{L^\infty(\Omega \cap B_x^2)} + \chi(2 - x_1)\|u_0\|_{C^{2+\beta}(\Omega_0 \cap B_x^2)})$$
$$\leqslant Q(\|u\|_{L^\infty(\Omega_+ \cap B_x^2)}) + C\chi(2 - x_1)\|u_0\|_{C^{2-\delta}(\Omega_0 \cap B_x^2)}, \quad x \in \Omega_+, \tag{3.12}$$

where the monotonic function Q and the constant C depend only on f and α, but are independent of $x \in \Omega$ and on the actual solution u, and $\chi(z)$ is a classical Heaviside function (which equals zero for $z \leqslant 0$ and one for $z > 0$).

Recall that we assume the first compatibility condition, $u_0|_{x_2=0} = 0$, to be valid. This assumption is necessary in order to obtain $C^{2-\delta}$-regularity in (3.12), in the case where x is close to the edge $\partial\Omega_0$).

Inserting now estimate (3.11) into the right-hand side of (3.12), we derive the analogue of estimate (3.3) for $C^{2-\delta}$-norm:

$$\|u\|_{C^{2-\delta}(B_x^1 \cap \Omega_+)} \leqslant Q(\|u_0\|_{C_b^{2-\delta}})e^{-\gamma x_1} + C_f. \tag{3.13}$$

In order to derive estimate (3.3), it is sufficient to use the elliptic interior estimate in the form

$$\|u\|_{C^{2+\beta}(\Omega_+ \cap B_x^1)}$$
$$\leqslant C(\|f(u)\|_{C^1(\Omega \cap B_x^1)} + \|u\|_{C(\Omega \cap B_x^2)} + \chi(2-x_1)\|u_0\|_{C^{2+\beta}(\Omega_0 \cap B_x^2)})$$
$$\leqslant Q(\|u\|_{C^1(\Omega_+ \cap B_x^2)}) + C\chi(2-x_1)\|u_0\|_{C^{2+\beta}(\Omega_0 \cap B_x^2)}, \quad x \in \Omega_+$$

(here we have implicitly used the second compatibility condition $\partial_{x_2}^2 u|_{x_2=0} = f(0)$, in order to obtain $C^{2+\beta}$-regularity near the edge $\partial\Omega_0$). Inserting estimate (3.13) into the last interior estimate, we derive inequality (3.3) for the $C^{2+\beta}$-norm.

Let us verify now the existence of a *positive* solution of problem (3.1). To this end, we consider a sequence of bounded domains Ω_+^N, $N \in \mathbb{N}$, defined via

$$\Omega_+^N := \Omega_+ \cap B_0^{N+1}$$

and a sequence of cut-off functions $\phi_N(x) \equiv 1$, if $x \in B_0^N$ and $\phi_N(x) \equiv 0$, if $x \notin B_0^{N+1}$, $0 \leqslant \phi \leqslant 1$. Consider also the family of auxiliary elliptic problems

$$\Delta_x u^N - f(u^N) = 0, \ x \in \Omega_+^N, \ u^N|_{\partial\Omega_+^N \cap \Omega_0} = u_0\phi_N, \ u^N|_{\partial\Omega_+^N \setminus \Omega_0} = 0. \tag{3.14}$$

Note that, according to our construction, $u^N|_{\partial\Omega_+^N} \geqslant 0$ and, according to assumptions (3.2), $w_-^N \equiv 0$ is a subsolution and $w_+^N \equiv R$ is a supersolution of (3.14), if R is large enough. Thus (see e.g. [104]), problem (3.14) has at least one *non-negative* solution $R \geqslant u^N(x) \geqslant 0$. Note that R is, in fact, independent of N. Consequently, applying again the interior regularity theorem, we derive that

$$\|u^N\|_{C^{2+\beta}(B_x^1 \cap \Omega_+^N)} \leqslant C, \tag{3.15}$$

with $C = C(f, u_0)$ independent of N and $x \in \Omega_+^N$. Having the uniform estimate (3.15), one can easily take the limit as $N \to \infty$ in Eq. (3.3) and construct a bounded *non-negative* solution $u(x)$ of the initial equation (3.1). *Theorem 3.1* is proved. □

3.3   The Attractor

Now we can apply the trajectory dynamical systems approach to (3.1). To this end, let us consider the union K^+ of all bounded positive solutions of (3.1), which correspond to every $u_0 \in C_b^{2+\beta}$. The set K^+ is nonempty due to *Theorem 3.1*. Then a semigroup of positive shifts,

$$(T_h u)(x_1, x_2, x_3) := u(x_1 + h, x_2, x_3),$$

acts on the set K^+:

$$T_h : K^+ \to K^+, \quad K^+ \subset C_b^{2+\beta}(\overline{\Omega_+}). \tag{3.16}$$

This semigroup acting on K^+ is called the trajectory dynamical system corresponding to (3.1). Our next task is to construct the global attractor for this system. Firstly, we note that the uniform topology of $C_b^{2+\beta}$ is too strong for our purposes. That is why we endow the space K^+ with a local topology, according to the embedding

$$K^+ \subset C_{loc}^{2+\beta}(\overline{\Omega_+}),$$

where, by definition, $\Phi := C_{loc}^{2+\beta}(\overline{\Omega_+})$ is a Fréchet space generated by the seminorms $\| \cdot \|_{C^{2+\beta}(B_{x_0}^1 \cap \Omega_+)}$, $x_0 \in \Omega_+$. Recall briefly the definition of the attractor from *Chap. 2* adapted in our case.

Definition 3.1 The set $\mathcal{A}_{tr} \subset K^+$ is called the global attractor for the trajectory dynamical system (3.16) (i.e., trajectory attractor for problem (3.1)), if the following conditions are valid:

1. The set $\mathcal{A}_{tr}$ is compact in $C_{loc}^{2+\beta}(\overline{\Omega_+})$;
2. It is strictly invariant with respect to T_h: $T_h \mathcal{A}_{tr} = \mathcal{A}_{tr}$;
3. $\mathcal{A}_{tr}$ attracts bounded subsets of solutions, when $x_1 \to \infty$. That means that, for every bounded (in the uniform topology of $C_b^{2+\beta}$) subset $B \subset K^+$ and for every neighborhood $\mathcal{O}(\mathcal{A}_{tr})$ in the $C_{loc}^{2+\beta}$ topology, there exists $H = H(B, \mathcal{O})$ such that

$$T_h B \subset \mathcal{O}(\mathcal{A}_{tr}) \text{ if } h \geqslant H.$$

Note that the first assumption of the definition claims that the restriction $\mathcal{A}_{tr}|_{\Omega_1}$ is compact in $C^{2+\beta}(\overline{\Omega_1})$, for every bounded $\Omega_1 \subset \Omega_+$, and the third one is equivalent to the following: For every bounded subdomain $\Omega_1 \subset \Omega_+$, for every B - bounded subset of K^+ and for every neighborhood $\mathcal{O}(\mathcal{A}_{tr}|_{\Omega_1})$ in the $C^{2+\beta}(\overline{\Omega_1})$-topology of the restriction of $\mathcal{A}_{tr}$ to this domain, there exists $H = H(\Omega_1, B, \mathcal{O})$ such that

$$(T_h B)|_{\Omega_1} \subset \mathcal{O}(\mathcal{A}_{tr}|_{\Omega_1}) \text{ if } h \geqslant H.$$

Theorem 3.2 *Let the assumptions of Theorem 3.1 hold. Then Eq. (3.1) possesses a trajectory attractor $\mathcal{A}_{tr}$ which has the following structure:*

$$\mathcal{A}_{tr} = \Pi_{\Omega_+} K(\Omega), \tag{3.17}$$

where $(x_1, x_2, x_3) \in \Omega := \mathbb{R} \times \mathbb{R}_+ \times \mathbb{R}^n$ and $K(\Omega)$ denotes the union of all bounded non-negative solutions $\hat{u}(x) \in C_b^{2+\beta}(\Omega)$ of

$$\Delta_x \hat{u} - f(\hat{u}) = 0, \quad x \in \Omega, \quad u|_{\partial\Omega} = 0, \quad \hat{u}(x) \geqslant 0, \tag{3.18}$$

i.e. the attractor $\mathcal{A}_{tr}$ consists of all bounded non-negative solutions u of (3.1) in Ω_+, which can be extended to bounded non-negative solution $\hat{u}$ in Ω.

Proof According to *Theorem 2.1*, in order to verify that a semigroup $T_h : K^+ \to K^+$ possesses an attractor, it should be verified that this semigroup is continuous for every fixed $h \geqslant 0$, and that this semigroup possesses a compact attracting (or absorbing) set in K^+. The continuity of the semigroup T_h on K^+ is obvious in our situation. Indeed, the semigroup T_h of positive shifts along the x_1 axis, is evidently continuous (for every fixed h) as a semigroup in $C_{loc}^{2+\beta}(\overline{\Omega_+})$, therefore, its restriction to K^+ is also continuous. Thus, it remains to construct a compact absorbing set for $T_h : K^+ \to K^+$. Let $\mathbb{B}_R$ be the R-ball centered in 0 in the space $C_b^{2+\beta}(\Omega_+)$. Then, estimate (3.3) implies that the set

$$\mathbb{M}_R := K^+ \cap \mathbb{B}_R$$

will be an absorbing set for the semigroup (3.16) (more precisely, for $R = 2C_f$, where C_f is defined in (3.3)). But this set is not compact in Φ. That is why we construct a new set,

$$\mathbb{V}_R := T_1 \mathbb{M}_R \subset \mathbb{M}_R \subset K^+.$$

Evidently, this set is also absorbing. We claim also that this set is precompact in Φ. Indeed, by definition, the set $\mathbb{V}_R$ consists of all bounded solutions u of Eq. (3.1), which can be extended to bounded solution $\hat{u}$, defined not for $x_1 \geqslant 0$, but for $x_1 \geqslant -1$, such that

$$\|\hat{u}\|_{C_b^{2+\beta}([-1,\infty] \times \Omega_0)} \leqslant R. \tag{3.19}$$

Note that, due to (3.2), $f \in C^1$; consequently, we may apply the interior estimate (see (3.12)) for the solution $\hat{u}$, not only with the exponent $2+\beta$, but with an arbitrary one, $2 + \beta'$, with $\beta' < 1$. In particular, if we fix $\beta' > \beta$, then the interior estimate together with (3.19), yields

$$\|u\|_{C_b^{2+\beta'}(T_1\Omega_+)} = \|\hat{u}\|_{C_b^{2+\beta'}(\Omega_+)} \leqslant R_1$$

where the constant R_1 depends only on R and f. Consequently, we have proved that

$$\mathbb{V}_R \subset C_b^{2+\beta'}(\Omega_+)$$

and is bounded in it. Now note that the embedding $C_b^{2+\beta'}(\Omega_+) \subset \Phi$ is compact if $\beta' > \beta$ and, consequently, $\mathbb{V}_R$ is indeed precompact in Φ. (This was the main reason to endow the trajectory phase space with the 'local' topology of Φ, not with the 'uniform' topology of $C_b^{2+\beta}(\Omega_+)$. Indeed, the embedding $C_b^{2+\beta'}(\Omega_+) \subset C_b^{2+\beta}(\Omega_+)$ is evidently non-compact and we cannot construct the compact absorbing set in this topology. Moreover, elementary examples show that problem (3.1) indeed may not possess an attractor in a 'uniform' topology—this we leave to the reader.) Thus, the precompact absorbing set $\mathbb{V}_R$ is already constructed and it remains to find the compact one. The simplest way is to take the compact absorbing set $\mathbb{V}_R' := [\mathbb{V}_R]_\Phi$, where $[\cdot]_\Phi$ denotes the closure in Φ. Indeed, since $\mathbb{V}_R \subset \mathbb{M}_R \subset K^+$ and $\mathbb{M}_R$ is evidently closed in Φ, $\mathbb{V}_R' \subset K^+$ and, consequently, it is a compact absorbing set for the semigroup (3.16). Thus, (due to the attractor existence *Theorem 2.1* for abstract semigroups), the semigroup (3.16) possesses an attractor $\mathcal{A}_{tr}$, which can be defined by the formula

$$\mathcal{A}_{tr} = \cap_{h \geqslant 0} \left[\cup_{s \geqslant h} T_s \mathbb{V}_R' \right]_\Phi. \tag{3.20}$$

It remains to prove representation (3.17). As we will see below, this follows from (3.20). Indeed, let $\widehat{u}(x)$, $x \in \Omega$, be a non-negative bounded solution of problem (3.18). Then in particular the sequence $\Pi_{\Omega_+}(T_{-h}\widehat{u})$, $h \in \mathbb{N}$, is uniformly bounded in $C_b^{2+\alpha}(\Omega_+)$ and, consequently, according to the attractor's definition,

$$T_h \Pi_{\Omega_+}(T_{-h}\widehat{u}) \to \mathcal{A}_{tr} \text{ in } \Phi, \text{ as } h \to \infty.$$

On the other hand, $T_h \Pi_{\Omega_+}(T_{-h}\widehat{u}) = \Pi_{\Omega_+}\widehat{u}$. Thus, $\Pi_{\Omega_+}\widehat{u} \in \mathcal{A}_{tr}$ and, consequently,

$$\Pi_{\Omega_+} K(\Omega) \subset \mathcal{A}^{tr}.$$

Let us prove the reverse inclusion. Let $u \in \mathcal{A}_{tr}$. Then, (3.20) implies that there exist a sequence $h_n \to +\infty$ and a sequence of solutions $u_n \in \mathbb{V}_R'$, such that

$$u = \Phi - \lim_{n \to \infty} T_{h_n} u_n. \tag{3.21}$$

Note that the solution $T_{h_n} u_n$ is defined not only in Ω_+, but also in the domain $T_{h_n}\Omega_+ := (-h_n, \infty) \times \Omega_0$, and

$$\|u_n\|_{C_b^{2+\beta}(T_{h_n}\Omega_+)} \leqslant R. \tag{3.22}$$

Consequently, arguing as in the proof of the compactness of $\mathbb{V}'_R$, we deduce that the sequence $T_{h_n} u_n$, $n \geqslant n_0$, is precompact in $C_{loc}^{2+\beta}(T_{h_{n_0}+1}\Omega_+)$, for every $n_0 \in \mathbb{N}$. Taking a subsequence, if necessary, and using Cantor's diagonal procedure and the fact that $h_n \to \infty$, we may assume that this sequence converges to some function $\hat{u} \in C_{loc}^{2+\beta}(\overline{\Omega})$, in the spaces $C_{loc}^{2+\beta}(T_{h_{n_0}+1}\Omega_+)$, for every $n_0 \in \mathbb{N}$. Then, (3.22) implies that $\hat{u} \in C_b^{2+\beta}(\Omega)$. Moreover, since $T_{h_n} u_n$ are non-negative solutions of (3.1), by letting $n \to \infty$, we easily obtain that $\hat{u}$ is a non-negative solution of Eq. (3.18) and formula (3.21) implies that $\Pi_{\Omega_+}\hat{u} = u$. Thus, $u \in \Pi_{\Omega_+}K(\Omega)$. *Theorem 3.2 is proved.*

$\square$

3.4 Symmetry and Stabilization

To obtain additional information on the behaviour of solutions of the initial problem (3.1), we use the description of non-negative bounded solutions in the half-space (see below).

Proposition 3.1 *Let assumptions* (3.2) *hold and let* $n = 1$, *that is,* $\Omega_+ = \mathbb{R}_+ \times \mathbb{R}_+ \times \mathbb{R}$. *Then any non-negative bounded solution* $\hat{u}(x)$ *of Eq.* (3.18) *depends only on the variable* x_2, *i.e.* $u(x) = V(x_2)$, *where* $V(z)$ *is a bounded solution of the following problem:*

$$V''(z) - f(V(z)) = 0, \quad z > 0, \quad V(0) = 0, \quad V(z) \geqslant 0. \tag{3.23}$$

The proof of this proposition is given in [21] (see also *Sect. 1.7*), for the case where the solution $\hat{u}(x)$ is strictly positive inside Ω. The general case can be reduced to the one above, using the following version of the strong maximum principle (see also *Sect. 1.4*).

Lemma 3.2 *[21] Let* $V \subset \mathbb{R}^n$ *be a (connected) domain with sufficiently smooth boundary and let* $w \in C^2(V) \cap C(\overline{V})$ *satisfy the following inequalities:*

$$\Delta_x w(x) - l(x)w(x) \leqslant 0, \quad x \in V, \quad w(x) \geqslant 0, \quad x \in V.$$

Assume also that $|l(x)| \leqslant K$ *for* $x \in V$. *Then either* $v(x) \equiv 0$, *or* $v(x) > 0$, *for every interior point* $x \in V$.

In order to apply the lemma to Eq. (3.18), we rewrite it in the following form:

$$\Delta_x \hat{u} - l(x)\hat{u} = f(0) \leqslant 0, \quad l(x) := \frac{f(\hat{u}(x)) - f(0)}{\hat{u}(x)}.$$

Since $f \in C^1$ and the solution $\hat{u}(x)$ is bounded, $l(x)$ is also bounded in Ω. Thus, according to *Lemma 3.2*, either $\hat{u}(x) \equiv 0$ (which is evidently symmetric), or $\hat{u}(x) > 0$, in the interior of Ω and, thus, *Proposition 3.1* follows from the result of [21] mentioned above. *Proposition 3.1* is proved.   □

Denote by $\mathcal{R}_V$ the set of all bounded non-negative solutions $V(z)$ of problem (3.23). Then *Proposition 3.1* implies that

$$\mathcal{A}_{tr} = \mathcal{R}_V.$$

Let us now study the positive solutions of problem (3.23). It is well known that every non-negative bounded solution of this problem is monotonically increasing, $V(z_1) \geqslant V(z_2)$, if $z_1 \geqslant z_2$, and, consequently, there exists the limit

$$z_0 = z_0(V) := \lim_{z \to +\infty} V(z), \quad f(z_0) = 0, \quad 0 \leqslant V(z) \leqslant z_0, \ z \geqslant 0. \qquad (3.24)$$

Moreover, it follows from *Lemma 3.2* that either $V(z) \equiv 0$, or $V'(z) > 0$, for every $z \geqslant 0$. Multiplying Eq. (3.23) by V' and integrating over $[0, z]$, we obtain the explicit expression for the derivative $V(z)$,

$$V'(z)^2 = -2F(V(z)) + C, \qquad (3.25)$$

where $F(V) := -\int_0^V f(V) \, dV$. Letting $z \to +\infty$ in (3.25) and taking into account (3.24), one can easily derive that $C = 2F(z_0)$. Therefore, we obtain the following equation for $V(z)$, stabilizing to z_0:

$$V'(z)^2 = 2(F(z_0) - F(V(z))). \qquad (3.26)$$

Assume now that $F(z_0) > 0$ (in the other case $V(z) \equiv 0$). Then, the solution $V_{z_0}(z)$ of (3.26), which satisfies (3.24), exists if and only if $F(z_0) - F(z) > 0$, for every $z \in (0, z_0)$. Moreover, such a solution is unique, because V_{z_0} satisfies (3.23) with the initial conditions

$$V_{z_0}(0) = 0, \quad V'_{z_0}(0) = \sqrt{2F(z_0)}. \qquad (3.27)$$

Denote by

$$\mathcal{R}_f^+ := \{z_0 \in \mathbb{R}_+ : f(z_0) = 0, \ F(z_0) - F(z) > 0 \text{ for every } z \in (0, z_0)\}. \qquad (3.28)$$

Note that the set (3.28) is totally disconnected in $\mathbb{R}$. Indeed, otherwise it should contain a segment $[\alpha, \beta] \in \mathcal{R}_f^+$, $\beta > \alpha \geqslant 0$. Then, $f(z_0) \equiv 0$, for $z_0 \in [\alpha, \beta]$ and, consequently, $F(z_0) = F(\beta)$, for every $z_0 \in [\alpha, \beta]$, which evidently contradicts the fact that $\beta \in \mathcal{R}_f^+$. Thus, we obtain the following proposition.

Proposition 3.2 *There exists a homeomorphism*

$$\tau : (\mathcal{R}_V, C_{loc}^{2+\beta}(\mathbb{R}_+)) \to (\mathcal{R}_f^+, \mathbb{R})$$

Moreover, the set $\mathcal{R}_f^+$ and, consequently, $\mathcal{R}_V$, are totally disconnected.

Indeed, (3.27) defines a homeomorphism between $\mathcal{R}_f^+$ and the set $\mathcal{R}_V(0) := \{(0, V'(0)) : V \in \mathcal{R}_V\}$ of values at $t = 0$, for functions from $\mathcal{R}_V$. Recall that $\mathcal{R}_V$ consists of solutions of the second order ODE (3.23) and, consequently, by a classical theorem on continuous dependence of solutions of ODE's, the set $\mathcal{R}_V$ is homeomorphic to $\mathcal{R}_V(0)$ and this homeomorphism is given by the solution operator $S : (V(0), V'(0)) \to V(t)$ of Eq. (3.23). *Proposition 3.2 is proved.*   □

Remark 3.1 Note that, although for generic functions f the sets $\mathcal{R}_f^+ \sim \mathcal{R}_V$ are finite, these sets may be uncountable, for some very special choices of the nonlinearity f. The simplest example of such an f is the following:

$$f(z) = - \operatorname{dist}(z, K),   \tag{3.29}$$

where K is a standard Cantor set on $[0, 1]$ and $\operatorname{dist}(z, K)$ denotes the distance from z to K. Indeed, it is easy to verify that, for this case, $\mathcal{R}^+ = K$ and, consequently, $\mathcal{R}_V$ consists of a continuum of elements. To be rigorous, function (3.29) is only Lipschitz continuous (but not in C^1) and does not satisfy the second assumption of (3.2), but by slightly modifying this function, one can construct the function $\tilde{f}$, which will satisfy all our assumptions and $\mathcal{R}_{\tilde{f}}^+ = \mathcal{R}_f^+ = K$. Indeed, for instance, define

$$\tilde{f}(z) = \begin{cases} 0 & \text{for } z \in K, \\ e^{-\frac{1}{(z-a)(z-b)}} & \text{for } z \in (a, b), \\ e^{-\frac{1}{z-1}} & \text{for } z > 1, \end{cases}   \tag{3.30}$$

where the numbers a and b are such that $0 < a < b < 1$, $a, b \in K$ and $(a, b) \subset (0, 1) \backslash K$.

We now state the main result of this section.

Theorem 3.3 *Let the assumptions of Theorem 3.1 hold. Then, for every non-negative bounded solution u of problem (3.1), there exists a solution $V(x_2) = V_u(x_2) \in \mathcal{R}_V$ of problem (3.23), such that, for every fixed R and $x = (x_1, x_2, x_3)$,*

$$\|u - V_u\|_{C^{2+\beta}(B_{x_h}^R \cap \Omega_+)} \to 0, \quad x_h := (x_1 + h, x_2, x_3),$$

when $h \to \infty$.

Proof Indeed, consider the ω-limit set of the solution $u \in K^+$, under the action of the semigroup T_h of shift in the x_1 direction,

$$\omega(u) = \cap_{h \geqslant 0} \left[\cup_{s \geqslant h} T_s u \right]_\Phi. \tag{3.31}$$

Recall that T_h possesses the attractor $\mathcal{A}_{tr}$ in K^+, and, consequently, the set (3.31) is nonempty. Thus,

$$\omega(u) \subset \mathcal{A}_{tr}.$$

It follows now from *Proposition 3.1* that $\omega(u) \subset \mathcal{R}_V$. Note that on the one hand, the set $\omega(u)$ must be connected (see e.g. [67]) and on the other, it is a subset of $\mathcal{R}_V$, which is totally disconnected (by *Proposition 3.2*). Therefore, $\omega(u)$ consists of a single point, $V_u \subset \mathcal{R}_V$:

$$\omega(u) = \{V_u\}.$$

The assertion of the theorem is a simple corollary of this fact and of our definition of the topology in K^+. *Theorem 3.3* is proved.

$\square$

Remark 3.2 We discuss assumptions (3.2) imposed on the nonlinear term $f(u)$, in order to obtain the results of *Chap. 3*. Note first that sign condition (3.2)(3) is, evidently, essential in order to prove the solvability of (3.1) in the class of positive bounded solutions, for every positive bounded initial data u_0 (and in fact, it is also essential for *Proposition 3.1* and *Lemma 3.2*, see e.g. [21]). It is easy to see that the assumption $f \in C^1$ is not necessary either for proving the existence of a positive solution of problem (3.1), or for applying the trajectory dynamical system approach to this problem, and can, therefore, be weakened to $f \in C(\mathbb{R}, \mathbb{R})$. Note, however, that the (local) Lipschitz continuity of the nonlinear term is very important for the symmetry result, formulated in *Proposition 3.1* (see [21]) and, consequently, for all results, obtained in *Chap. 3*. Thus, assumptions (3.2)(1) and (3.2)(3) seem to be close to optimal in order to derive the results of *Chap. 3*. In contrast, the dissipativity assumption (3.2)(2) is far from optimal and has been imposed in such a form only in order to avoid additional technicalities and to make the trajectory approach for the study of the behavior of positive solutions clearer. In fact, it can be proved, using the standard sub- and super-solutions technique and some monotonicity results for positive solutions of elliptic equations, that under assumptions (3.2)(1) and (3.2)(3), problem (3.1) has at least one positive bounded solution, for every positive bounded initial value u_0, if and only if its one dimensional analogue,

$$V''(z) - f(V(z)) = 0, \quad z > 0, \quad V(0) = M, \tag{3.32}$$

is solvable in the class of bounded non-negative solutions, for every $M \geqslant 0$. Recall that (3.32) is a second order ODE of Newtonian type and can be easily analyzed using e.g. the phase portrait technique.

In the case where $n = 1$, that is, $\Omega_+ = \mathbb{R}_+ \times \mathbb{R}_+ \times \mathbb{R}$, using the explicit description of the set of bounded positive solutions of the Eq. (3.1) in Ω, one can easily show that the attractor $\mathcal{A}_{tr}$ exists if and only if the set K of all bounded positive solutions of the problem

$$V''(z) - f(V(z) = 0, \ z > 0, \ V(0) = 0, \tag{3.33}$$

is globally bounded in $C(\mathbb{R}_+)$. Combining (3.32) and (3.33), we derive, after straightforward analysis of the corresponding phase portrait, that, under the assumptions (3.2) (1) as well as (3.2) (3), problem (3.1) possesses the trajectory attractor $\mathcal{A}_{tr}$ if and only if the potential $F(v) := -\int_0^v f(u)\,du$ achieves its global maximum on $[0, \infty)$, i.e. if there exists $v_0 \geqslant 0$, such that

$$F(z_0) = \max_{v \in \mathbb{R}_+} F(v).$$

Hence, all results of *Chap. 3* remain valid, in the case when condition (3.2)(2) is replaced by (3.2). Evidently, condition (3.2)(2) is sufficient, but not necessary for (3.2).

Remark 3.3 Note that neither our concrete choice of of the domain $\Omega_+ = \mathbb{R} \times \mathbb{R}_+ \times \mathbb{R}^n$ nor the concrete choice of the 'time' direction x_1 are essential for the trajectory dynamical system approach. Indeed, let us replace the 'time' direction x_1 by any fixed direction $\vec{l} \in \mathbb{R}^n$ and (and correspondingly $(T_h u)(x) := u(x + h\vec{l})$. Then the above construction seems to be applicable if the domain Ω_+ satisfies the following assumptions:

1. $T_h \Omega_+ \subset \Omega_+$ (it is necessary in order to define the restriction T_h to the trajectory phase space K_+).
2. $\Omega = \cup_{h \leqslant 0} T_{-h} \Omega_+$ (it is required in order to obtain representation (3.17)).

Further generalizations of the result of *Theorem 3.3* to higher dimensions, as well as other classes of equations, will be discussed in the following chapters.

Chapter 4
Symmetry and Attractors: The Case $N \leqslant 4$

4.1 Introduction

In the previous chapter, we showed that nonnegative solutions of elliptic equations in "asymptotically symmetric" domains are "asymptotically symmetric" as well (see *Theorem 3.3*). However, in order to prove *Theorem 3.3*, we imposed a restriction on the dimension (less or equal 3) of the underlying domain, which was crucial for our proof. The goal of this chapter is to extend *Theorem 3.3* for higher dimensions. As we will see below, this is not straightforward and requires both new ideas and stronger assumptions on the nonlinearity. Moreover, even under these stronger assumptions, we manage to extend the results of the previous *Chap. 3* only up to dimension 4 of the underlying domain (see below and [50]).

We consider the following elliptic boundary problem in an unbounded domain $\Omega_+ := \{x = (x_1, x_2, x_3, x_4) \mid 0 \leqslant x_1, x_4 < +\infty, x_2, x_3 \in \mathbb{R}\}$,

$$\begin{cases} \Delta_x u - f(u) = 0; \\ u|_{x_1=0} = u_0; \quad u|_{x_4=0} = 0. \end{cases} \tag{4.1}$$

Here Δ_x denotes the four-dimensional Laplacian. It is assumed that the nonlinear term satisfies

$$\begin{cases} 1. \ f \in C^2(\mathbb{R}, \mathbb{R}), \\ 2. \ f(v).v \geqslant -C + \alpha |v|^2, \quad \alpha > 0, \\ 3. \ \text{for some } \mu > 0, \ f(u) \leqslant 0 \text{ in } [0, \mu] \text{ and } f(u) \geqslant 0 \text{ in } [\mu, \infty[\end{cases} \tag{4.2}$$

We suppose also that $u_0(x_2, x_3, x_4) \geqslant 0$ and will consider only nonnegative solutions of (4.1):

© Springer Nature Switzerland AG 2018
M. Efendiev, *Symmetrization and Stabilization of Solutions
of Nonlinear Elliptic Equations*, Fields Institute Monographs 36,
https://doi.org/10.1007/978-3-319-98407-0_4

$$u(x) \geqslant 0 \quad x \in \Omega_+$$

It is assumed also that

$$u_0 \in C_b^{2+\beta}(\Omega_0), \quad (x_2, x_3, x_4) \in \Omega_0 := \mathbb{R} \times \mathbb{R} \times \mathbb{R}_+$$

for some $0 < \beta < 1$. Here and below we write

$$C_b^{2+\beta}(\Omega_0) := \{u_0 : \|u_0\|_{C_b^{2+\beta}} = \sup_{\xi \in \Omega_0} \|u_0\|_{C^{2+\beta}(B_\xi^1 \cap \Omega_0)} < \infty\}$$

where B_ξ^r means a ball of radius r centered at ξ.

A bounded solution of (4.1) is understood to be a function $u \in C_b^{2+\beta}(\overline{\Omega}_+)$ which satisfies (4.1) in a classical sense. (In fact due to the interior estimates this assumption is equivalent to $u \in C_b(\overline{\Omega}_+)$). The main goal of this chapter is to study the behaviour of a bounded solution of (4.1) when $x_1 \to \infty$. We are particularly interested in the symmetrization and stabilization properties of a bounded solution of (4.1) as $x_1 \to \infty$. As we mentioned above, we apply to (4.1) the dynamical systems approach, which was initiated in the seminal work of Kirchgässner [72] where a local center manifold for a semilinear elliptic equation on a strip was constructed.

As we already mentioned both in the Preface and in *Chap. 3*, one of the main difficulties arising in a dynamical study of elliptic equations is that the corresponding Cauchy problem is not well posed and thus, in general (4.1) can only be rigorously defined as a semigroup of multivalued maps (see [8]). The use of multivalued maps can be overcome by using a new approach, the so-called trajectory dynamical one (see *Chaps. 2–3* for details). Under this approach one considers the set K^+ of all bounded solutions of (4.1), endowed with a suitable topology (in our case it will be $C_{loc}^{2+\beta}(\Omega_+)$ for some $0 < \beta < 1$), as a (trajectory) phase space for the semigroup defined by the translation

$$(T_h u)(x_1, x_2, x_3, x_4) := u(x_1 + h, x_2, x_3, x_4), \quad h \geqslant 0$$

If an associated attractor exists it is called a trajectory attractor for (4.1).

For the case of second order elliptic systems there is another way to avoid the use of multivalued maps. In [31] the problem under consideration was studied with Cauchy initial conditions

$$u|_{t=0} = u_0, \quad \partial_t u|_{t=0} = u_1$$

In this case under certain assumptions a solution will be unique and one can define a semigroup

$$S_t : (u(0), \partial_t u(0)) \mapsto (u(t), \partial_t u(t))$$

But the set on which it is defined is not described in explicit form. As for the application of the trajectory dynamical approach to evolution problems we refer to [12, 12, 32, 52, 92]. In the previous *Chap. 3*, we considered (4.1) from the viewpoint of dynamical systems, when $\Omega_+ = \{x = (x_1, x_2, x_3) \in \mathbb{R}^3_+ | 0 \leqslant x_1, x_2 < \infty, x_3 \in \mathbb{R}\}$. Such a restriction on the dimension of Ω_+ (dim $\Omega_+ \leqslant 3$) was crucial because in *Chap. 3* symmetrization and stabilization to the attractor for solutions of (4.1) among others, were based on the paper [21], where a surprising link with a problem concerning the Schrödinger operator led to the restriction on dimension of the underlying domain Ω_+, namely $n \leqslant 3$. Whether or not the symmetry result in $\mathbb{R}^n_+ = \{x = (x_1, \ldots, x_n) | x_n \geqslant 0\}$ holds for $n \geqslant 4$ under the assumptions (4.2) was open at that time.

In this section, using recent development on the De Giorgi conjecture (see [16, 22] and references therein, for symmetry results for parabolic equations we refer to [11, 68]), especially the paper [5] and a personal communication with X. Cabre, we prove the existence of the trajectory attractor for (4.1) in

$$\Omega_+ := \{x = (x_1, x_2, x_3, x_4) \mid 0 \leqslant x_1, x_4 < +\infty, x_2, x_3 \in \mathbb{R}\}$$

and analyse the symmetrization and stabilization of solutions (4.1) to the attractor.

The chapter is organised as follows. The dissipative a priori estimate with respect to $x_1 \to \infty$ for the positive solutions of (4.1) is derived in *Sect. 4.2*, which allows us to apply the trajectory approach to our situation, and, in particular, gives the existence of at least one nonnegative solution.

In *Sect. 4.3*, we construct the trajectory dynamical system (T_h, K^+) associated with problem (4.1) and prove existence of a global attractor for this dynamical system.

Section 4.4 is devoted to a more comprehensive study of the four dimensional case. Following [5] we establish that the trajectory attractor $\mathcal{A}_{tr}$ consists of functions $u(x_1, x_2, x_3, x_4) := V(x_4)$ which satisfy the ordinary differential equation

$$V''(z) - f(V(z)) = 0, \quad z > 0, \quad V(0) = 0, \quad V(z) \geqslant 0 \tag{4.3}$$

In this section stabilisation of solutions of (4.1) to $\mathcal{A}_{tr}$ is also discussed.

4.2 A Priori Estimates and Solvability Results

In this section, we prove that the problem (4.1) possesses at least one non-negative bounded solution u and derive the so-called dissipative estimate for such solutions, which is of fundamental significance in order to apply the dynamical approach to the elliptic equation (4.1). For the convenience of the reader, we present a sketch of the proof. It uses the same techniques and ideas and leads to the same results, as in *Theorem 3.1*. As was mentioned in *Remark 3.3*, the dimension of Ω_+ doesn't play an important role in the proof of existence of trajectory attractor for Eq. (4.1). The difference between the assumptions (4.2) and (3.2) on the nonlinearity is related to the symmetry result which is formulated in *Sect. 4.4*.

The main result of this section is the following theorem.

Theorem 4.1 *Let $u_0 \in C_b^{2+\beta}(\Omega_0)$ and let the first and second compatibility conditions at $\partial\Omega_0$ be valid (i.e. $u_0(x_2, x_3, 0) = 0$ and $\partial_{x_4}^2 u_0(x_2, x_3, 0) = f(0)$). Then (4.1) possesses at least one nonnegative bounded solution and every such solution u satisfies the estimate*

$$\|u\|_{C^{2+\beta}(B_x^1 \cap \Omega_+)} \leqslant Q(\|u_0\|_{C_b^{2+\beta}})e^{-\gamma x_1} + C_f. \tag{4.4}$$

Here $x = (x_1, x_2, x_3, x_4) \in \Omega_+$, $\gamma > 0$, Q is an appropriate monotonic function, and C_f is independent of u_0.

Sketch of the Proof First we verify the a priori estimate (4.4). To this end we consider the function $w(t, x) = u^2(t, x)$ which evidently satisfies the equation

$$\Delta_x w = 2f(u).u + 2\nabla_x u.\nabla_x u \geqslant -2C + 2\alpha w, \quad w|_{x_1=0} = u_0^2, \quad w|_{x_2=0} = 0 \tag{4.5}$$

Consider also the auxiliary linear problem with the same boundary conditions as for the function w,

$$\Delta_x w_1 = -2C + 2\alpha w_1, \quad w_1|_{x_1=0} = w|_{x_1=0} = u_0^2, \quad w_1|_{x_2=0} = 0. \tag{4.6}$$

Then applying the comparison principle to the solutions w and w_1 of (4.5) and (4.6) respectively and the evident fact that $w = u^2$ is non-negative, we derive

$$\|u\|_{C(B_x^1)}^2 \leqslant \|w\|_{C(B_x^1)} \leqslant \|w_1\|_{C(B_x^1)} \leqslant C_1 \|u_0\|_{C_b^{2+\beta}(\Omega_0)} e^{-\alpha x_1} + C_2 \tag{4.7}$$

The latter inequality in (4.7) is due to the fact that a solution of linear equation (4.6) admits the following estimate

$$\|w_1\|_{C(B_x^1)} \leqslant C_1 \|u_0\|_{C_b(\Omega_0)}^2 e^{-\alpha x_1} + C_2$$

Consequently, the estimate (4.4) can be obtained using a classical elliptic interior estimate (see [72], as well as *Lemma 3.1*). A proof of the existence of at least one nonnegative solution for problem (4.1) is based on the approximation of Ω_+ by a sequence of bounded domains and the sub-super solution techniques and follows exactly the same arguments as in the proof of *Theorem 3.1*. This proves *Theorem 4.1*. $\square$

4.3 The Attractor

In this section, we study the behaviour of the non-negative solutions of problem (4.1) when $x_1 \to \infty$ applying the dynamical system approach to the elliptic boundary value problem (4.1) for general unbounded domains Ω_+. This approach we already applied in the *Chaps. 2* and *3* for several class of semilinear elliptic equations in unbounded domains taking into account both different geometries of underlying domains and different class of their solutions.

Recall that in such a case we fix some direction in our unbounded domain Ω_+ and interpret it as the 'time' direction. In our case this will be the x_1-direction, the x_1 variable will then play the role of 'time' variable and we (formally) consider (4.1) as an 'evolutionary' equation in an unbounded domain Ω_0. The main difficulty which arises here is the fact that the solution of (4.1) may be not unique and consequently we cannot construct the semigroup corresponding to the 'evolutionary' equation (4.1) in the standard way.

One of the possible ways to overcome this difficulty, as was mentioned in the beginning of this chapter, is to use the trajectory approach which takes into account the dynamical system for (4.1) in another way. Namely, let us consider the union K^+ of all bounded positive solutions of (4.1) which corresponds to every $u_0 \in C_b^{2+\beta}$. Then a semigroup of positive shifts

$$(T_h u)(x_1, x_2, x_3, x_4) := u(x_1 + h, x_2, x_3, x_4)$$

acts on the set K^+:

$$T_h : K^+ \to K^+, \quad K^+ \subset C_b^{2+\beta}(\overline{\Omega}_+) \tag{4.8}$$

This semigroup acting on K^+ is called the trajectory dynamical system, corresponding to (4.1). Our next task is to construct the attractor for this system. Firstly we note that the uniform topology of $C_b^{2+\beta}$ is too strong for our purposes. That is why we endow the space K^+ with a local topology according to the embedding

$$K^+ \subset C_{loc}^{2+\beta}(\overline{\Omega}_+)$$

where by definition $\Phi := C_{loc}^{2+\beta}(\overline{\Omega}_+)$ is a Fréchet space generated by the seminorms $\| \cdot \|_{C^{2+\beta}(B_{x_0}^1 \cap \Omega_+)}$, $x_0 \in \Omega_+$.

For the convenience of the reader we recall briefly the definition of the attractor adapted to our case.

Definition 4.1 The set $\mathcal{A}_{tr} \subset K^+$ is called the attractor for the trajectory dynamical system (4.8) (= trajectory attractor for the problem (4.1)) if the following conditions are valid.

1. The set $\mathcal{A}_{tr}$ is compact in $C_{loc}^{2+\beta}(\overline{\Omega}_+)$.
2. The set $\mathcal{A}_{tr}$ is strictly invariant with respect to T_h: $T_h \mathcal{A}_{tr} = \mathcal{A}_{tr}$
3. $\mathcal{A}_{tr}$ attracts bounded subsets of solutions when $x_1 \to \infty$. This means that for every bounded subset $B \subset K^+$ and for every neighbourhood $\mathcal{O}(\mathcal{A}_{tr})$ in the $C_{loc}^{2+\beta}$ topology there exists $H = H(B, \mathcal{O})$ such that

$$T_h B \subset \mathcal{O}(\mathcal{A}_{tr}) \text{ if } h \geqslant H$$

Note that the first condition of the definition asserts that the restriction $\mathcal{A}_{tr}|_{\Omega_1}$ is compact in $C^{2+\beta}(\overline{\Omega_1})$ for every bounded $\Omega_1 \subset \Omega_+$ and the third one is equivalent to the following:

For every bounded subdomain $\Omega_1 \subset \Omega_+$, for every B – bounded subset of K^+ and for every neighbourhood $\mathcal{O}(\mathcal{A}_{tr}|_{\Omega_1})$ in the $C^{2+\beta}(\overline{\Omega_1})$-topology of the restriction $\mathcal{A}_{tr}$ to this domain, there exists $H = H(\Omega_1, B, \mathcal{O})$ such that

$$(T_h B)|_{\Omega_1} \subset \mathcal{O}(\mathcal{A}_{tr}|_{\Omega_1}) \text{ if } h \geqslant H$$

Theorem 4.2 *Let the assumptions of Theorem 4.1 hold. Then the Eq. (4.1) possesses the trajectory attractor $\mathcal{A}_{tr}$ which has the following structure:*

$$\mathcal{A}_{tr} = \Pi_{\Omega_+} K(\Omega) \tag{4.9}$$

where $(x_1, x_2, x_3, x_4) \in \Omega := \mathbb{R} \times \mathbb{R} \times \mathbb{R} \times \mathbb{R}_+$ *and the symbol $K(\Omega)$ means the union of all bounded nonnegative solutions* $\hat{u}(x) \in C_b^{2+\beta}(\Omega)$ *of*

$$\Delta_x \hat{u} - f(\hat{u}) = 0, \quad x \in \Omega, \quad \hat{u}|_{\partial\Omega} = 0, \quad \hat{u}(x) \geqslant 0 \tag{4.10}$$

That is, the attractor $\mathcal{A}_{tr}$ consists of all bounded nonnegative solutions u of (4.1) in Ω_+ which can be extended to a bounded nonnegative solution $\hat{u}$ in Ω.

The proof of this theorem repeats word by word the proof of *Theorem 3.2*, we omit it.

Remark 4.1 Note that neither our concrete choice of the domain $\Omega_+ = \mathbb{R}_+ \times \mathbb{R} \times \mathbb{R} \times \mathbb{R}_+$ nor the concrete choice of the 'time' direction x_1 are essential for the use of the trajectory dynamical system approach. Indeed, let us replace the 'time' direction x_1 by any fixed direction $\vec{l} \in \mathbb{R}^4$ and (and correspondingly $(T_h u)(x) := u(x + h\vec{l})$). Then the above construction seems to be applicable if the domain Ω_+ satisfies the following assumptions:

1. $T_h \Omega_+ \subset \Omega_+$ (this is necessary in order to define the restriction T_h to the trajectory phase space K^+).
2. $\Omega = \cup_{h \leqslant 0} T_{-h} \Omega_+$ (this is required in order to obtain the representation (4.9)).

4.4 Symmetry and Stabilization

In this section based on a recent development on a conjecture of E. DeGiorgi, especially the paper [5], we give the main result.

Theorem 4.3 *Let assumptions (4.2) hold. Then for every nonnegative bounded solution u of the problem (4.1) there is a solution $V(x_4) = V_u(x_4)$ of the problem (4.3) such that for every fixed R and $x = (x_1, x_2, x_3, x_4)$*

$$\|u - V_u\|_{C^{2+\beta}(B^R_{x_h} \cap \Omega_+)} \to 0, \quad x_h := (x_1 + h, x_2, x_3, x_4)$$

when $h \to \infty$.

The proof of *Theorem 4.3* will be carried out in three steps, based on the following *Propositions 4.1, 4.2, 4.3.*

Proposition 4.1 ([21]) *Let $Q \subset \mathbb{R}^n$ be a (connected) domain with sufficiently smooth boundary and let $w \in C^2(Q) \cap C(\overline{Q})$ satisfy the following inequalities*

$$\Delta_x w(x) - l(x)w(x) \leqslant 0, \quad x \in Q, \quad w(x) \geqslant 0, \quad x \in Q$$

Assume also that $|l(x)| \leqslant K$ for $x \in Q$. Then either $v(x) \equiv 0$ or $v(x) > 0$ for every interior point $x \in Q$.

Proposition 4.2 *Let the assumptions (4.2) hold. Then any non-negative bounded solution $\hat{u}(x)$ of the Eq. (4.10) is symmetric, that is depends only on the variable x_4, $\hat{u}(x) = V(x_4)$ where $V(z)$ is a bounded solution of the following problem:*

$$V''(z) - f(V(z)) = 0, \quad z > 0, \quad V(0) = 0, \quad V(z) \geqslant 0 \tag{4.11}$$

Proof We rewrite the Eq. (4.10) in the following form

$$\Delta_x \hat{u} - l(x)\hat{u} = f(0) \leqslant 0, \quad l(x) := \frac{f(\hat{u}(x)) - f(0)}{\hat{u}(x)}$$

Since $f \in C^2$ and the solution $\hat{u}(x)$ is bounded then $l(x)$ is also bounded in Ω. Thus, according to Proposition *Proposition 4.1* either $\hat{u}(x) \equiv 0$ (which is evidently symmetric) or $\hat{u}(x) > 0$ in the interior of Ω. Hence it remains to prove *Proposition 4.2* in the case where $\hat{u}(x) > 0$ in Ω. We emphasize again that we cannot apply the arguments of *Chap. 3*, which leads to the restrictions for the dimension of the underlying domain $\Omega_+ \subset \mathbb{R}^n$, $n \leqslant 3$. Here we mainly follow [5].

Following [21], to show the symmetry result in $\mathbb{R}^4_+$ we have to prove that $f(\sup \hat{u}) \geqslant 0$. To this end, we note that from the assumptions (4.2) it follows that $\partial_{x_4} \hat{u} > 0$ (see [21]). We define in $\mathbb{R}^3$

$$\bar{u}(x_1, x_2, x_3) := \lim_{x_4 \to +\infty} \hat{u}(x_1, x_2, x_3, x_4), \quad (x_1, x_2, x_3) \in \mathbb{R}^3.$$

Obviously, this limit exists. Since $\sup \bar{u} = \sup \hat{u} = M$, it suffices to show that $f(\sup \bar{u}) \geqslant 0$. For this purpose we proceed as follows. We remark that, $\partial_{x_4} u$ satisfies

$$\Delta_x \partial_{x_4} \hat{u} - f'(\hat{u}) \partial_{x_4} \hat{u} = 0 \text{ in } \mathbb{R}^4_+ \tag{4.12}$$

Multiplying (4.12) by $\frac{\xi^2(x)}{\partial_{x_4} \hat{u}}$, where $\xi \in C_0^\infty(\mathbb{R}^4_+)$ and integrating by parts, we obtain

$$\int \left(\frac{2\xi(x)}{\partial_{x_4} \hat{u}} \nabla_x \partial_{x_4} \hat{u} \cdot \nabla \xi + f'(\hat{u})\xi^2 \right) dx = \int \frac{\xi^2(x)}{(\partial_{x_4} \hat{u})^2} |\nabla_x \partial_{x_4} \hat{u}|^2 dx$$

Using the Cauchy-Schwarz inequality the last equality leads to

$$\int (|\nabla_x \xi|^2 + f'(\hat{u})\xi^2) dx \geqslant 0 \text{ for all } \xi \in C_0^\infty(\mathbb{R}^4) \tag{4.13}$$

As was shown in [5] from (4.13) one can deduce (here for the first time we need $f \in C^2$)

$$\int_{\mathbb{R}^3} \left(|\nabla_x \eta|^2 + f'(\bar{u})\eta^2 \right) dx \geqslant 0 \quad \text{for all} \quad \eta \in C_0^\infty(\mathbb{R}^3)$$

and as a result (see [5]) we obtain the existence of a strictly positive solution φ such that

$$\Delta_x \varphi - f'(\bar{u})\varphi = 0. \tag{4.14}$$

Note that the function $\bar{u}(x_1, x_2, x_3)$ defined by (4.4) also satisfies

$$\Delta_x \bar{u} - f(\bar{u}) = 0. \tag{4.15}$$

Next, we prove that, $\bar{u}$ defined by (4.4) satisfies

$$\int_{B_R} |\nabla \bar{u}|^2 dx \leqslant CR^2 \tag{4.16}$$

for every ball $B_R \subset \mathbb{R}^3$ of radius R and centered at the origin. To this end, we first prove,

$$\Delta_x \hat{u} \leqslant 0 \text{ in } \mathbb{R}^4_+. \tag{4.17}$$

Indeed, if $f(\sup \hat{u}) \geqslant 0$, then it leads immediately to a symmetry result; that is $\hat{u}(x_1, x_2, x_3, x_4) = V(x_4)$ (see [21]). Assume that $f(\sup \hat{u}) < 0$. Since $\Delta_x \hat{u} =$

$f(\hat{u})$, it follows from conditions (4.2) that $\Delta_x \hat{u} < 0$. Hence (4.17) holds. Now using (4.17) we show that

$$\int_{Q_R} |\nabla_x \hat{u}|^2 dx \leqslant C R^3 \tag{4.18}$$

for every cylinder $Q_R = B_R \times (b, b+R) \subset \mathbb{R}_+^4$, where C is a constant independent of R and b. Indeed, since $-\Delta_x \hat{u} \geqslant 0$, we have $\hat{u}(-\Delta_x \hat{u}) \leqslant M(-\Delta_x \hat{u})$ and hence

$$\int_{Q_R} |\nabla_x \hat{u}|^2 dx = \int_{Q_R} \hat{u}(-\Delta_x \hat{u})dx + \int_{\partial Q_R} \hat{u} \frac{\partial \hat{u}}{\partial \nu} d\sigma$$

$$\leqslant -M \int_{Q_R} \Delta_x \hat{u} dx + \int_{\partial Q_R} \hat{u} \frac{\partial \hat{u}}{\partial \nu} d\sigma$$

$$= \int_{\partial Q_R} (\hat{u} - M) \frac{\partial \hat{u}}{\partial \nu} d\sigma \leqslant C R^3,$$

that is (4.18) holds. By letting $b \to \infty$ in (4.18), it is not difficult to see that

$$\int_{B_R} |\nabla_x \bar{u}|^2 dx \leqslant C R^2,$$

that is (4.16) holds. To show $f(\sup \bar{u}) \geqslant 0$ we define $\sigma_i(x_1, x_2, x_3) := \frac{\partial_{x_i} \bar{u}}{\varphi}$, $i = 1, 2, 3$. Then, due to (4.14), (4.15) and (4.16), σ_i satisfies the following conditions

$$\sigma_i \mathrm{div}(\varphi^2 \nabla \sigma_i) = 0 \text{ in } \mathbb{R}^3$$

and

$$\int (\varphi \sigma_i)^2 dx \leqslant C R^2, \ \forall R > 1.$$

Then a Liouville-type theorem due to [21] implies that σ_i is necessarily constant, hence each partial derivative of $\bar{u}$ is constant, that is $\partial_{x_i} \bar{u} = c_i \varphi$, $i = 1, 2, 3$. In particular, $\bar{u}$ is either constant, which leads to $f'(M) \geqslant 0$, or $\bar{u}(x_1, x_2, x_3) = \tilde{u}(c_1 x_1 + c_2 x_2 + c_3 x_3)$. Then, the bounded function $\tilde{u}(\theta)$, which is a monotone function of only one variable satisfies the ODE $\tilde{u}'' + f(\tilde{u}(\theta)) = 0$. It is then a classical result that $f(\sup \bar{u}) = 0$.

This proves *Proposition 4.2*. $\square$

Denote by $\mathcal{R}_V$ the set of all bounded non-negative solutions $V(z)$ of problem (4.11). Then *Proposition 4.2* implies that

$$\mathcal{A}_{tr} = \mathcal{R}_V.$$

Proposition 4.3 *The set $\mathcal{A}_{tr}$ (trajectory attractor) endowed with the local topology $C_{loc}^{2+\beta}(\overline{\Omega}_+)$ is totally disconnected.*

The proof of *Proposition 4.3* repeats word by word the proof of *Proposition 3.2* and arguments (3.23)–(3.27). We omit it.

Now we are in the position to prove *Theorem 4.3*.

Proof Indeed, consider the ω-limit set of the solution $u \in K^+$ under the action of the semigroup T_h of shifts in the x_1 direction:

$$\omega(u) = \bigcap_{h \geqslant 0} \left[\cup_{s \geqslant h} T_s u \right]_\Phi \tag{4.19}$$

Recall that T_h possesses the attractor $\mathcal{A}_{tr}$ in K^+, consequently the set (4.19) is non-empty and

$$\omega(u) \subset \mathcal{A}_{tr}$$

It follows now from the *Proposition 4.2* that $\omega(u) \subset \mathcal{R}_V$.

Note that the set $\omega(u)$ must be connected [1] but also it is a subset of the set $\mathcal{R}_V$ which is totally disconnected (due to *Proposition 4.2*). Therefore $\omega(u)$ consists of a single point $V_u \subset \mathcal{R}_V$:

$$\omega(u) = \{V_u\}$$

The assertion of the theorem is a simple corollary of this fact and of our definition of the topology in K^+. *Theorem 4.3* is proved.   □

Remark 4.2 We would like to emphasize especially that, in order to prove the main result of *Chap. 4*, that is, *Theorem 4.3*, we were forced to impose in (4.2) a strong condition on the nonlinearity f. This assumption on f we used in order to prove the symmetry result (see *Proposition 4.2*). Note that, in the case when the underlying domain Ω_+ has dimension less or equal 3, weaker assumptions on f, such as (3.2), are sufficient (see *Remark 3.2*).

Chapter 5
Symmetry and Attractors

5.1 Introduction

In this chapter, symmetry results in the half-space and in $\mathbb{R}^N$ will be used towards the characterization of the asymptotic profiles of solutions in the quarter-space and in the half-space, respectively. As we have seen in the previous *Chaps. 3 and 4*, here the dimension of the underlying domain plays an important role and to extend the results on the symmetrization and stabilization of solutions of semilinear elliptic equations for dimensions less or equal 3 to the case of dimensions less or equal 4 requires nontrivial arguments and assumptions on the nonlinearities. The goal of this chapter is to extend the results from *Chaps. 3 and 4* to the case of dimensions less or equal 5. As we will see below, to this end we need new arguments and we cannot use the techniques from *Chaps. 3 and 4*. Similar to the previous chapters, we will apply the trajectory dynamical systems approach in order to study the asymptotic profiles of solutions for this new case of dimension 5 or higher. Moreover, in contrast to the previous chapters, we will also study the case when the asymptotic profile is a constant.

5.1.1 Statement of Results

For $N \in \mathbb{N}$, $N \geqslant 2$, we consider the half-space

$$\mathbb{R}^N_+ := \{x = (x_1, \ldots, x_N) \in \mathbb{R}^N \text{ s.t. } x_1 > 0\}$$

and the quarter-space

$$\mathbb{R}^N_{++} := \{x = (x_1, \ldots, x_N) \in \mathbb{R}^N \text{ s.t. } x_1 > 0 \text{ and } x_N > 0\}.$$

© Springer Nature Switzerland AG 2018
M. Efendiev, *Symmetrization and Stabilization of Solutions of Nonlinear Elliptic Equations*, Fields Institute Monographs 36,
https://doi.org/10.1007/978-3-319-98407-0_5

The purpose of this chapter is to develop asymptotic results, as $x_1 \to +\infty$, of solutions of elliptic PDEs on either $\mathbb{R}_+^N$ or $\mathbb{R}_{++}^N$. These results extend some previous work of [15, 50] by making use of some symmetry results and classifications of [45, 60].

Following are the main results of this chapter. First, we show that the solutions of elliptic equations on the quarter-space are asymptotic to one-dimensional solutions, up to dimension 5, according to our next result:

Theorem 5.1 *Let* $f \in C^1(\mathbb{R})$. *Let us assume that there exist* $c_1 \geqslant 0$, $c_2 > 0$ *and* $r \geqslant 2$ *such that*

$$f(s)s \geqslant -c_1 + c_2 s^r, \text{ for any } s \geqslant 0, \tag{5.1}$$

and that there exists $\mu > 0$ *such that*

$$f(s) \leqslant 0 \text{ for any } s \in [0, \mu] \text{ and } f(s) \geqslant 0 \text{ for any } s \in [\mu, +\infty).$$

Let $N \leqslant 5$ *and* $u \in C^2(\mathbb{R}_{++}^N) \cap C^0(\overline{\mathbb{R}_{++}^N})$ *be a non-negative solution of*

$$\begin{cases} \Delta u = f(u) \text{ in } \mathbb{R}_{++}^N, \\[2mm] u|_{\{x_1=0\}} = u_0, \\[2mm] u|_{\{x_N=0\}} = 0. \end{cases} \tag{5.2}$$

Then, there exists uniquely defined $w : \mathbb{R} \to \mathbb{R}$ *such that, as* $x_1 \to +\infty$, u *converges in* $C_{loc}^2(\mathbb{R}_{++}^N)$ *to* $w(x_N)$. *In particular, we have that either* w *vanishes identically, or it is a bounded solution of*

$$\begin{cases} w''(t) = f(w(t)) \text{ for any } t \in \mathbb{R}, \\[2mm] w(0) = 0, \\[2mm] w'(t) > 0 \text{ for any } t \in \mathbb{R}. \end{cases}$$

Remark 5.1 Nonlinearities satisfying (5.1) have been extensively studied in the dynamical system framework, and they are sometimes called "dissipative" in the literature.

The next result shows that the stable solutions on the half-spaces are asymptotic to constants, up to dimension 4:

Theorem 5.2 *Let* $f \in C^1(\mathbb{R})$. *Let us assume that there exist* $c_1 \geqslant 0$, $c_2 > 0$ *and* $r \geqslant 2$ *such that*

$$f(s)s \geqslant (-c_1 + c_2 s^r)^+, \text{ for any } s \geqslant 0, \qquad (5.3)$$

and that

$$\text{the set } E := \{s \geqslant 0 \text{ s.t. } f(s) = 0\} \text{ is non-empty.} \qquad (5.4)$$

Let $N \leqslant 4$ and $u \in C^2(\mathbb{R}^N_+) \cap C^0(\overline{\mathbb{R}^N_+})$ be a non-negative stable solution of

$$\begin{cases} \Delta u = f(u) \text{ in } \mathbb{R}^N_+, \\[2mm] u|_{\{x_1=0\}} = u_0. \end{cases} \qquad (5.5)$$

Then, as $x_1 \to +\infty$, u converges in $C^2_{loc}(\mathbb{R}^N_+)$ to a constant $c \in E$.

Remark 5.2 In the statement of *Theorem 5.2* we used the standard notation

$$v^+(x) := \max\{v(x), 0\}, \qquad (5.6)$$

and the classical language of the calculus of variation, according to which a solution u of (5.5) is stable if

$$\int_{\mathbb{R}^N_+} |\nabla \psi(x)|^2 + f'(u(x))\psi^2(x)\, dx \geqslant 0 \qquad (5.7)$$

for any $\psi \in C^\infty_c(\mathbb{R}^N_+)$. Such stability condition is classical and widely studied (see, e.g., [44], Sect. 7 in [59], and the references therein). We observe that solutions minimizing the associated energy functional are always stable. Moreover, *Theorem 5.2* remains valid (with the same proof) if the stability assumption is weakened to the so-called "stability outside a compact set" $K \subset \mathbb{R}^N_+$, namely if one requires (5.7) to hold for any $\psi \in C^\infty_c(\mathbb{R}^N_+)$ whose support does not intersect K (these kind of assumptions play a role in the study of solutions with finite Morse index, which are always stable outside a compact set, see [55, 57]).

Remark 5.3 From (5.6), we have that condition (5.3) implies (5.1). Condition (5.4) is also necessary to make sense of *Theorem 5.2*, because if E is empty there cannot be any c to which u is asymptotic, since such c should satisfy $0 = \Delta c = f(c)$. In fact, if (5.3) holds but (5.4) does not hold, there does not exist any non-negative solution of $\Delta u = f(u)$ in $\mathbb{R}^N_+$ (see *Remark 5.9*).

A particular case comprised by *Theorem 5.2* is when f has the special form

$$f(r) = r \prod_{j=1}^m (r - p_j)^2, \qquad (5.8)$$

with $0 \leqslant p_1 \leqslant \cdots \leqslant p_m$. In this case, our *Theorem 5.2* provides a positive answer, at least for stable solutions and in dimension up to 4. In fact, in this case, *Theorem 5.2* gets even stronger, since it is not necessary to assume that u is non-negative: more precisely, for the nonlinearity in (5.8) we have:

Theorem 5.3 *Let $N \leqslant 4$, $0 \leqslant p_1 \leqslant \cdots \leqslant p_m$, and $u \in C^2(\mathbb{R}_+^N) \cap C^0(\overline{\mathbb{R}_+^N})$ be a stable solution of*

$$
\begin{cases}
\Delta u = u \displaystyle\prod_{j=1}^{m} (u - p_j)^2 \text{ in } \mathbb{R}_+^N, \\[2em]
u\big|_{\{x_1=0\}} = u_0.
\end{cases}
$$

Then, as $x_1 \to +\infty$, u converges in $C_{loc}^2(\mathbb{R}_+^N)$ to a constant $c \in E$.

Remark 5.4 We think that our *Theorems 5.1* and *5.2* leaves open some very intriguing questions. For instance, we wonder whether the results remain true in higher dimension or not. In particular, related to the proof of *Theorem 5.2*, we wonder if there exists positive and bounded, non-constant, or non-stable, solutions of $\Delta u = f(u)$ in the whole of $\mathbb{R}^N$, with $f \geqslant 0$, or, in particular, with f as in (5.8).

5.2 The Dynamical System Approach

The goal of this Section is to apply the dynamical systems approach in order to study the symmetrization and stabilization (as $|x| \to \infty$) properties of nonnegative solutions we consider non-negative solutions of problem (5.2) (in the same manner one can handle (5.5)).

$$
u(x) \geqslant 0, \ x \in \mathbb{R}_{++}^N
$$

and study their behavior when $x_1 \to +\infty$. Thus, in this case, the x_1-axis will play the role of time ($\vec{l} := (1, 0, \ldots, 0) \in \mathbb{R}^N$). Moreover, we restrict our consideration to bounded (with respect to $x \to \infty$) solutions of (5.2). More precisely, a bounded solution of (5.2) is understood to be a function $u \in C_b^{2+\beta}(\overline{\mathbb{R}_{++}^N})$, for some fixed $0 < \beta < 1$, which satisfies (5.2) in a classical sense (in fact, due to interior estimates, this assumption is equivalent to $u \in C_b(\overline{\mathbb{R}_{++}^N})$, but we prefer to work with classical solutions). Therefore, the boundary data is assumed to be non-negative $u_0(x_2 \ldots x_N) \geqslant 0$ and to belong to the space

$$
u_0 \in C_b^{2+\beta}(\Omega_0), \ (x_2 \ldots x_N) \in \Omega_0 := \mathbb{R}_{++}^N \cap \{x_1 = 0\}.
$$

Here and below, we use the notation

$$C_b^{2+\beta}(V) := \{u_0 \ : \ \|u_0\|_{C_b^{2+\beta}} := \sup_{\xi \in V} \|u_0\|_{C^{2+\beta}(B_\xi^1 \cap V)} < \infty\},$$

where B_ξ^r denotes the ball of radius r, centered in ξ.

The dissipative estimate for nonnegative solutions of (5.2), which allows us to apply the trajectory dynamical systems approach, can be obtained in the same manner as in the previous Sections. We omit these details.

Theorem 5.4 *Let $u_0 \in C_b^{2+\beta}(\Omega_0)$ and let the first and second compatibility conditions be valid on $\partial\Omega_0$ (i.e. $u_0(x_2 \ldots x_{N-1}, 0) = 0$ and $\frac{\partial^2 u_0}{\partial_{x_N}^2}(x_2 \ldots x_{N-1}, 0) = f(0)$). Then (5.2) has at least one non-negative bounded solution, every such solution u satisfies the estimate*

$$\|u\|_{C^{2+\beta}(B_x^1 \cap \mathbb{R}_{++}^N)} \leq Q(\|u_0\|_{C_b^{2+\beta}})e^{-\gamma x_1} + C_f, \tag{5.9}$$

$x \in \mathbb{R}_{++}^N$, *where $\gamma > 0$, Q is an appropriate monotonic function and C_f is independent of u_0.*

Remark 5.5 It is interesting enough to note that, if $r > 2$, then one can show that without sub- and supersolution technique (without sign condition on f at the origin) that any weak solution of (5.2) in an appropriately defined space is necessarily bounded (see *Sect. 2.6*).

Now we can apply the trajectory dynamical systems approach to (5.2). To this end, as in the previous chapters, we consider the union K^+ of all bounded positive solutions of (5.2), which correspond to every $u_0 \in C_b^{2+\beta}$. Then a semigroup of positive shifts,

$$(T_h u)(x_1, x', x_N) := u(x_1 + h, x', x_N), \ x' = (x_2, \ldots, x_{N-1}) \in \mathbb{R}^{N-2}, \tag{5.10}$$

acts on the set K^+:

$$T_h : K^+ \to K^+, \quad K^+ \subset C_b^{2+\beta}(\overline{\mathbb{R}_{++}^N}).$$

This semigroup acting on K^+ is called the trajectory dynamical system corresponding to (5.2). Our next task is to construct the global attractor for this system. To this end, we endow the space K^+ with a local topology, according to the embedding

$$K^+ \subset C_{loc}^{2+\beta}(\overline{\mathbb{R}_{++}^N}),$$

where, by definition, $\Phi := C_{loc}^{2+\beta}(\overline{\mathbb{R}_{++}^N})$ is a Fréchet space generated by the seminorms $\|\cdot\|_{C^{2+\beta}(B_{x_0}^1 \cap \mathbb{R}_{++}^N)}$, $x_0 \in \mathbb{R}_{++}^N$.

For convenience of the reader, we recall briefly the definition of the attractor adapted in our case.

Definition 5.1 The set $\mathcal{A}_{tr} \subset K^+$ is called the global attractor for the trajectory dynamical system (5.10) (= trajectory attractor for problem (5.2)), if the following conditions are valid.

1. The set $\mathcal{A}_{tr}$ is compact in $C_{loc}^{2+\beta}(\overline{\mathbb{R}_{++}^N})$.
2. It is strictly invariant with respect to T_h: $T_h \mathcal{A}_{tr} = \mathcal{A}_{tr}$
3. $\mathcal{A}_{tr}$ attracts bounded subsets of solutions, when $x_1 \to \infty$. That means that, for every bounded (in the uniform topology of $C_b^{2+\beta}$) subset $B \subset K^+$ and for every neighborhood $\mathcal{O}(\mathcal{A}_{tr})$ in the $C_{loc}^{2+\beta}$ topology, there exists $H = H(B, \mathcal{O})$ such that

$$T_h B \subset \mathcal{O}(\mathcal{A}_{tr}) \text{ if } h \geqslant H.$$

Note that the first assumption of the definition claims that the restriction $\mathcal{A}_{tr}|_{\Omega_1}$ is compact in $C^{2+\beta}(\overline{\Omega_1})$, for every bounded $\Omega_1 \subset \mathbb{R}_{++}^N$, and the third one is equivalent to the following:

For every bounded subdomain $\Omega_1 \subset \mathbb{R}_{++}^N$, for every B – bounded subset of K^+ and for every neighborhood $\mathcal{O}(\mathcal{A}_{tr}|_{\Omega_1})$ in the $C^{2+\beta}(\overline{\Omega_1})$-topology of the restriction of $\mathcal{A}_{tr}$ to this domain, there exists $H = H(\Omega_1, B, \mathcal{O})$ such that

$$(T_h B)|_{\Omega_1} \subset \mathcal{O}(\mathcal{A}_{tr}|_{\Omega_1}) \text{ if } h \geqslant H.$$

Theorem 5.5 *Let the assumptions of Theorem 5.4 hold. Then Eq. (5.2) possesses a trajectory attractor $\mathcal{A}_{tr}$ which has the following structure:*

$$\mathcal{A}_{tr} = \Pi_{\mathbb{R}_{++}^N} K(\mathbb{R}_+^N), \tag{5.11}$$

where $(x_1, x', x_N) \in \mathbb{R}_+^N$ and $K(\mathbb{R}_+^N)$ denotes the union of all bounded non-negative solutions $\hat{u}(x) \in C_b^{2+\beta}(\mathbb{R}_+^N)$ of

$$\Delta_x \hat{u} - f(\hat{u}) = 0, \quad x \in \mathbb{R}_+^N, \quad u|_{x_N=0} = 0, \quad \hat{u}(x) \geqslant 0, \tag{5.12}$$

i.e. the attractor $\mathcal{A}_{tr}$ consists of all bounded non-negative solutions u of (5.2) in $\mathbb{R}_{++}^N$, which can be extended to bounded non-negative solution $\hat{u}$ in $\mathbb{R}_+^N$.

The proof of this theorem is identical to *Theorem 3.2*. We omit the details.

Remark 5.6 Note that neither our concrete choice of $\mathbb{R}_{++}^N$ as the domain, on which we study our problem, nor the concrete choice of the 'time' direction x_1 are not essential for the of the trajectory dynamical system approach. Indeed, let us replace the 'time' direction x_1 by any fixed direction $\vec{l} \in \mathbb{R}^N$ and (and correspondingly $(T_h u)(x) := u(x + h\vec{l})$. Then the above construction is applicable if the domain Ω_+ satisfies the following assumptions:

1. $T_h \Omega_+ \subset \Omega_+$ (it is necessary in order to define the restriction T_h to the trajectory phase space K_+).
2. $\mathbb{R}_+^N = \cup_{h \leqslant 0} T_{-h} \Omega_+$ (it is required in order to obtain representation (5.11)).

5.3   Proof of *Theorem 5.1*

To obtain additional information on the behaviour of solutions of the initial problem (5.2), we use the description of non-negative bounded solutions in the half-space (see below).

Proposition 5.1 *Let assumptions of Theorem 5.1 hold and let $N \leqslant 5$. Then any non-negative bounded solution $\hat{u}(x)$ of Eq. (5.12) depends only on the variable x_N, i.e. $u(x) = V(x_N)$, where $V(z)$ is a bounded solution of the following problem:*

$$V''(z) - f(V(z)) = 0, \quad z > 0, \quad V(0) = 0, \quad V(z) \geqslant 0. \qquad (5.13)$$

The proof of this proposition is given in [56], for the case where the solution $\hat{u}(x)$ is strictly positive inside Ω. The general case can be reduced to the one above, using the following version of the strong maximum principle.

Lemma 5.1 *Let $V \subset \mathbb{R}^n$ be a (connected) domain with sufficiently smooth boundary and let $w \in C^2(V) \cap C(\overline{V})$ satisfy the following inequalities:*

$$\Delta_x w(x) - l(x)w(x) \leqslant 0, \quad x \in V, \quad w(x) \geqslant 0, \quad x \in V.$$

Assume also that $|l(x)| \leqslant K$ for $x \in V$. Then either $v(x) \equiv 0$, or $v(x) > 0$, for every interior point $x \in V$.

In order to apply the lemma to Eq. (5.12), we rewrite it in the following form:

$$\Delta_x \hat{u} - l(x)\hat{u} = f(0) \leqslant 0, \quad l(x) := \frac{f(\hat{u}(x)) - f(0)}{\hat{u}(x)}.$$

Since $f \in C^1$ and the solution $\hat{u}(x)$ is bounded, $l(x)$ is also bounded in $\mathbb{R}_+^N$. Thus, according to *Lemma 5.1*, either $\hat{u}(x) \equiv 0$ (which is evidently symmetric), or $\hat{u}(x) > 0$, in the interior of $\mathbb{R}_+^N$ and, thus, *Proposition 5.1* follows from the result of [56] mentioned above. *Proposition 5.1* is proved.

Denote by $\mathcal{R}_V$ the set of all bounded non-negative solutions $V(z)$ of problem (5.13). Then *Proposition 5.1* implies that

$$\mathcal{A}_{tr} = \mathcal{R}_V.$$

In the same manner as in the previous sections, one can obtain:

Proposition 5.2 *There exists a homeomorphism*

$$\tau : (\mathcal{R}_V, C_{loc}^{2+\beta}(\mathbb{R}_+)) \to (\mathcal{R}_f^+, \mathbb{R})$$

Moreover, the set $\mathcal{R}_f^+$ and, consequently, $\mathcal{R}_V$, are totally disconnected.

We now state the main result of this section.

Theorem 5.6 *Let the assumptions of Theorem 5.1 hold. Then, for every non-negative bounded solution u of problem* (5.2), *there exists a solution $V(x_N) = V_u(x_N) \in \mathcal{R}_V$ of problem* (5.13), *such that, for every fixed R and $x = (x_1, x', x_N)$,*

$$\|u - V_u\|_{C^{2+\beta}(B_{x_h}^R \cap \mathbb{R}_{++}^N)} \to 0, \quad x_h := (x_1 + h, x', x_N),$$

when $h \to \infty$.

Proof The proof of this theorem repeats word by word the proof of *Theorem 3.3*. Still, since it is short, we give it here for the convenience of the reader. Indeed, consider the ω-limit set of the solution $u \in K^+$, under the action of the semigroup T_h of shift in the x_1 direction,

$$\omega(u) = \cap_{h \geqslant 0}[\cup_{s \geqslant h} T_s u]_\Phi. \tag{5.14}$$

Recall that T_h possesses the attractor $\mathcal{A}_{tr}$ in K^+, and, consequently, the set (5.14) is nonempty. Thus,

$$\omega(u) \subset \mathcal{A}_{tr}.$$

It follows now from *Proposition 5.1* that $\omega(u) \subset \mathcal{R}_V$.

Note that on the one hand, the set $\omega(u)$ must be connected (see e.g. [1]) and on the other, it is a subset of $\mathcal{R}_V$, which is totally disconnected (by *Proposition 5.2*). Therefore, $\omega(u)$ consists of a single point, $V_u \subset \mathcal{R}_V$:

$$\omega(u) = \{V_u\}.$$

The assertion of the theorem is a simple corollary of this fact and of our definition of the topology in K^+. *Theorem 5.6* is proved, which in turn implies the assertion of *Theorem 5.1*.   □

5.4   Proof of *Theorem 5.2*

5.4.1   *Symmetry of the Profiles*

We recall a result of [45]:

Theorem 5.7 *Let $N \leqslant 4$ and $f \in C^1(\mathbb{R})$. Suppose that $u \in C^2(\mathbb{R}^N) \cap L^\infty(\mathbb{R}^N)$ is a stable solution of $\Delta u = f(u)$ in the whole of $\mathbb{R}^N$. Assume that*

$$f(u(x)) \geqslant 0 \text{ for almost any } x \in \mathbb{R}^N. \qquad (5.15)$$

Then u is constant.

Proof This is Theorem 1.1 of [45]. In fact, there it was assumed that

$$f(r) \leqslant 0 \text{ for any } r \in \mathbb{R}, \qquad (5.16)$$

but of course one can replace u with $-u$ and so change (5.16) into

$$f(r) \geqslant 0 \text{ for any } r \in \mathbb{R}. \qquad (5.17)$$

In fact, the same proof works by requiring (5.17) in the domain of u, that is (5.15).

$\square$

Remark 5.7 It is worth noticing that, when $N \leqslant 2$ the stability condition in *Theorem 5.7* may be dropped. Indeed, if $N = 1$, it just follows by an argument on convex functions, and if $N = 2$, one defines $w := u^2$ and checks by a computation that w is bounded and its Laplacian has a sign (hence is constant by the Liouville theorem in $\mathbb{R}^2$, see *Theorem 2.26* and *Lemma 2.20*).

Then, we have:

Theorem 5.8 *Let $f \in C^1(\mathbb{R})$. Let us assume that*

$$f(s) \geqslant 0 \text{ for any } s \geqslant 0, \qquad (5.18)$$

and that

$$\text{the set } E := \{s \geqslant 0 \text{ s.t. } f(s) = 0\} \text{ is non-empty.} \qquad (5.19)$$

Let $N \leqslant 4$ and u be a bounded and non-negative stable solution $u \in C^2(\mathbb{R}^N_+) \cap C^0(\overline{\mathbb{R}^N_+})$ of

$$
\begin{cases}
\Delta u = f(u) \text{ in } \mathbb{R}^N_+, \\[2mm]
u|_{\{x_1=0\}} = u_0.
\end{cases}
$$

Then, as $x_1 \to +\infty$, u converges in $C^2_{loc}(\mathbb{R}^N_{++})$ to a constant $c \in E$.

Proof Let $\mathcal{K}^+_s$ be the set of all non-negative stable solutions $v \in C^2(\mathbb{R}^N_+) \cap C^0(\overline{\mathbb{R}^N_+}) \cap L^\infty(\mathbb{R}^N_+)$ of $\Delta v = f(v)$ in $\mathbb{R}^N_+$. Notice that $E \subseteq \mathcal{K}^+_s$, so, by (5.19), we have that $\mathcal{K}^+_s$ is non-empty. For any $v \in \mathcal{K}^+_s$ and any $h \geqslant 0$, we define

$$\mathcal{T}_h v(x_1, x_2, \ldots, x_N) := v(x_1 + h, x_2, \ldots, x_N).$$

Since the above PDE and the stability condition (5.7) are translation invariant, we
have that $T_h(\mathcal{K}_s^+) \subseteq \mathcal{K}_s^+$. Also, T_h is continuous in the $C_{loc}^{2+\beta}(\overline{\mathbb{R}_{++}^N})$ topology, due
to elliptic estimates. Therefore, by *Theorem 5.5*, there exists a global stable attractor
$\mathcal{A}_s$, that is, as $x_1 \to +\infty$, any $v \in \mathcal{K}^+$ converges to an element of $\mathcal{A}_s$. On the other
hand, if $w \in \mathcal{A}_s$, it is a bounded and non-negative stable solution of $\Delta w = f(w)$ in
the whole of $\mathbb{R}^N$. Since we know from *Theorem 5.7* that w has to be constant, we
are done.   □

5.4.2 Completion of the Proof of Theorem 5.2

Notice that condition (5.3) implies (5.18) and (5.1) (recall *Remark 5.3*). Hence,
thanks to *Theorem 5.8*, we have that all the solutions of (5.2) are bounded, and
so *Theorem 5.2* is a consequence of *Theorem 5.8*.   □

Remark 5.8 Notice that the assumption that u is non-negative has been used only at
the end of *Theorem 5.8* to say that w is non-negative (so to fulfill (5.15) and be able
to use *Theorem 5.7*). If one knows that such a w is non-negative for other reasons,
the assumption that u is non-negative in *Theorems 5.2* and *5.8* can be dropped. This
observation will be important in the forthcoming proof of *Theorem 5.3*.

Remark 5.9 If (5.3) holds but (5.4) does not hold, we have that:

there does not exist any non-negative solution of $\Delta u = f(u)$ in $\mathbb{R}_+^N$,   (5.20)

hence condition (5.4) is natural and cannot be avoided. To prove this, we argue
as follows. If (5.4) is violated, there exists $c_3 > 0$ such that $f(s) > c_3$ for any
$s \in [0, A]$, where $A := (2(1 + c_1)/c_2)^{1/r}$. Let

$$c_4 := \min\left\{ \frac{c_3}{(1 + A)^r}, \frac{c_2}{2}\left(\frac{A}{1 + A}\right)^r \right\}.$$

We claim that, for any $s \geq 0$,

$$f(s) \geq c_4(1 + s)^r \tag{5.21}$$

To prove (5.21), we distinguish two cases. If $s \in [0, A]$, we have that

$$c_4(1 + s)^r \leqslant c_4(1 + A)^r \leqslant c_3 \leqslant f(s)$$

which proves (5.21) in this case. On the other hand, if $s \geqslant A$, we have that $1 \leqslant s/A$, hence

$$c_4(1 + s)^r \leqslant c_4 \left(\frac{s}{A} + s \right)^r = c_4 \left(\frac{1 + A}{A} \right)^r s^r$$

$$= -c_1 + c_1 + c_4 \left(\frac{1 + A}{A} \right)^r s^r \leqslant -c_1 + \frac{c_1}{A^r} s^r + \frac{c_2}{2} s^r$$

$$= -c_1 + \frac{c_1 c_2}{2(1 + c_1)} s^r + \frac{c_2}{2} s^r \leqslant -c_1 + c_2 s^r \leqslant f(s),$$

which completes the proof of (5.21). Now, if u were a non-negative solution of $\Delta u = f(u)$ in $\mathbb{R}_+^N$, by *Theorem 5.8*, we know that, as $x_1 \to +\infty$, u converges to some v which is a non-negative solution of $\Delta v = f(v)$ in the whole of $\mathbb{R}^N$. Let now $w := 1 + v$. By (5.21), we have

$$\Delta w^+ = \Delta w = \Delta v = f(v) \geqslant c_4(1 + v)^r = c_4(w^+)^r.$$

Accordingly, by Lemma 2 of [28], we have that $0 \geqslant w^+ = w = 1 + v$, that is $v \leqslant -1$. Since we knew v to be non-negative, this contradiction proves (5.20).

5.5 Proof of *Theorem 5.3*

5.5.1 *Positivity of Solutions*

Global solutions of the PDE with the special nonlinearity in (5.8) are always non-negative, as pointed out by the following result:

Lemma 5.2 *Let $N \in \mathbb{N}$. Let w be a solution of $\Delta w = w \prod_{j=1}^{m} (w - p_j)^2$ in the whole of $\mathbb{R}^N$. Then $0 \leqslant w \leqslant p_m$.*

Proof The proof is inspired by some arguments in [54]. Let $v := -w$. We have

$$\Delta v = v \prod_{j=1}^{m} (v + p_j)^2$$

and so, by Kato's Inequality (see Lemma A.1 in [28]),

$$\Delta v^+ \geq v \prod_{j=1}^{m}(v + p_j)^2 \chi_{\{v \geq 0\}}$$

$$\geq v^{2m+1}\chi_{\{v \geq 0\}} = (v^+)^{2m+1}$$

in the sense of distribution. Therefore, by Lemma 2 of [28], $v^+ \leq 0$ almost everywhere, that is $w \geq 0$, as desired. Now, let $q_j := p_m - p_j$ and $w := u - p_m$. Notice that $q_j \geq 0$ and that

$$\Delta w = \psi(w + p_m)(w + p_m) \prod_{j=1}^{m}(w + q_j)^2.$$

Accordingly, by Kato's Inequality,

$$\Delta w^+ \geq \psi(w + p_m)(w + p_m) \prod_{j=1}^{m}(w + q_j)\chi_{\{w \geq 0\}}$$

$$\geq w^{2m+1}\chi_{\{w \geq 0\}} = (w^+)^{2m+1}.$$

Once again, by Lemma 2 of [28], we have that $w^+ \leq 0$ almost everywhere, that is $u \leq p_m$, as desired.   □

5.5.2   Completion of the Proof of Theorem 5.3

We proceed as in the proof of *Theorem 5.2*. The only difference here is that the function w involved in the proof of *Theorem 5.8* is non-negative, thanks to *Lemma 5.2* (recall *Remark 5.8*).   □

Chapter 6
Symmetry and Attractors: Arbitrary Dimension

6.1 Introduction

Let Ω be the domain of $\mathbb{R}^N$ ($N \geqslant 2$) defined by

$$\Omega = (0, +\infty) \times \mathbb{R}^{N-2} \times (0, +\infty)$$
$$= \{x = (x_1, x', x_N) \in \mathbb{R}^N \mid x_1 > 0, \ x' = (x_2, \ldots, x_{N-1}) \in \mathbb{R}^{N-2}, \ x_N > 0\}.$$

This chapter is devoted to the study of the large space behavior, that is as $x_1 \to +\infty$, of the nonnegative bounded classical solutions u of the equation

$$\begin{cases} \Delta u + f(u) = 0 & \text{in } \Omega, \\ u(x_1, x', 0) = 0 & \text{for all } x_1 > 0 \text{ and } x' \in \mathbb{R}^{N-2}, \\ u(0, x', x_N) = u_0(x', x_N) & \text{for all } x' \in \mathbb{R}^{N-2} \text{ and } x_N > 0, \end{cases} \tag{6.1}$$

where the function

$$u_0 : \mathbb{R}^{N-2} \times (0, +\infty) \to \mathbb{R}_+ = [0, +\infty)$$

is given, continuous and bounded. The solutions u are understood to be bounded, of class $C^2(\Omega)$ and to be continuous on $\overline{\Omega} \setminus (\{0\} \times \mathbb{R}^{N-2} \times \{0\})$. From standard elliptic estimates, they are then automatically of class $C_b^{2,\beta}([\epsilon, +\infty) \times \mathbb{R}^{N-2} \times \mathbb{R}_+)$ for all $\epsilon > 0$ and $\beta \in [0, 1)$. Here and below, for any closed set $F \subset \mathbb{R}^N$ and $\beta \in [0, 1)$, we write

$$C_b^{2,\beta}(F) := \left\{ u : F \to \mathbb{R} \mid \|u\|_{C_b^{2+\beta}(F)} = \sup_{x \in F} \|u\|_{C^{2,\beta}(\overline{B(x,1)} \cap F)} < \infty \right\}, \tag{6.2}$$

© Springer Nature Switzerland AG 2018
M. Efendiev, *Symmetrization and Stabilization of Solutions
of Nonlinear Elliptic Equations*, Fields Institute Monographs 36,
https://doi.org/10.1007/978-3-319-98407-0_6

where $B(x, 1)$ means the open Euclidean ball of radius 1 centered at x. Problems sets in the half-space

$$\Omega' = (0, +\infty) \times \mathbb{R}^{N-1}$$

will also be considered in this chapter, see (6.11) below. The value of u at $x_1 = 0$ is then given and the goal is to describe, as it was in the previous chapters, the limiting profiles of u as $x_1 \to +\infty$. If the equation were parabolic in the variable x_1, we would then be reduced to characterize the ω-limit set of the initial condition u_0. However, problem (6.1) is an elliptic equation in all variables, including x_1, and the "Cauchy" problem (6.1) with the "initial value" u_0 at $x_1 = 0$ is ill-posed. There might indeed be several solutions u with the same value u_0 at $x_1 = 0$. Nevertheless, under some assumptions on the nonlinearity f, we will see that the behavior as $x_1 \to +\infty$ of any solution u of (6.1) or of similar problems in the half-space $\Omega' = (0, +\infty) \times \mathbb{R}^{N-1}$ is well-defined and unique (roughly speaking, no oscillation occur). In some cases, we will prove that all solutions u converge as $x_1 \to +\infty$ to the same limiting one-dimensional profile, irrespectively of u_0. To do so, we will use two different approaches. The first one is a pure PDE approach based on comparisons with suitable sub-solutions and on Liouville type results. This chapter indeed contains new Liouville type results of independent interest for the solutions of some elliptic equations in half-spaces $\mathbb{R}^{N-1} \times (0, +\infty)$ with homogeneous Dirichlet boundary conditions, or in the whole space $\mathbb{R}^N$ (see Sect. 6.2 for more details). The second approach is a dynamical systems' approach which says that x_1 can all the same be viewed as a time variable for a suitably defined dynamical system whose global attractor can be proved to exist and can be characterized. Let us now describe more precisely the types of assumptions we make on the functions f, which are always assumed to be locally Lipschitz-continuous from $\mathbb{R}_+$ to $\mathbb{R}$. The first class of functions we consider corresponds to functions f such that

$$\begin{cases} \exists \mu > 0, \quad f > 0 \text{ on } (0, \mu), \quad f \leqslant 0 \text{ on } [\mu, +\infty), \\ \exists 0 < \mu' < \mu, \quad f \text{ is nonincreasing on } [\mu', \mu], \\ \text{either } \left[f(0) > 0 \right] \text{ or } \left[f(0) = 0 \text{ and } \liminf_{s \to 0^+} \frac{f(s)}{s} > 0 \right]. \end{cases} \quad (6.3)$$

Under assumption (6.3) on f, it is immediate to see that there exists a unique solution $V \in C^2(\mathbb{R}_+)$ of the one-dimensional equation

$$\begin{cases} V''(\xi) + f(V(\xi)) = 0 \text{ for all } \xi \geqslant 0, \\ V(0) = 0 < V(\xi) < \mu = V(+\infty) \text{ for all } \xi > 0. \end{cases} \quad (6.4)$$

Furthermore, $V'(\xi) > 0$ for all $\xi \geqslant 0$.

Under assumption (6.3), the behavior of the nontrivial solutions u of (6.1) as $x_1 \to +\infty$ is uniquely determined, as the following theorem shows.

Theorem 6.1 *Let N be any integer such that $N \geqslant 2$ and assume that f satisfies (6.3). Let u be any nonnegative and bounded solution of (6.1), where $u_0 : \mathbb{R}^{N-2} \times (0, +\infty) \to \mathbb{R}_+$ is any continuous and bounded function such that $u_0 \neq 0$ in $\mathbb{R}^{N-2} \times (0, +\infty)$. Then*

$$\lim_{R \to +\infty} \left(\inf_{(R,+\infty) \times \mathbb{R}^{N-2} \times (R,+\infty)} u \right) \geqslant \mu \tag{6.5}$$

and

$$u(x_1 + h, x', x_N) \to V(x_N) \text{ as } h \to +\infty \text{ in } C_b^{2,\beta}([A, +\infty) \times \mathbb{R}^{N-2} \times [0, B]) \tag{6.6}$$

for all $A \in \mathbb{R}$, $B > 0$ and $\beta \in [0, 1)$, where $V \in C^2(\mathbb{R}_+)$ is the unique solution of (6.4).

Notice that property (6.5) means that the non-trivial nonnegative solutions u of (6.1) are separated from 0, irrespectively of u_0, far away from the boundary $\partial\Omega$. If $u_0 \leqslant \mu$, then since $f \leqslant 0$ on $[\mu, +\infty)$ and $u = \mu$ on $\partial\Omega^+$, where $\Omega^+ = \Omega \cap \{u > \mu\}$, it follows that from the maximum principle applied in Ω^+ (see [13], since $\mathbb{R}^N \backslash \overline{\Omega^+}$ contains the closure of an infinite open connected cone), that actually $u \leqslant \mu$ in Ω^+ whence $\Omega^+ = \varnothing$ and $u \leqslant \mu$ in Ω. In this case, it also follows from *Theorem 6.1* and standard elliptic estimates that the convergence (6.6) holds not only locally in x_N, but in $C_b^{2,\beta}([A, +\infty) \times \mathbb{R}^{N-2} \times \mathbb{R}_+)$ for all $A \in \mathbb{R}$ and $\beta \in [0, 1)$. However, without the assumption $u_0 \leqslant \mu$, it is not clear that this last convergence property holds globally with respect to x_N in general. The second class of functions f we consider corresponds to the following assumption:

$$\begin{cases} f \geqslant 0 \text{ on } \mathbb{R}_+, \\ \forall z \in E, \ \liminf_{s \to z^+} \dfrac{f(s)}{s - z} > 0, \end{cases} \tag{6.7}$$

where

$$E = \{z \in \mathbb{R}_+; \ f(z) = 0\} \tag{6.8}$$

denotes the set of zeroes of f. A typical example of such a function f is $f(s) = |\sin s|$ for all $s \geqslant 0$, with $E = \pi\mathbb{N}$. More generally speaking, under the assumption (6.7), it follows immediately that the set E is at most countable. Furthermore, it is easy to check that, for each $z \in E\backslash\{0\}$, there exists a unique solution $V_z \in C^2(\mathbb{R}_+)$ of the one-dimensional equation

$$\begin{cases} V_z''(\xi) + f(V_z(\xi)) = 0 \text{ for all } \xi \geqslant 0, \\ V_z(0) = 0 < V_z(\xi) < z = V_z(+\infty) \text{ for all } \xi > 0. \end{cases} \tag{6.9}$$

Furthermore, $V_z'(\xi) > 0$ for all $\xi \geqslant 0$.

The following theorem states any solution of (6.1) is asymptotically one-dimensional as $x_1 \to +\infty$.

Theorem 6.2 *Let N be any integer such that $N \geqslant 2$ and assume that f satisfies (6.7). Let u be any nonnegative and bounded solution of (6.1), where $u_0 : \mathbb{R}^{N-2} \times (0, +\infty) \to \mathbb{R}_+$ is any continuous and bounded function such that $u_0 \not\equiv 0$ in $\mathbb{R}^{N-2} \times (0, +\infty)$. Then there exists $R > 0$ such that*

$$\inf_{(R,+\infty) \times \mathbb{R}^{N-2} \times (R,+\infty)} u > 0$$

and there exists $z \in E \backslash \{0\}$ such that

$$u(x_1 + h, x', x_N) \to V_z(x_N) \ \text{as } h \to +\infty \text{ in } C_b^{2,\beta}([A, +\infty) \times \mathbb{R}^{N-2} \times \mathbb{R}_+)$$
$$(6.10)$$

for all $A \in \mathbb{R}$ and $\beta \in [0, 1)$, where $V_z \in C^2(\mathbb{R}_+)$ is the unique solution of (6.9) with the limit $V_z(+\infty) = z$.

This result shows that any non-trivial bounded solution u of (6.1) converges to a single one-dimensional profile as $x_1 \to +\infty$. More precisely, given u, the real number z defined by (6.10) is unique and, in the proof of *Theorem 6.2*, the explicit expression of z will be provided. Observe that the asymptotic profile may now depend on the solution u (unlike in *Theorem 6.1*) but *Theorem 6.2* says that the oscillations in the x_1 variable are excluded at infinity, for any solution u. The last two results are concerned with the analysis of the asymptotic behavior, as $x_1 \to +\infty$, of the nonnegative bounded classical solutions u of

$$\begin{cases} \Delta u + f(u) = 0 & \text{in } \Omega' = (0, +\infty) \times \mathbb{R}^{N-1}, \\ u(0, x_2, \ldots, x_N) = u_0(x_2, \ldots, x_N) & \text{for all } (x_2, \ldots, x_N) \in \mathbb{R}^{N-1}, \end{cases} \quad (6.11)$$

in the half-space Ω', where the function $u_0 : \mathbb{R}^{N-1} \to \mathbb{R}_+$ is given, continuous and bounded. The solutions u of (6.11) are understood to be bounded, of class $C^2(\Omega')$ and to be continuous on $\overline{\Omega'}$. They are then automatically of class $C_b^{2,\beta}([\epsilon, +\infty) \times \mathbb{R}^{N-1})$ for all $\epsilon > 0$ and $\beta \in [0, 1)$. Firstly, under the same assumptions (6.7) as in the previous theorem, the behavior as $x_1 \to +\infty$ of any non-trivial solution u of (6.11) is well-defined:

Theorem 6.3 *Let N be any integer such that $N \geqslant 2$ and assume that f satisfies (6.7). Let u be any nonnegative and bounded solution of (6.11), where $u_0 : \mathbb{R}^{N-1} \to \mathbb{R}_+$ is any continuous and bounded function such that $u_0 \not\equiv 0$ in $\mathbb{R}^{N-1}$. Then there exists $z \in E \backslash \{0\}$ such that*

$$u(x_1 + h, x_2, \ldots, x_N) \to z \ \text{as } h \to +\infty \text{ in } C_b^{2,\beta}([A, +\infty) \times \mathbb{R}^{N-1}) \qquad (6.12)$$

for all $A \in \mathbb{R}$ and $\beta \in [0, 1)$.

Notice that the conclusion implies in particular that u is separated from 0 far away from the boundary $\{0\} \times \mathbb{R}^{N-1}$ of Ω'. Furthermore, as in *Theorem 6.2*, the real number z in (6.12) is uniquely determined by u and its explicit value will be given during the proof. In *Theorem 6.3*, if instead of (6.7) the function f now satisfies assumption (6.3), then u may not converge in general to a constant as $x_1 \to +\infty$. Furthermore, even if u does converge to a constant as $x_1 \to +\infty$, that constant may not be equal to the real number μ given in (6.3). For instance, if there exists $\rho \in (\mu, +\infty)$ such that $f(\rho) = 0$, then the constant function $u = \rho$ solves (6.1) with $u_0 = \rho$. Therefore, under assumption (6.3), the asymptotic profile of a solution u of problem (6.11) in the half-space Ω' depends on u and is even not clearly well-defined in general. The situation is thus very different from *Theorem 6.1* about the existence and uniqueness of the asymptotic behavior of the solutions of problem (6.1) in the quarter-space Ω. However, under (6.3) and an additional appropriate assumption on f, the following result holds:

Theorem 6.4 *Let N be any integer such that $N \geqslant 2$ and assume that, in addition to (6.3), f is such that*

$$\liminf_{s \to z^-} \frac{f(s)}{s - z} > 0 \ \text{ for all } z > \mu \text{ such that } f(z) = 0. \tag{6.13}$$

Let u be any nonnegative and bounded solution of (6.11), where $u_0 : \mathbb{R}^{N-1} \to \mathbb{R}_+$ is any continuous and bounded function such that $u_0 \not\equiv 0$ in $\mathbb{R}^{N-1}$. Then there exists $z \geqslant \mu$ such that $f(z) = 0$ and

$$u(x_1 + h, x_2, \ldots, x_N) \to z \ \text{ as } h \to +\infty \text{ in } C_b^{2,\beta}([A, +\infty) \times \mathbb{R}^{N-1})$$

for all $A \in \mathbb{R}$ and $\beta \in [0, 1)$.

Remark 6.1 It is worth noticing that all above results hold in any dimension $N \geqslant 2$.

6.2 The PDE Approach

In this section, we use a pure PDE approach to prove the main results announced in *Sect. 6.1*. In *Sect. 6.2.1*, we deal with the case of problem (6.1) set in the quarter-space $\Omega = (0, +\infty) \times \mathbb{R}^{N-2} \times (0, +\infty)$, while *Sect. 6.2.2* is concerned with problem (6.11) set in the half-space $\Omega' = (0, +\infty) \times \mathbb{R}^{N-1}$.

6.2.1   Problem (6.1) in the Quarter-Space $\Omega = (0, +\infty) \times \mathbb{R}^{N-2} \times (0, +\infty)$

Let us first begin with the

Proof of Theorem 6.1 The proof is divided into three main steps: we first prove that u is bounded from below away from 0 when x_1 and x_N are large, uniformly with respect to $x' \in \mathbb{R}^{N-2}$. Then, we pass to the limit as $x_1 \to +\infty$ and use a classification result, which finally leads to the uniqueness and one-dimensional symmetry of the limiting profiles of u as $x_1 \to +\infty$.

Step 1 It consists in the following result, which we state as a lemma since it will be used several times in the chapter. Notice that only the third line of (6.3) is needed.

Lemma 6.1 *Let N be any integer such that $N \geqslant 2$ and assume that either $f(0) > 0$, or $f(0) = 0$ and $\liminf_{s \to 0^+} f(s)/s > 0$. Let u be any nonnegative and bounded solution of (6.1), where $u_0 : \mathbb{R}^{N-2} \times (0, +\infty) \to \mathbb{R}_+$ is any continuous and bounded function such that $u_0 \not\equiv 0$ in $\mathbb{R}^{N-2} \times (0, +\infty)$. Then there exist $R > 0$ and $\epsilon > 0$ such that*

$$u \geqslant \epsilon \ \ in \ [R, +\infty) \times \mathbb{R}^{N-2} \times [R, +\infty). \tag{6.14}$$

Proof First of all, observe that, since $f(0) \geqslant 0$, the strong maximum principle implies that the function u is either positive in Ω, or identically equal to 0 in Ω. But since u is continuous up to $\{0\} \times \mathbb{R}^{N-2} \times (0, +\infty)$ and since $u_0 \not\equiv 0$, it follows that $u > 0$ in Ω.

In the sequel, for any $x \in \mathbb{R}^N$ and $R > 0$, denote $B(x, R)$ the open Euclidean ball of centre x and radius R. For each $R > 0$, let λ_R be the principal eigenvalue of the Laplace operator in $B(0, R)$ with Dirichlet boundary condition on $\partial B(0, R)$, and let φ_R be the normalized principal eigenfunction, that is

$$\begin{cases} \Delta \varphi_R + \lambda_R \, \varphi_R = 0 & \text{in } B(0, R), \\ \varphi_R > 0 & \text{in } B(0, R), \\ \|\varphi_R\|_{L^\infty(B(0,R))} = \varphi_R(0) = 1, \\ \varphi_R = 0 & \text{on } \partial B(0, R). \end{cases} \tag{6.15}$$

Notice that $\lambda_R \to 0$ as $R \to +\infty$. If $f(0) = 0$, we can then choose $R > 0$ large enough so that

$$\lambda_R < \liminf_{s \to 0^+} \frac{f(s)}{s}.$$

If $f(0) > 0$, we simply choose $R = 1$. Then, fix a point $x_0 \in \Omega$ in such a way that $\overline{B(x_0, R)} \subset \Omega$. Since u is continuous and positive on $\overline{B(x_0, R)}$, there holds

$\min_{\overline{B(x_0,R)}} u > 0$. Therefore, it follows from the choice of R that there exists $\epsilon > 0$ small enough, such that the function

$$\underline{u}(x) = \epsilon\, \varphi_R(x - x_0)$$

is a subsolution in $\overline{B(x_0, R)}$, that is

$$\Delta \underline{u} + f(\underline{u}) \geqslant 0 \text{ and } \underline{u} < u \text{ in } \overline{B(x_0, R)}. \tag{6.16}$$

Next, let $\tilde{x}_0$ be any point in Ω such that

$$\overline{B(\tilde{x}_0, R)} \subset \overline{\Omega},$$

that is $\tilde{x}_0 = (\tilde{x}_{0,1}, \tilde{x}_0', \tilde{x}_{0,N})$ with $\tilde{x}_{0,1} \geqslant R$ and $\tilde{x}_{0,N} \geqslant R$. For all $t \in [0, 1]$, call

$$y_t = x_0 + t\,(\tilde{x}_0 - x_0)$$

and observe that $\overline{B(y_t, R)} \subset \Omega$ for all $t \in [0, 1)$. Define

$$\underline{u}_t(x) = \underline{u}(x - y_t + x_0) = \epsilon\, \varphi_R(x - y_t)$$

for all $t \in [0, 1]$ and $x \in \overline{B(y_t, R)}$. By continuity and from (6.16), there holds $\underline{u}_t < u$ in $\overline{B(y_t, R)}$ for $t \in [0, t_0]$, where $t_0 > 0$ is small enough. On the other hand, for all $t \in [0, 1]$, the function $\underline{u}_t$ is a subsolution of the equation satisfied by u, that is

$$\Delta \underline{u}_t + f(\underline{u}_t) \geqslant 0 \text{ in } \overline{B(y_t, R)}.$$

We shall now use a sliding method (see *Chap. 1* and [13, 18]) to conclude that

$$\underline{u}_t < u \text{ in } \overline{B(y_t, R)} \text{ for all } t \in [0, 1).$$

Indeed, if this were not true, there would then exist a real number $t^* \in (0, 1)$ such that the inequality $\underline{u}_{t^*} \leqslant u$ holds in $\overline{B(y_{t^*}, R)}$ with equality at some point $x^* \in \overline{B(y_{t^*}, R)}$. Since $\overline{B(y_{t^*}, R)} \subset \Omega$, $u > 0$ in Ω and $\underline{u}_{t^*} = 0$ on $\partial B(y_{t^*}, R)$, one has $x^* \in B(y_{t^*}, R)$. But since $\underline{u}_{t^*}$ is a subsolution of the equation satisfied by u, the strong maximum principle yields $\underline{u}_{t^*} = u$ in $B(y_{t^*}, R)$ and also on the boundary by continuity, which is impossible. One has then reached a contradiction. Hence, $\underline{u}_t < u$ in $\overline{B(y_t, R)}$ for all $t \in [0, 1)$. By continuity, one also gets that $\underline{u}_1 \leqslant u$ in $\overline{B(y_1, R)}$.

Therefore, for any $\tilde{x}_0 = (\tilde{x}_{0,1}, \tilde{x}_0', \tilde{x}_{0,N})$ with $\tilde{x}_{0,1} \geqslant R$ and $\tilde{x}_{0,N} \geqslant R$, there holds

$$u(\tilde{x}_0) \geqslant \underline{u}_1(\tilde{x}_0) = \epsilon\, \varphi_R(0) = \epsilon.$$

In other words, (6.14) holds and the proof of *Lemma 6.1* is complete. □

Step 2 Let $(x_{1,n})_{n\in\mathbb{N}}$ be any sequence of positive numbers such that $x_{1,n} \to +\infty$ as $n \to +\infty$, and let $(x'_n)_{n\in\mathbb{N}}$ be any sequence in $\mathbb{R}^{N-2}$. From standard elliptic estimates, there exists a subsequence such that the functions

$$u_n(x) = u(x_1 + x_{1,n}, x' + x'_n, x_N)$$

converge in $C^{2,\beta}_{loc}(\mathbb{R}^N_+)$, for all $\beta \in [0, 1)$, to a bounded classical solution u_∞ of

$$\begin{cases} \Delta u_\infty + f(u_\infty) = 0 & \text{in } \mathbb{R}^N_+, \\ \qquad\qquad u_\infty = 0 & \text{on } \partial\mathbb{R}^N_+, \end{cases}$$

where $\mathbb{R}^N_+ = \mathbb{R}^{N-1} \times [0, +\infty)$. Furthermore, $u_\infty \geqslant 0$ and $u_\infty \neq 0$ in $\mathbb{R}^N_+$ from Step 1, since

$$u_\infty \geqslant \epsilon > 0 \quad \text{in } \mathbb{R}^{N-1} \times [R, +\infty)$$

from *Lemma 6.1*. Thus, $u_\infty > 0$ in $\mathbb{R}^{N-1} \times (0, +\infty)$ from the strong maximum principle. It follows from Theorems 1.1 and 1.2 of Berestycki, Caffarelli and Nirenberg [13][1] (see also [6, 37]) that u_∞ is unique and has one-dimensional symmetry. By uniqueness of the problem (6.4), one gets that $u_\infty(x) = V(x_N)$ for all $x \in \mathbb{R}^N_+$, and the limit does not depend on the sequences $(x_{1,n})_{n\in\mathbb{N}}$ or $(x'_n)_{n\in\mathbb{N}}$. Property (6.6) of *Theorem 6.1* then follows from the uniqueness of the limit.

Step 3 Let us now prove formula (6.5). One already knows from *Lemma 6.1* that

$$m := \lim_{R\to+\infty} \left(\inf_{(R,+\infty)\times\mathbb{R}^{N-2}\times(R,+\infty)} u \right) \geqslant \epsilon > 0.$$

Let $(x_n)_{n\in\mathbb{N}} = (x_{1,n}, x'_n, x_{N,n})_{n\in\mathbb{N}}$ be a sequence in Ω such that $(x_{1,n}, x_{N,n})_{n\in\mathbb{N}} \to (+\infty, +\infty)$ and $u(x_n) \to m$ as $n \to +\infty$. Up to extraction of a subsequence, the functions

$$v_n(x) = u(x + x_n)$$

converge in $C^2_{loc}(\mathbb{R}^N)$ to a classical bounded solution v_∞ of

$$\Delta v_\infty + f(v_\infty) = 0 \quad \text{in } \mathbb{R}^N$$

such that $v_\infty \geqslant m$ in $\mathbb{R}^N$ and $v_\infty(0) = m > 0$. Thus, $f(m) \leqslant 0$, whence $m \geqslant \mu$ due to (6.3). The proof of *Theorem 6.1* is thereby complete. $\qquad\square$

---

[1]In [13], the function f was assumed to be globally Lipschitz-continuous. Here, f is just assumed to be locally Lipschitz-continuous. However, since u is bounded, it is always possible to find a Lipschitz-continuous function $\tilde{f} : \mathbb{R}^+ \to \mathbb{R}$ satisfying (6.3) and such that $\tilde{f}$ and f coincide on the range of u.

The proof of *Theorem 6.2* is based on two Liouville type results for the bounded nonnegative solutions u of the elliptic equation

$$\Delta u + f(u) = 0$$

in the whole space $\mathbb{R}^N$ or in the half-space $\mathbb{R}^{N-1} \times \mathbb{R}_+$ with Dirichlet boundary condition on $\mathbb{R}^{N-1} \times \{0\}$.

Theorem 6.5 *Let N be any integer such that $N \geqslant 1$ and assume that the function f satisfies (6.7). Let u be a bounded nonnegative solution of*

$$\Delta u + f(u) = 0 \ \ in \ \mathbb{R}^N. \tag{6.17}$$

Then u is constant.

The following result is concerned with the one-dimensional symmetry of non-negative bounded solutions in a half-space with Dirichlet boundary conditions.

Theorem 6.6 *Let N be any integer such that $N \geqslant 1$ and assume that the function f satisfies (6.7). Let u be a bounded nonnegative solution of*

$$\begin{cases} \Delta u + f(u) = 0 \ \ in \ \mathbb{R}^N_+ = \mathbb{R}^{N-1} \times \mathbb{R}_+, \\ \qquad\qquad u = 0 \ \ on \ \partial\mathbb{R}^N_+ = \mathbb{R}^{N-1} \times \{0\}. \end{cases} \tag{6.18}$$

Then u is a function of x_N only. Furthermore, either $u = 0$ in $\mathbb{R}^{N-1} \times \mathbb{R}_+$ or there exists $z > 0$ such that $f(z) = 0$ and $u(x) = V_z(x_N)$ for all $x \in \mathbb{R}^{N-1} \times \mathbb{R}_+$, where the function V_z satisfies equation (6.9).

These results are of independent interest and will be proved in *Sect. 6.3*. Notice that one of the main points is that they hold in any dimensions $N \geqslant 1$ without any other assumption on u than its boundedness. In low dimensions $N \leqslant 4$, and under the additional assumption that u is stable, the conclusion of *Theorem 6.5* holds for any nonnegative function f of class $C^1(\mathbb{R}_+)$, see Dupaigne and Farina [43]. Consequently, because of the monotonicity result in the direction x_N due to Berestycki, Caffarelli and Nirenberg [13] and Dancer [39] (since $f(0) \geqslant 0$), it follows that the conclusion of *Theorem 6.6* holds for any nonnegative function f of class $C^1(\mathbb{R}_+)$, provided that $N \leqslant 5$, see Farina and Valdinoci [60]. However, observe that the nonnegativity and the C^1 character of f are incompatible with (6.7) for any positive zero z of f. Furthermore, assumption (6.7) is crucially used in the proof of *Theorem 6.5* and 6.6. It is actually not true that these theorems stay valid in general when f is just assumed to be nonnegative and locally Lipschitz-continuous. For instance, non-constant solutions of (6.17), which are even stable, exist for power-like nonlinearities f in high dimensions (see [57] and the references therein).

With these results in hand, let us turn to the

Proof of Theorem 6.2 Observe that, from (6.7), either $f(0) > 0$, or $f(0) = 0$ and $\liminf_{s \to 0^+} f(s)/s > 0$. Therefore, *Lemma 6.1* gives the existence of $R > 0$ and $\epsilon > 0$ such that (6.14) holds. Set

$$M = \lim_{A \to +\infty} \left(\sup_{[A,+\infty) \times \mathbb{R}^{N-2} \times [0,+\infty)} u \right). \tag{6.19}$$

Our goal is to prove that the conclusion of *Theorem 6.2* holds with $z = M$.

Since u is bounded and satisfies (6.14), M is such that $\epsilon \leqslant M < +\infty$. Furthermore, there exists a sequence $(x_n)_{n \in \mathbb{N}} = (x_{1,n}, x_n', x_{N,n})_{n \in \mathbb{N}}$ of points in $\overline{\Omega}$ such that $x_{1,n} \to +\infty$ and $u(x_n) \to M$ as $n \to +\infty$.

Assume first, up to extraction of a subsequence, that the sequence $(x_{N,n})_{n \in \mathbb{N}}$ converges to a nonnegative real number $x_{N,\infty}$ as $n \to +\infty$. From standard elliptic estimates, the functions

$$u_n(x) = u(x_1 + x_{1,n}, x' + x_n', x_N)$$

converge in $C_{loc}^{2,\beta}(\mathbb{R}^{N-1} \times \mathbb{R}_+)$ for all $\beta \in [0, 1)$, up to extraction of another subsequence, to a bounded nonnegative solution u_∞ of the problem (6.18) in the half-space $\mathbb{R}^{N-1} \times [0, +\infty)$, such that

$$u_\infty(0, 0, x_{N,\infty}) = M = \sup_{\mathbb{R}^{N-1} \times [0,+\infty)} u_\infty > 0.$$

It follows from *Theorem 6.6* that $u_\infty(x) = V_z(x_N)$ is a one-dimensional increasing solution of (6.9), whence $z = M$. But since V_M is (strictly) increasing, it cannot reach its maximum M at the finite point $x_{N,\infty}$. This case is then impossible.

Therefore, one can assume without loss of generality that $x_{N,n} \to +\infty$ as $n \to +\infty$. From standard elliptic estimates, the functions

$$u_n(x) = u(x + x_n)$$

converge in $C_{loc}^{2,\beta}(\mathbb{R}^N)$ for all $\beta \in [0, 1)$, up to extraction of another subsequence, to a bounded nonnegative solution u_∞ of the problem (6.17) in $\mathbb{R}^N$, such that

$$u_\infty(0) = M = \sup_{\mathbb{R}^N} u_\infty > 0.$$

Theorem 6.5 implies that $u_\infty = M$ in $\mathbb{R}^N$, whence $f(M) = 0$. Furthermore, since the limit M is unique, the convergence of the functions u_n to the constant M holds for the whole sequence.

Now, in order to complete the proof of *Theorem 6.2*, we shall make use of the following lemma of independent interest:

Lemma 6.2 *Let* $g : \mathbb{R}_+ \rightarrow \mathbb{R}$ *be a locally Lipschitz-continuous nonnegative function. Then, for each* $z > 0$ *such that* $g(z) = 0$ *and for each* $\epsilon \in (0, z]$, *there exist* $R' = R'_{g,z,\epsilon} > 0$ *and a classical solution* v *of*

$$
\begin{cases}
\Delta v + g(v) = 0 \ in \ \overline{B(0, R')}, \\
0 \leqslant v < z \ in \ \overline{B(0, R')}, \\
v = 0 \ on \ \partial B(0, R'), \\
v(0) = \max_{\overline{B(0,R')}} v \geqslant z - \epsilon.
\end{cases}
\tag{6.20}
$$

The proof of this lemma is postponed at the end of this section. Let us now finish the proof of *Theorem 6.2*. Fix an arbitrary ϵ in $(0, M]$. Let $R'(\epsilon) = R'_{f,M,\epsilon}$ be as in *Lemma 6.2* and let v be a solution of (6.20) with $g = f$ and $z = M$. Since the functions $u_n(x) = u(x + x_n)$ converge locally uniformly in $\mathbb{R}^N$ to the constant M and since

$$
\max_{\overline{B(0,R'(\epsilon))}} v < M,
$$

there exists $n_0 \in \mathbb{N}$ large enough so that $\overline{B(x_{n_0}, R'(\epsilon))} \subset \Omega$ and

$$
v(x - x_{n_0}) < u(x) \ for \ all \ x \in \overline{B(x_{n_0}, R'(\epsilon))}.
\tag{6.21}
$$

But since v solves the same elliptic equation as u, the same sliding method as in *Theorem 6.1* implies that

$$
u \geqslant v(0) \geqslant M - \epsilon \ in \ [R'(\epsilon), +\infty) \times \mathbb{R}^{N-2} \times [R'(\epsilon), +\infty).
\tag{6.22}
$$

Lastly, choose any sequence $(\tilde{x}_{1,n})_{n \in \mathbb{N}}$ converging to $+\infty$, and any sequence $(\tilde{x}'_n)_{n \in \mathbb{N}}$ in $\mathbb{R}^{N-2}$. Up to extraction of a subsequence, the functions

$$
\tilde{u}_n(x) = u(x_1 + \tilde{x}_{1,n}, x' + \tilde{x}'_n, x_N)
$$

converge in $C^{2,\beta}_{loc}(\mathbb{R}^{N-1} \times \mathbb{R}_+)$ for all $\beta \in [0, 1)$ to a bounded nonnegative solution $\tilde{u}_\infty$ of problem (6.18) in the half-space $\mathbb{R}^{N-1} \times \mathbb{R}_+$. The function $\tilde{u}_\infty$ satisfies $\tilde{u}_\infty \leqslant M$ in $\mathbb{R}^{N-1} \times \mathbb{R}_+$ by definition of M, while

$$
\lim_{A \rightarrow +\infty} \left(\inf_{\mathbb{R}^{N-1} \times [A, +\infty)} \tilde{u}_\infty \right) \geqslant M
$$

because $\epsilon > 0$ in (6.22) can be arbitrarily small. *Theorem 6.6* and the above estimates imply that

$$
\tilde{u}_\infty(x) = V_M(x_N) \ for \ all \ x \in \mathbb{R}^{N-1} \times \mathbb{R}_+.
$$

Since this limit does not depend on any subsequence, and due to (6.19), (6.22) and standard elliptic estimates, it follows in particular that

$$u(x_1 + h, x', x_N) \rightarrow V_M(x_N) \text{ in } C_b^{2,\beta}([A, +\infty) \times \mathbb{R}^{N-2} \times \mathbb{R}_+) \text{ as } h \rightarrow +\infty,$$

for all $A \in \mathbb{R}$ and $\beta \in [0, 1)$. The proof of *Theorem 6.2* is thereby complete. □

Remark 6.2 Instead of *Theorem 6.5*, if f is just assumed to be nonnegative and locally Lipschitz-continuous on $\mathbb{R}_+$ and if u_∞ is a solution of (6.17) which reaches its maximum and is such that $f(\max_{\mathbb{R}^N} u_\infty) = 0$ (as in some assumptions of [14, 21]), then u_∞ is constant, from the strong maximum principle. Therefore, it follows from similar arguments as in the above proof that if, instead of (6.7) in *Theorem 6.2*, the function f is just assumed to be nonnegative, locally Lipschitz-continuous and positive almost everywhere on $\mathbb{R}_+$ and if u is a bounded nonnegative solution of (6.1) such that $f(M) = 0$, where M is defined by (6.19), then either $M = 0$ and $u(x_1 + h, x', x_N) \rightarrow 0$ in $C^2([A, +\infty) \times \mathbb{R}^{N-2} \times \mathbb{R}_+)$ as $h \rightarrow +\infty$ for all $A \in \mathbb{R}$, or the conclusion (6.10) holds with $z = M$.

Proof of Lemma 6.2 The proof uses classical variational arguments, which we sketch here for the sake of completeness (see also e.g. [24] for applications of this method). Let g and z be as in *Lemma 6.2* and let $\tilde{g}$ be the function defined in $\mathbb{R}$ by

$$\tilde{g}(s) = \begin{cases} g(0) & \text{if } s < 0, \\ g(s) & \text{if } 0 \leqslant s \leqslant z, \\ 0 & \text{if } s > z. \end{cases}$$

The function $\tilde{g}$ is nonnegative, bounded and Lipschitz-continuous on $\mathbb{R}$. Set

$$G(s) = \int_s^z \tilde{g}(\tau) \, d\tau \geqslant 0$$

for all $s \in \mathbb{R}$. The function G is nonnegative and Lipschitz-continuous on $\mathbb{R}$.

Let r be any positive real number. Define

$$I_r(v) = \frac{1}{2} \int_{B(0,r)} |\nabla v|^2 + \int_{B(0,r)} G(v)$$

for all $v \in H_0^1(B(0, r))$. The functional I_r is well-defined in $H_0^1(B(0, r))$ and it is coercive, from Poincaré's inequality and the nonnegativity of G. From Rellich's and Lebesgue's theorems, the functional I_r has a minimum v_r in $H_0^1(B(0, r))$. The function v_r is a weak and hence, from the elliptic regularity theory, a classical $C^2(\overline{B(0, r)})$ solution of the equation

$$\begin{cases} \Delta v_r + \tilde{g}(v_r) = 0 \text{ in } \overline{B(0, r)}, \\ \qquad\qquad v_r = 0 \text{ on } \partial B(0, r). \end{cases}$$

Since $\widetilde{g} \geqslant 0$ on $(-\infty, 0]$, it follows from the strong maximum principle that $v_r \geqslant 0$ in $\overline{B(0, r)}$. Furthermore, either $v_r = 0$ in $\overline{B(0, r)}$, or $v_r > 0$ in $B(0, r)$. Similarly, since $\widetilde{g} = 0$ on $[z, +\infty)$, one gets that $v_r < z$ in $\overline{B(0, r)}$. Consequently, $\widetilde{g}(v_r) = g(v_r)$ in $\overline{B(0, r)}$. It also follows from the method of moving planes and Gidas, Ni and Nirenberg [65] that v_r is radially symmetric and decreasing with respect to $|x|$ (provided that $v_r \not\equiv 0$ in $\overline{B(0, r)}$). In all cases, there holds

$$0 \leqslant v_r(0) = \max_{\overline{B(0,r)}} v_r < z.$$

In order to complete the proof of *Lemma 6.2*, it is sufficient to prove that, given ϵ in $(0, z]$, there exists $r > 0$ such that $v_r(0) \geqslant z - \epsilon$. Let $\epsilon \in (0, z]$ and assume that $\max_{\overline{B(0,r)}} v_r = v_r(0) < z - \epsilon$ for all $r > 0$. Observe that the function G is nonincreasing in $\mathbb{R}$, and actually decreasing and positive on the interval $[0, z)$, from (6.7). Therefore,

$$I_r(v_r) \geqslant \alpha_N r^N G(z - \epsilon) \tag{6.23}$$

for all $r > 0$, where $\alpha_N > 0$ denotes the Lebesgue measure of the unit Euclidean ball in $\mathbb{R}^N$. For $r > 1$, let w_r be the test function defined in $\overline{B(0, r)}$ by

$$w_r(x) = \begin{cases} z & \text{if } |x| < r - 1, \\ z\,(r - |x|) & \text{if } r - 1 \leqslant |x| \leqslant r. \end{cases}$$

This function w_r belongs to $H_0^1(B(0, r))$ and $|\nabla w_r|^2$ and $G(w_r)$ are supported on the shell $\overline{B(0, r)} \backslash B(0, r - 1)$. Thus, there exists a constant C independent of r such that

$$I_r(w_r) \leqslant C\,(r^n - (r - 1)^n) \tag{6.24}$$

for all $r > 1$. But since $I_r(v_r) \leqslant I_r(w_r)$, by definition of v_r, and since $G(z - \epsilon) > 0$, inequalities (6.23) and (6.24) lead to a contradiction as $r \to +\infty$. Therefore, there exists a radius $R' > 0$ such that $v_{R'}(0) \geqslant z - \epsilon$, and $v_{R'}$ solves (6.20). The proof of *Lemma 6.2* is now complete. □

6.2.2 Problem (6.11) in the Half-Space $\Omega' = (0, +\infty) \times \mathbb{R}^{N-1}$

Let us now turn to problem (6.11) set in the half-space $\Omega' = (0, +\infty) \times \mathbb{R}^{N-1}$. This section is devoted to the proof of *Theorems 6.3* and *6.4*. A useful ingredient in the proof of these two theorems is the following result, which is the analogue of

Lemma 6.1 in the case of half-spaces. Its proof is very similar to that of *Lemma 6.1* and it is left to the reader.

Lemma 6.3 *Let N be any integer such that $N \geqslant 2$ and assume that either $f(0) > 0$, or $f(0) = 0$ and $\liminf_{s \to 0^+} f(s)/s > 0$. Let u be any nonnegative and bounded solution of (6.11), where $u_0 : \mathbb{R}^{N-1} \to \mathbb{R}_+$ is any continuous and bounded function such that $u_0 \not\equiv 0$ in $\mathbb{R}^{N-1}$. Then there exist $R > 0$ and $\epsilon > 0$ such that*

$$u \geqslant \epsilon \ \text{in} \ [R, +\infty) \times \mathbb{R}^{N-1}.$$

Proof of Theorem 6.3 Assume here that f satisfies (6.7). First of all, it follows from *Lemma 6.3* that there exists $R > 0$ such that $\inf_{[R,+\infty) \times \mathbb{R}^{N-1}} u > 0$. Call

$$M = \lim_{A \to +\infty} \left(\sup_{[A,+\infty) \times \mathbb{R}^{N-1}} u \right) > 0.$$

We shall now prove that the conclusion of *Theorem 6.3* holds with $z = M$. Choose a sequence $(x_n)_{n \in \mathbb{N}} = (x_{1,n}, \ldots, x_{N,n})_{n \in \mathbb{N}}$ in Ω' such that $x_{1,n} \to +\infty$ and $u(x_n) \to M$ as $n \to +\infty$. As in the proof of *Theorem 6.2*, it follows from *Theorem 6.5* that the functions

$$u_n(x) = u(x + x_n)$$

converge in $C^{2,\beta}_{loc}(\mathbb{R}^N)$ for all $\beta \in [0, 1)$ to the constant M, whence $f(M) = 0$.

Then, for any $\epsilon \in (0, M]$, let $R'(\epsilon) = R'_{f,M,\epsilon}$ be as in *Lemma 6.2* and let v be a solution of (6.20) with $g = f$ and $z = M$. As in the proof of *Theorem 6.2*, there exists $n_0 \in \mathbb{N}$ large enough so that $\overline{B(x_{n_0}, R'(\epsilon))} \subset \Omega'$ and (6.21) holds. The sliding method yields

$$u \geqslant v(0) \geqslant M - \epsilon \ \text{in} \ [R'(\epsilon), +\infty) \times \mathbb{R}^{N-1}.$$

Since $\epsilon > 0$ is arbitrarily small, the definition of M implies that, for all $A \in \mathbb{R}$,

$$u(x_1 + h, x_2, \ldots, x_N) \to M \ \text{as} \ h \to +\infty$$

uniformly with respect to $(x_1, x_2, \ldots, x_N) \in [A, +\infty) \times \mathbb{R}^{N-1}$. The convergence also holds in $C^{2,\beta}_b([A, +\infty) \times \mathbb{R}^{N-1})$ for all $\beta \in [0, 1)$ from standard elliptic estimates. The proof of *Theorem 6.3* is thereby complete. $\square$

In the case when f satisfies (6.3) and (6.13), the conclusion is similar to that of *Theorem 6.3, as the following proof of Theorem 6.4* will show. As a matter of fact, it is also based on a Liouville type result for the bounded nonnegative solutions of (6.17), which is the counterpart of *Theorem 6.5* under assumptions (6.3) and (6.13).

Theorem 6.7 *Let N be any integer such that $N \geqslant 1$ and assume that the function f satisfies (6.3) and (6.13). Then any bounded nonnegative solution u of (6.17) is constant.*

The proof is postponed in *Sect. 6.3* and we now complete the

Proof of Theorem 6.4 First of all, it follows from *Lemma 6.3* that there exist $\epsilon > 0$ and $R > 0$ such that $u \geqslant \epsilon$ in $[R, +\infty) \times \mathbb{R}^{N-1}$. Call now

$$m = \lim_{A \to +\infty} \left(\inf_{[A,+\infty) \times \mathbb{R}^{N-1}} u \right).$$

One has $m \in [\epsilon, +\infty)$. Let $(x_n)_{n \in \mathbb{N}} = (x_{1,n}, \ldots, x_{N,n})_{n \in \mathbb{N}}$ be a sequence in Ω' such that $x_{1,n} \to +\infty$ and $u(x_n) \to m$ as $n \to +\infty$. Up to extraction of a subsequence, the functions

$$u_n(x) = u(x + x_n)$$

converge in $C^{2,\beta}_{loc}(\mathbb{R}^N)$ for all $\beta \in [0, 1)$ to a classical bounded solution u_∞ of (6.17) such that $u_\infty \geqslant m > 0$ in $\mathbb{R}^N$ and $u_\infty(0) = m$. *Theorem 6.7* then implies that u_∞ is constant in $\mathbb{R}^N$, whence it is identically equal to m and $f(m) = 0$.

Call now

$$M' = \sup_{\Omega'} u$$

and let $g : [0, +\infty) \to \mathbb{R}$ be the function defined by

$$g(s) = \begin{cases} -f(M' + 1 - s) & \text{if } 0 \leqslant s \leqslant M' + 1 - m, \\ 0 & \text{if } s > M' + 1 - m. \end{cases}$$

The function g is Lipschitz-continuous and nonnegative. The real number

$$z = M' + 1 - m$$

is positive and fulfills $g(z) = f(m) = 0$. Choose any ϵ in $(0, z]$. From *Lemma 6.2*, there exist $R' > 0$ and a classical solution v of (6.20) in $\overline{B(0, R')}$, that is

$$\begin{cases} \Delta v + g(v) = 0 \text{ in } \overline{B(0, R')}, \\ \quad\quad 0 \leqslant v < z \text{ in } \overline{B(0, R')}, \\ \quad\quad\quad\quad v = 0 \text{ on } \partial B(0, R'), \\ v(0) = \max_{\overline{B(0,R')}} v \geqslant z - \epsilon. \end{cases}$$

The function $V = M' + 1 - v$ then satisfies

$$\begin{cases} \Delta V + f(V) = 0 \text{ in } \overline{B(0, R')}, \\ m < V \leqslant M' + 1 \text{ in } \overline{B(0, R')}, \\ \phantom{\Delta V + f(V) m < } V = M' + 1 \text{ on } \partial B(0, R'), \\ V(0) = \min_{\overline{B(0,R')}} V \leqslant m + \epsilon. \end{cases}$$

Since $u_n(x) = u(x+x_n) \to u_\infty(x) = m$ as $n \to +\infty$ in $C_{loc}^{2,\beta}(\mathbb{R}^N)$ for all $\beta \in [0, 1)$, it follows that there exists $n_0 \in \mathbb{N}$ large enough so that $\overline{B(x_{n_0}, R')} \subset \Omega'$ and

$$V(x - x_{n_0}) > u(x) \text{ for all } x \in \overline{B(x_{n_0}, R')}.$$

Since $V = \sup_{\Omega'} u + 1 > \sup_{\Omega'} u$ on $\partial B(0, R')$, it follows from the elliptic maximum principle and the sliding method that

$$u(x) \leqslant V(0) \leqslant m + \epsilon \text{ for all } x \in [R', +\infty) \times \mathbb{R}^{N-1}.$$

Owing to the definition of m, one concludes that, for all $A \in \mathbb{R}$,

$$u(x_1 + h, x_2, \dots, x_N) \to m \text{ as } h \to +\infty$$

uniformly with respect to $(x_1, x_2, \dots, x_N) \in [A, +\infty) \times \mathbb{R}^{N-1}$ and the convergence holds in $C_b^{2,\beta}([A, +\infty) \times \mathbb{R}^{N-1})$ for all $\beta \in [0, 1)$ from standard elliptic estimates. The proof of *Theorem 6.4* is thereby complete. $\qquad\qquad\square$

6.3 Classification Results in the Whole Space $\mathbb{R}^N$ or in the Half-Space $\mathbb{R}^{N-1} \times (0, +\infty)$ with Dirichlet Boundary Conditions

This section is devoted to the proof of the Liouville type results for the bounded nonnegative solutions u of problems (6.17) or (6.18). *Theorems 6.6* and *6.7* are actually corollaries of *Theorem 6.5*. We then begin with the proof of the latter.

Proof of Theorem 6.5 Let u be a bounded nonnegative solution of (6.17) under assumption (6.7). Denote

$$m = \inf_{\mathbb{R}^N} u \geqslant 0.$$

Since $f(m) \geqslant 0$, the constant m is a subsolution for (6.17). It follows from the strong elliptic maximum principle that either $u = m$ in $\mathbb{R}^N$, or $u > m$ in $\mathbb{R}^N$.

Let us prove that the second case, that is $u > m$, is impossible. That will give the desired conclusion. Assume that $u > m$ in $\mathbb{R}^N$ and let us get a contradiction. Let us first check that

$$f(m) = 0. \qquad (6.25)$$

This could be done by considering a sequence along which u converges to its minimum; after changing the origin, the limiting function would be identically equal to m from the strong maximum principle, which would yield (6.25). Let us choose an alternate elementary parabolic argument. Assume that $f(m) > 0$ and let $\xi : [0, T) \to \mathbb{R}$ be the maximal solution of

$$\begin{cases} \xi'(t) = f(\xi(t)) & \text{for all } t \in [0, T), \\ \xi(0) = m. \end{cases}$$

The maximal existence time T satisfies $0 < T \leqslant +\infty$ (and $T = +\infty$ if f is globally Lipschitz-continuous). Since $\xi(0) \leqslant u$ in $\mathbb{R}^N$, it follows from the parabolic maximum principle for the equation $v_t = \Delta v + f(v)$, satisfied by both ξ and u in $[0, T) \times \mathbb{R}^N$, that

$$\xi(t) \leqslant u(x) \text{ for all } x \in \mathbb{R}^N \text{ and } t \in [0, T).$$

But $\xi'(0) = f(\xi(0)) = f(m) > 0$. Hence, there exists $\tau \in (0, T)$ such that $\xi(\tau) > m$, whence $u(x) \geqslant \xi(\tau) > m$ for all $x \in \mathbb{R}^N$, which contradicts the definition of m.

Therefore, (6.25) holds. Now, as in the proof of *Lemma 6.1*, because of property (6.7) at $z = m$, there exist $R > 0$ and $\epsilon > 0$ such that

$$\Delta(m + \epsilon \, \varphi_R) + f(m + \epsilon \, \varphi_R) \geqslant 0 \text{ in } B(0, R)$$

and $m + \epsilon \, \varphi_R < u$ in $\overline{B(0, R)}$, where φ_R solving (6.15) is the principal eigenfunction of the Dirichlet-Laplace operator in $B(0, R)$. Since $m + \epsilon \, \varphi_R = m$ on $\partial B(0, R)$ and $u > m$ in $\mathbb{R}^N$, the same sliding method as in the proof of *Lemma 6.1* implies that

$$m + \epsilon \, \varphi_R(x - y) \leqslant u(x) \text{ for all } x \in \overline{B(y, R)}$$

and for all $y \in \mathbb{R}^N$. Therefore, $u \geqslant m + \epsilon \, \varphi_R(0) > m$ in $\mathbb{R}^N$, which contradicts the definition of m.

As a conclusion, the assumption $u > m$ is impossible and, as already emphasized, the proof of *Theorem 6.5* is thereby complete. □

The proof of *Theorem 6.6* also uses the sliding method and *Theorem 6.5*, combined with limiting arguments as $x_N \to +\infty$ and comparison with non-small subsolutions.

Proof of Theorem 6.6 Let u be a bounded nonnegative solution of (6.18) under assumption (6.7). Since $f(0) \geqslant 0$, it follows from the strong maximum principle that either $u = 0$ in $\mathbb{R}^{N-1} \times \mathbb{R}_+$, or $u > 0$ in $\mathbb{R}^{N-1} \times (0, +\infty)$. Let us then consider the second case. Since $f(0) \geqslant 0$, it follows from Corollary 1.3 of Berestycki, Caffarelli and Nirenberg [21] that u is increasing in x_N. Denote

$$M = \sup_{\mathbb{R}^{N-1} \times [0, +\infty)} u.$$

There exists a sequence $(x_n)_{n \in \mathbb{N}} = (x_{1,n}, \dots, x_{N,n})_{n \in \mathbb{N}}$ in $\mathbb{R}^{N-1} \times \mathbb{R}_+$ such that $x_{N,n} \to +\infty$ and $u(x_n) \to M$ as $n \to +\infty$. From standard elliptic estimates, the functions

$$u_n(x) = u(x + x_n),$$

which satisfy the same equation as u, converge in $C_{loc}^{2,\beta}(\mathbb{R}^N)$ for all $\beta \in [0, 1)$, up to extraction of a subsequence, to a solution u_∞ of (6.17) such that $u_\infty(0) = M$. Theorem 6.5 implies that

$$u_\infty = M \text{ in } \mathbb{R}^N,$$

whence $f(M) = 0$. It follows then from Berestycki, Caffarelli and Nirenberg [14] (see also Theorem 1.4 in [21]) that u depends on x_N only. In other words, the function $u(x)$ is equal to $V_M(x_N)$, which completes the proof of *Theorem 6.6*. □

Let us complete this section with the

Proof of Theorem 6.7 Assume that the function f satisfies (6.3) and (6.13) and let u be a bounded nonnegative solution of (6.17). As already underlined, it first follows from the strong maximum principle that either $u \equiv 0$ in $\mathbb{R}^N$, or $u > 0$ in $\mathbb{R}^N$. Let us then consider the second case. Applying the sliding method and using the same notations as in the proof of *Lemma 6.1*, one gets the existence of $R > 0$ and $\epsilon > 0$ such that $u(x) \geqslant \epsilon \varphi_R(x - y)$ for all $x \in \overline{B(y, R)}$ and for all $y \in \mathbb{R}^N$, whence

$$m = \inf_{\mathbb{R}^N} u > 0.$$

Let $(x_n)_{n \in \mathbb{N}}$ be a sequence in $\mathbb{R}^N$ such that $u(x_n) \to m$ as $n \to +\infty$. Up to extraction of a subsequence, the functions

$$u_n(x) = u(x + x_n)$$

converge in $C_{loc}^{2,\beta}(\mathbb{R}^N)$ for all $\beta \in [0, 1)$ to a classical bounded solution u_∞ of (6.17) in $\mathbb{R}^N$ such that $u_\infty \geqslant m$ in $\mathbb{R}^N$ and $u_\infty(0) = m$. Hence,

$$f(m) \leqslant 0,$$

whence $m \geqslant \mu$ from (6.3) and since $m > 0$.

Call now

$$M = \sup_{\mathbb{R}^N} u.$$

If $M = m$, then u is constant, which is the desired result. Assume now that $M > m$. The function

$$v = M - u$$

is a nonnegative bounded solution of

$$\Delta v + g(v) = 0 \ \text{ in } \mathbb{R}^N,$$

where the function $g : [0, +\infty) \to \mathbb{R}$ is defined by

$$g(s) = \begin{cases} -f(M - s) \ \text{if } 0 \leqslant s \leqslant M - m, \\ -f(m) \quad\ \ \text{if } s > M - m. \end{cases}$$

Because of (6.3) and (6.13), the function g is Lipschitz-continuous and fulfills property (6.7). *Theorem 6.5* applied to g and v implies that the function v is actually constant. Hence, u is also constant, which actually shows that the assumption $M > m$ is impossible. As a conclusion, $M = m$ and u is then constant. $\quad\square$

6.4   The Dynamical Systems' Approach

The goal of this section is to apply the dynamical systems' (shortly DS) approach to study the symmetrization and stabilization (as $x_1 \to +\infty$) properties of the nonnegative solutions (6.1) in

$$\Omega = (0, +\infty) \times \mathbb{R}^{N-2} \times (0, +\infty)$$

and (6.11) in

$$\Omega' = (0, +\infty) \times \mathbb{R}^{N-1}.$$

To this end we apply as aforementioned the DS approach which we developed in *Chaps. 2–5*. One of the main difficulties which arises in the dynamical study of (6.1) in Ω or (6.11) in Ω' is the fact that the corresponding Cauchy problem is not well posed for (6.1) in Ω and for (6.11) in Ω', and consequently the straightforward interpretation of (6.1) and (6.11) as an evolution equation leads to semigroups of multivalued maps even in the case of cylindrical domains, see [8]. The usage of multivalued maps can be overcome using the so-called trajectory dynamical approach (see *Chaps.* 2–5 and [12, 35, 92] and the references therein). Under this approach, one fixes a signed direction $\vec{l}$ in $\mathbb{R}^N$, which will play role of time. Then the space K^+ of all bounded nonnegative classical solutions of (6.1) in Ω or (6.11) in Ω' (in the sense described in *Sect. 6.1*) is considered as a trajectory phase space for the semi-flow $(T_h^{\vec{l}})_{h \in \mathbb{R}_+}$ of translations along the direction $\vec{l}$ defined via

$$\left(T_h^{\vec{l}} u\right)(x) = u(x + h\vec{l}), \quad h \in \mathbb{R}_+, \ u \in K^+.$$

As already mentioned in a previous chapter, the trajectory dynamical system $\left(T_h^{\vec{l}}, K^+\right)$ to be well defined, one needs the domains Ω and Ω' to be invariant with respect to positive translations along the $\vec{l}$ directions, that is

$$T_h^{\vec{l}}(\Omega) \subset \Omega \ \left(\text{resp. } T_h^{\vec{l}}(\Omega') \subset \Omega'\right), \quad T_h^{\vec{l}}(x) := x + h\vec{l},$$

for all $h \geqslant 0$. In our case, the x_1-axis will play the role of time, that is $\vec{l} = (1, 0, \ldots 0, 0)$. For the sake of simplicity of the notation, we then set

$$T_h^{\vec{l}} = T_h.$$

To apply the DS approach for our purposes, we apply the following *Lemma 6.4* (see below), which also has an independent interest. For that purpose, let us introduce a few more notations. For any locally Lipschitz-continuous function f from $\mathbb{R}_+$ to $\mathbb{R}$, such that $f(0) \geqslant 0$, let Z_f be defined by

$$Z_f = \{z_0 \in \mathbb{R}_+ \mid f(z_0) = 0 \text{ and } F(z) < F(z_0) \text{ for all } z \in [0, z_0)\}, \tag{6.26}$$

where

$$F(z) = \int_0^z f(\sigma)d\sigma.$$

The set Z_f is then a subset of the set E of zeroes of f, defined in (6.8). Lastly, by R_f we denote the set of all bounded, nonnegative solutions $V \in C^2(\mathbb{R}_+)$ of

$$\begin{cases} V''(\xi) + f(V(\xi)) = 0 \text{ for all } \xi \geqslant 0, \\ V(0) = 0, \quad V \geqslant 0, \quad V \text{ is bounded.} \end{cases} \tag{6.27}$$

Lemma 6.4 *Let f be a locally Lipschitz-continuous function from $\mathbb{R}_+$ to $\mathbb{R}$, such that $f(0) \geqslant 0$. Then the set R_f is homeomorphic to Z_f and as a consequence is totally disconnected.*

The proof of *Lemma 6.4* is identical to the proof of *Proposition 3.2*. We omit it.

Below we state the main results of this *Sect. 6.4*, that is *Theorems 6.8–6.11*, which are obtained by the dynamical systems' approach. To this end we define a class of functions K^+ to which the solutions of (6.1) as well as (6.11) belong to. Namely, a bounded nonnegative solution of (6.1) (resp. (6.11)) is understood to be a solution u of class $C^2(\Omega)$ (resp. $C^2(\Omega')$) and continuous on $\overline{\Omega} \setminus \{0\} \times \mathbb{R}^{N-2} \times \{0\}$ (resp. on $\overline{\Omega'}$). The set K^+ is endowed with the local topology according to the embedding of K^+ in $C^{2,\beta}_{loc}(\overline{\Omega} \setminus \{x_1 = 0\})$ (resp. $C^{2,\beta}_{loc}(\overline{\Omega'} \setminus \{x_1 = 0\})$) for all $\beta \in [0, 1)$. Actually, as already emphasized in *Sect. 6.1*, all solutions $u \in K^+$ are automatically in $C^{2,\beta}_b([\epsilon, +\infty) \times \mathbb{R}^{N-2} \times [0, +\infty))$ (resp. $C^{2,\beta}_b([\epsilon, +\infty) \times \mathbb{R}^{N-1}))$ for all $\beta \in [0, 1)$ and for all $\epsilon > 0$, where we refer to (6.2) for the definition of the sets $C^{2,\beta}_b(F)$.

The first two theorems are concerned with the case of functions f fulfilling the condition (6.3).

Theorem 6.8 *Let N be any integer such that $N \geqslant 2$ and let f be a locally Lipschitz-continuous function from $\mathbb{R}_+$ to $\mathbb{R}$, satisfying (6.3). Then the trajectory dynamical system (T_h, K^+) associated to (6.1) possesses a global attractor A_{tr} in K^+ which is bounded in $C^{2,\beta}_b(\overline{\Omega})$ and then compact in $C^{2,\beta}_{loc}(\overline{\Omega})$ for all $\beta \in [0, 1)$. Moreover A_{tr} has the following structure*

$$A_{tr} = \Pi_{\overline{\Omega}} K^+(\widetilde{\Omega})$$

where $\widetilde{\Omega} = \mathbb{R}^N_+$, $K^+(\widetilde{\Omega})$ is the set of all bounded nonnegative solutions of (6.18) in $\widetilde{\Omega} = \mathbb{R}^N_+$, and $\Pi_{\overline{\Omega}}$ denotes the restriction to $\overline{\Omega}$. Hence,

$$A_{tr} \subset \{x \mapsto 0, \ x \mapsto V(x_N)\}$$

and $A_{tr} = \{x \mapsto V(x_N)\}$ if $f(0) > 0$, where V is the unique solution of (6.4). Lastly, for any bounded nonnegative solution u of (6.1) in Ω, the functions $T_h u$ converge as $h \to +\infty$ in $C^{2,\beta}_{loc}(\overline{\Omega})$ for all $\beta \in [0, 1)$ either to 0 or to $x \mapsto V(x_N)$, and they do converge to the function $x \mapsto V(x_N)$ if $f(0) > 0$.

The next theorem deals with the analysis of the asymptotic behavior as $x_1 \to +\infty$ of the nonnegative bounded classical solutions $u \in K^+$ of Eq. (6.11) in the half-space $\Omega' = (0, +\infty) \times \mathbb{R}^{N-1}$. For any $M \geqslant 0$, we define

$$K^+_M = K^+ \cap \{0 \leqslant u \leqslant M\}.$$

Theorem 6.9 *Let N be any integer such that $N \geqslant 2$ and assume that, in addition to (6.3), the given locally Lipschitz-continuous function f from $\mathbb{R}_+$ to $\mathbb{R}$ satisfies (6.13). Then, for every $M \geqslant \mu$, the trajectory dynamical system (T_h, K^+_M) associated to (6.11) possesses a global attractor A_{tr}, which is bounded in $C^{2,\beta}_b(\overline{\Omega'})$ and then compact in $C^{2,\beta}_{loc}(\overline{\Omega'})$ for all $\beta \in [0, 1)$, and satisfies*

$$A_{tr} = \Pi_{\overline{\Omega'}} K_M^+ (\widetilde{\Omega'}) \tag{6.28}$$

where $\widetilde{\Omega'} = \mathbb{R}^N$, $K_M^+ (\widetilde{\Omega'}) = K^+ (\widetilde{\Omega'}) \cap \{0 \leqslant u \leqslant M\}$, $K^+ (\widetilde{\Omega'})$ is the set of all bounded nonnegative solutions of (6.17) in $\widetilde{\Omega'} = \mathbb{R}^N$, and $\Pi_{\overline{\Omega'}}$ denotes the restriction to $\overline{\Omega'}$. Hence,

$$A_{tr} = \{z \in [0, M] \mid f(z) = 0\} = E \cap [0, M].$$

Lastly, for any bounded nonnegative solution u of (6.11) in Ω' such that $0 \leqslant u \leqslant M$, the functions $T_h u$ converge as $h \rightarrow +\infty$ in $C_{loc}^{2,\beta}(\overline{\Omega'})$ for all $\beta \in [0, 1)$ to some $z \in E \cap [0, M]$ which is uniquely defined by u.

The next two theorems, which are concerned with the case of functions f fulfilling the condition (6.7), are based on the new Liouville type *Theorems 6.5* and *6.6* which were already stated in *Sect. 6.2*.

Theorem 6.10 *Let N be any integer such that $N \geqslant 2$, let f be any locally Lipschitz-continuous function from $\mathbb{R}_+$ to $\mathbb{R}$ satisfying (6.7) and assume that, for problem (6.1) in the quarter-space Ω, the set K^+ is not empty. Then, for every sufficiently large $M \geqslant 0$, the trajectory dynamical system (T_h, K_M^+) associated to (6.1) possesses a global attractor A_{tr}, which is bounded in $C_b^{2,\beta}(\overline{\Omega})$ and then compact in $C_{loc}^{2,\beta}(\overline{\Omega})$ for all $\beta \in [0, 1)$ and has the following structure*

$$A_{tr} = \Pi_{\overline{\Omega}} K_M^+ (\widetilde{\Omega}) \tag{6.29}$$

where $K_M^+ (\widetilde{\Omega}) = K^+ (\widetilde{\Omega}) \cap \{0 \leqslant u \leqslant M\}$. Hence,

$$A_{tr} = \{x \mapsto V_z(x_N) \mid z \in [0, M], \ f(z) = 0\}.$$

Lastly, for any bounded nonnegative solution u of (6.1) in Ω such that $0 \leqslant u \leqslant M$, the functions $T_h u$ converge as $h \rightarrow +\infty$ in $C_{loc}^{2,\beta}(\overline{\Omega})$ for all $\beta \in [0, 1)$ to some function $x \mapsto V_z(x_N)$, where $z \in E \cap [0, M]$ is uniquely defined by u and E denotes the set of zeroes of the function f.

Analogously to *Theorem 6.9* we have the following *Theorem 6.11* in the case of the half-space $\Omega' = (0, +\infty) \times \mathbb{R}^{N-1}$.

Theorem 6.11 *Let N be any integer such that $N \geqslant 2$, let f be any locally Lipschitz-continuous function from $\mathbb{R}_+$ to $\mathbb{R}$ satisfying (6.7), and assume that, for problem (6.11) in the half-space Ω', the set K^+ is not empty. Then, for every sufficiently large $M \geqslant 0$, the trajectory dynamical system (T_h, K_M^+) associated to (6.11) possesses a global attractor A_{tr}, which is bounded in $C_b^{2,\beta}(\overline{\Omega'})$ and then compact in $C_{loc}^{2,\beta}(\overline{\Omega'})$ for all $\beta \in [0, 1)$, and satisfies (6.28). Lastly, for any bounded nonnegative solution u of (6.11) in Ω' such that $0 \leqslant u \leqslant M$, the functions $T_h u$*

converge as $h \to +\infty$ *in* $C_{loc}^{2,\beta}(\overline{\Omega'})$ *for all* $\beta \in [0, 1)$ *to some* $z \in E \cap [0, M]$ *which is uniquely defined by* u.

In what follows we prove *Theorem 6.8*. A proof of *Theorems 6.9–6.11* can be done in the same manner as in *Theorem 6.8* with some minor modifications (see *Remark 6.4*).

Proof of Theorem 6.8 Let K^+ be the set of all bounded nonnegative solutions of (6.1) in Ω. Due to the assumptions (6.3), the set K^+ is not empty (the function $x \mapsto V(x_N)$, where V is the unique solution of (6.4), belongs to K^+) and due to the translation invariance of (6.1), it follows that $T_h : K^+ \to K^+$ is well defined for all $h \geqslant 0$, where $(T_h u)(x_1, x', x_N) := u(x_1 + h, x', x_N)$.

To show that (T_h, K^+) possesses a global attractor, it suffices to show (see [8] and the references therein) that

- for any fixed $h > 0$, T_h is a continuous map in K^+ (we recall that K^+ is endowed with local topology according to the embedding of K^+ in $C_{loc}^{2,\beta}(\overline{\Omega}\backslash\{x_1 = 0\})$) for all $\beta \in [0, 1)$;
- the semi-flow $(T_h)_{h \geqslant 0}$ possesses a compact attracting (absorbing) set in $C_{loc}^{2,\beta}(\overline{\Omega})$, which is even bounded in $C_b^{2,\beta}(\overline{\Omega})$, for all $\beta \in [0, 1)$.

Note that, the continuity of T_h in $C_{loc}^{2,\beta}(\overline{\Omega}\backslash\{x_1 = 0\})$ is obvious, because the shift operator is continuous in this topology, as well as its restriction to K^+. As for the existence of compact attracting (absorbing) set for the semi-flow $(T_h)_{h \geqslant 0}$, it follows from the fact that the set of all bounded nonnegative solutions of (6.18) in $\cup_{h \geqslant 0} T_{-h}(\overline{\Omega}) = \widetilde{\Omega} = \mathbb{R}_+^N$ under the assumption (6.3) on f is uniformly bounded. Indeed, as already recalled in *Sect. 6.2.1* and according to a result of [13], under the assumption (6.3), any bounded solution of (6.18) in $\mathbb{R}_+^N$ which is positive in $\mathbb{R}^{N-1} \times (0, +\infty)$ has one-dimensional symmetry, that is

$$u(x_1, x', x_N) = V(x_N)$$

where $0 \leqslant V < \mu$ is the unique solution of (6.4). On the other hand, since $f(0) \geqslant 0$, any bounded nonnegative solution of (6.18) is either positive in $\mathbb{R}^{N-1} \times (0, +\infty)$, or identically 0 in $\mathbb{R}_+^N$, and it cannot be 0 if $f(0) > 0$. Thus, the set $K^+(\widetilde{\Omega})$ of all bounded nonnegative solutions of (6.18) in $\widetilde{\Omega} = \mathbb{R}_+^N$ is bounded in $L^\infty(\widetilde{\Omega})$, namely

$$\sup_{u \in K^+(\widetilde{\Omega})} \|u\|_\infty \leqslant \mu.$$

Then the existence of a compact absorbing set for $(T_h)_{h \geqslant 0}$ in $C_{loc}^{2,\beta}(\overline{\Omega})$, which is even bounded in $C_b^{2,\beta}(\overline{\Omega})$, for all $\beta \in [0, 1)$ is a consequence of the uniform boundedness of all solutions of (6.18) in $\widetilde{\Omega} = \mathbb{R}_+^N$ and of standard elliptic estimates. Hence the semigroup (T_h, K^+) possesses a global attractor A_{tr} in K^+ which is bounded in $C_b^{2,\beta}(\overline{\Omega})$ and compact in $C_{loc}^{2,\beta}(\overline{\Omega})$ for all $\beta \in [0, 1)$.

To prove the convergence part of *Theorem 6.8*, as we will see below, it is sufficient to show that $A_{tr} = \Pi_{\overline{\Omega}} K^+(\tilde{\Omega})$. Assuming for a moment that this representation is true, we obtain from the previous considerations that

$$A_{tr} \subset \{x \mapsto 0, \; x \mapsto V(x_N)\}, \qquad (6.30)$$

and A_{tr} is then equal to the singleton $\{x \mapsto V(x_N)\}$ if $f(0) > 0$. Hence, for any bounded nonnegative solution u of (6.1), since $\{T_h u, \, h \geqslant 1\}$ is bounded in $C_b^{2,\beta}(\overline{\Omega})$ and compact in $C_{loc}^{2,\beta}(\overline{\Omega})$ for all $\beta \in [0, 1)$, the ω-limit set $\omega(u)$ of u is not empty and it is an invariant and connected subset of A_{tr}. Since A_{tr} is totally disconnected,[2] it follows that either $\omega(u) = \{x \mapsto 0\}$ or $\omega(u) = \{x \mapsto V(x_N)\}$, the latter being necessarily true if $f(0) > 0$.

To complete the proof of *Theorem 6.8*, it remains to show that $A_{tr} = \Pi_{\overline{\Omega}} K^+(\tilde{\Omega})$. First we prove that $\Pi_{\overline{\Omega}} \hat{u} \in A_{tr}$ for any bounded nonnegative solution $\hat{u}$ of (6.18) in $\tilde{\Omega} = \mathbb{R}_+^N$, that is $\hat{u} \in K^+(\tilde{\Omega})$. Indeed, for such a $\hat{u}$, the family $(\Pi_{\overline{\Omega}}(T_{-h}\hat{u}))_{h \geqslant 0}$ is uniformly bounded in $C_b^{2,\beta}(\overline{\Omega})$ and compact in $C_{loc}^{2,\beta}(\overline{\Omega})$ for all $\beta \in [0, 1)$ and, according to definition of the attractor, there holds

$$T_h \Pi_{\overline{\Omega}}(T_{-h}\hat{u}) \longrightarrow A_{tr} \text{ in } C_{loc}^{2,\beta}(\overline{\Omega}) \text{ as } h \to +\infty,$$

for all $\beta \in [0, 1)$. On the other hand, $T_h \Pi_{\overline{\Omega}}(T_{-h}\hat{u}) = \Pi_{\overline{\Omega}}\hat{u}$. Hence $\Pi_{\overline{\Omega}}\hat{u} \in A_{tr}$.

Next we prove the reverse inclusion. To this end, let us recall that (T_h, K^+) possesses an absorbing set which is bounded in $C_b^{2,\beta}(\overline{\Omega})$ and then compact in $C_{loc}^{2,\beta}(\overline{\Omega})$ for all $\beta \in [0, 1)$, say $\mathbb{B}_* \subset K^+$, and, as a consequence,

$$A_{tr} = \omega(\mathbb{B}_*) = \bigcap_{h \geqslant 0} \left[\bigcup_{s \geqslant h} T_s \mathbb{B}_* \right],$$

where [ ] means the closure in $C_{loc}^{2,\beta}(\overline{\Omega})$ (see [8, 35] and the references therein). Let now $u \in A_{tr}$. The property $A_{tr} = \omega(\mathbb{B}_*)$ implies that there exist an increasing sequence $(h_k)_{k \in \mathbb{N}} \to +\infty$ and a sequence of solutions $(u_k)_{k \in \mathbb{N}}$ in $\mathbb{B}_*$, such that

$$u = \lim_{k \to +\infty} T_{h_k} u_k \qquad (6.31)$$

in $C_{loc}^{2,\beta}(\overline{\Omega})$ for all $\beta \in [0, 1)$. Note that the solution $T_{h_k} u_k$ is defined not only in Ω, but also in the domain $(-h_k, +\infty) \times \mathbb{R}^{N-2} \times \mathbb{R}_+$, and that

---

[2]This property is obvious here due to (6.30). See *Remark 6.4* for a comment about the other situations, corresponding to *Theorems 6.9, 6.10* and *6.11*.

$$\sup_{k \in \mathbb{N}} \|T_{h_k} u_k\|_{C_b^{2,\beta}\left([-h_k+\epsilon,\infty)\times\mathbb{R}^{N-2}\times\mathbb{R}_+\right)} < +\infty \tag{6.32}$$

for all $\epsilon > 0$ and $\beta \in [0,1)$, from standard elliptic estimates. Consequently, for every $k_0 \in \mathbb{N}$ and $\beta \in [0,1)$, the sequence $(T_{h_k} u_k)_{k>k_0}$ is precompact in $C_{loc}^{2,\beta}\left([-h_{k_0},\infty)\times\mathbb{R}^{N-2}\times\mathbb{R}_+\right)$. Taking a subsequence, if necessary, and using Cantor's diagonal procedure and the fact $h_k \to \infty$, we can say that this sequence converges to $\hat{u} \in C_{loc}^{2,\beta}(\mathbb{R}_+^N)$ in the spaces $C_{loc}^{2,\beta}\left([-h_{k_0},\infty)\times\mathbb{R}^{N-2}\times\mathbb{R}_+\right)$ for every $k_0 \in \mathbb{N}$ and for every $\beta \in [0,1)$. Then (6.32) implies that $\hat{u} \in C_b^{2,\beta}(\mathbb{R}_+^N)$ for every $\beta \in [0,1)$. Lastly, the functions $T_{h_k} u_k$ are nonnegative solutions of (6.1) in $(-h_k,+\infty)\times\mathbb{R}^{N-2}\times\mathbb{R}_+$ and by letting $k \to +\infty$, we easily obtain that $\hat{u}$ is a bounded nonnegative solution of (6.18) in $\tilde{\Omega} = \mathbb{R}_+^N$. Finally, formula (6.31) implies that

$$\Pi_{\overline{\Omega}}\hat{u} = u.$$

Thus $u \in \Pi_{\overline{\Omega}} K^+(\tilde{\Omega})$ and the representation formula $A_{tr} = \Pi_{\overline{\Omega}} K^+(\tilde{\Omega})$ is proved. The proof of *Theorem 6.8* is thereby complete.   □

Remark 6.3 As far as *Theorems 6.1* and *6.2* on the one hand, and *Theorems 6.8* and *6.10* on the other hand, are concerned, we especially emphasize that the DS approach simplified in a very elegant way most of the computations regarding the asymptotic behavior of the solutions of (6.1) as $x_1 \to +\infty$. However, the DS approach and *Theorem 6.8* (resp. *Theorem 6.10*) do not provide as in PDE approach the fact that only the limiting profile $V(x_N)$ (resp. $V_z(x_N)$ for some $z \in E\backslash\{0\}$) is selected, even if $f(0) = 0$, as soon as $u_0 \not\equiv 0$ on $\{0\} \times \mathbb{R}^{N-2} \times (0,+\infty)$. In the PDE proof of *Theorems 6.1* and *6.2*, it is indeed shown that the condition $u_0 \not\equiv 0$ implies that u is separated from 0 for large enough x_1 and x_N. This property is not shown in the DS proof of *Theorems 6.8* and *6.10*. Similar comments also hold for *Theorems 6.3, 6.4, 6.9* and *6.11*, where the PDE proof provides the convergence to a *non-zero* zero of f, what the DS proof does not. Lastly, in some of the results obtained through the PDE approach, the convergence of the solutions as $x_1 \to +\infty$ is proved to be uniform with respect to the variables (x', x_N), while the DS approach only provides local convergence, due to the necessity of using the local topology to get the existence of a global attractor.

Remark 6.4 In *Theorem 6.10* (resp. *Theorem 6.11*) under assumption (6.7), the phase space K_M^+, which is invariant under the semigroup T_h, is not empty for any sufficiently large M, because K^+ is assumed to be not empty (one can take M as any nonnegative real number such that $M \geq \|U\|_\infty$, where U is any fixed element in K^+). Then, in the same manner as in the proof of *Theorem 6.8*, using both the representation formula $A_{tr} = \Pi_{\overline{\Omega}} K_M^+(\tilde{\Omega})$ (resp. $A_{tr} = \Pi_{\overline{\Omega'}} K_M^+(\mathbb{R}^N)$), the Liouville theorems of *Sect. 6.2* and *Lemma 6.4*, one obtains the desired conclusions. In particular, for *Theorem 6.10* (resp. *Theorem 6.11*) about problem (6.1) in Ω

(resp. (6.11) in Ω'), the total disconnectedness of A_{tr} follows from the representation formula (6.29) (resp. (6.28)), from *Theorem 6.6* (resp. *Theorem 6.5*) and from *Lemma 6.4* with, here, $Z_f = E$ (resp. condition (6.7) again). Note that for *Theorem 6.9* in the case of assumptions (6.3) and (6.13), the total disconnectedness of A_{tr} follows from *Theorem 6.7* and assumption (6.13) again.

Remark 6.5 Note that neither the concrete choice of the domain Ω (or Ω') nor the concrete choice of the "time" direction x_1 are essential for the use of the trajectory dynamical system' approach. Indeed, let us replace the "time" direction x_1 by any fixed direction $\vec{l} \in \mathbb{R}^N$ and correspondingly $T_h^{\vec{l}} u = u(\cdot + h\vec{l})$ for $h \in \mathbb{R}_+$ and $u \in K^+$. Then the above construction seems to be applicable if the domain Ω satisfies the following assumptions:

- $T_h\Omega \subset \Omega$ (this is necessary in order to define the restriction of T_h to the trajectory phase space K^+ or K_M^+).
- $\cup_{h \geqslant 0} T_{-h}\overline{\Omega} = \mathbb{R}_+^N$, or $\mathbb{R}^N$ (this is required in order to obtain representation formulas of the type $A_{tr} = \Pi_{\overline{\Omega}} K^+(\mathbb{R}_+^N)$ or $A_{tr} = \Pi_{\overline{\Omega}} K^+(\mathbb{R}^N)$, with possibly K_M^+ instead of K^+).

Chapter 7
The Case of p-Laplacian Operator

7.1 Introduction

We are interested in quasilinear elliptic problems over a half-space of the form

$$\begin{cases} \Delta_p u + f(u) = 0 \text{ in } \mathbb{R}_+ \times \mathbb{R}^{N-1}, \\ u(0, x_2, \dots, x_N) = u_0(x_2, \dots, x_N), \end{cases} \tag{7.1}$$

and similar problems over a quarter-space

$$\begin{cases} \Delta_p u + f(u) = 0 \text{ in } \mathbb{R}_+ \times \mathbb{R}^{N-2} \times \mathbb{R}_+, \\ u(0, x_2, \dots, x_N) = u_0(x_2, \dots, x_N), \\ u(x_1, x_2, \dots, x_{N-1}, 0) = 0. \end{cases} \tag{7.2}$$

Here $u_0 \geq 0$ may be regarded as a given bounded continuous function, $\mathbb{R}_+ = [0, +\infty)$, and $\Delta_p u = div(|\nabla u|^{p-2}\nabla u)$ is the usual p-Laplacian operator, and we always assume that $p > 1$. If $u \not\equiv 0$ is a bounded nonnegative solution to either (7.1) or (7.2), we want to understand the behavior of $u(x_1, x_2, \dots, x_N)$ as $x_1 \to \infty$. For some general classes of nonlinearities f, we show that, in the half-space case, $\lim_{x_1 \to \infty} u(x_1, x_2, \dots, x_N)$ always exists and is a positive zero of f; and in the quarter-space case,

$$\lim_{x_1 \to \infty} u(x_1, x_2, \dots, x_N) = V(x_N),$$

where V is a solution of the one-dimensional problem

$$\Delta_p V + f(V) = 0 \text{ in } \mathbb{R}_+, \quad V(0) = 0, \quad V(t) > 0 \text{ for } t > 0, \quad V(+\infty) = z, \tag{7.3}$$

© Springer Nature Switzerland AG 2018
M. Efendiev, *Symmetrization and Stabilization of Solutions
of Nonlinear Elliptic Equations*, Fields Institute Monographs 36,
https://doi.org/10.1007/978-3-319-98407-0_7

where z is a positive zero of f. These features of (7.1) and (7.2) were previously investigated and shown in *Chap. 6* (see also [51]) for the special case $p = 2$. The nonlinearities f considered in *Chap. 6* are mainly of two types. The first type consists of functions which are nonnegative (see (6.7) in *Chap. 6* for details), and the second type are sign-changing functions satisfying $f(s) > 0$ in $(0, a)$ and $f(s) \leqslant 0$ in $(a, +\infty)$ for some constant $a > 0$ (see (6.2) in *Chap. 6*). All the functions considered in *Chap. 6* are locally Lipschitz continuous. Among other techniques, the arguments in *Chap. 6* rely on various forms of the comparison principles, and in particular, the strong comparison principle plays a key role. These rely crucially on the assumption that $p = 2$. In this chapter, we show that most of the results of *Chap. 6* continue to hold for the corresponding general p-Laplacian problems. Since there is no strong comparison principle in general for the p-Laplacian problem when $p \neq 2$, we have to take rather different approaches in many key steps. A crucial ingredient here is a simple weak sweeping principle, which is a consequence of the weak comparison principle for the p-Laplacian, and is a variant of Serrin's famous sweeping principle for the Laplacian equations. We will show that in many situations, it is possible to use the weak sweeping principle to replace the moving plane or sliding method, which are based on the strong comparison principle for the Laplacian case and frequently used in *Chap. 6*. The weak sweeping principle was used, for example, in [38, 41] and [42] for related problems. The techniques in these papers are further developed here to treat (7.1) and (7.2). As in *Chap. 6*, to obtain a good understanding of the asymptotic behavior of the solutions to (7.1) and (7.2), we need a thorough classification of the solutions of (7.3), and also some Liouville type results over the entire $\mathbb{R}^N$. These preparations will be done in *Sect. 7.2*. The results in this section are mostly of interests on their own. The Liouville theorem here (see *Theorem 7.5*) improves the corresponding result of *Chap. 6* (see *Remark 7.3* for details) when $p = 2$. For convenience of the reader, we also present a version of the weak sweeping principle in this section in a form that is easy to use. In *Sect. 7.3*, we make use of the results in *Sect. 7.2* to study (7.1) and (7.2). The main results for the half-space problem are *Theorems 7.6* and *7.7*, which imply, respectively, *Theorems 6.3* and *6.4* of *Chap. 6* in the special case $p = 2$, except that we require the extra condition that the zeros of f are isolated, but on the other hand, our conditions on f near its zeros are less restrictive then those in *Chap. 6* due to our better Liouville theorem. Our main results for the quarter-space problem are *Theorems 7.10* and *7.11*. Our *Theorem 7.11* extends *Theorem 6.1* of *Chap. 6*. In proving these results, apart from the techniques developed for treating the half-space problems, we also need two one-dimensional symmetry results for half-space problems of the form

$$\Delta_p u + f(u) = 0 \text{ in } \mathbb{R}^{N-1} \times \mathbb{R}_+, \ u = 0 \text{ on } \{x_N = 0\}. \tag{7.4}$$

For sign-changing nonlinearities considered in *Theorem 7.11*, the one-dimensional symmetry of positive solutions of (7.4) follows from a modification of the arguments used in [42] (see *Theorem 7.9* below). However, for nonnegative nonlinearities considered in *Theorem 7.10*, we are able to prove such symmetry only under the extra assumption that $\partial_{x_N} u \geqslant 0$ (see *Theorem 7.8* and *Remark 7.4*). As a consequence our *Theorem 7.10* also requires the solution u to satisfy $\partial_{x_N} u \geqslant 0$. Such a monotonicity condition for u is not needed in *Chap. 6* for the case $p = 2$. We believe that here this monotonicity condition for u is also unnecessary. If the half-space $\{x \in \mathbb{R}^N : x_1 > 0\}$ in (7.1) is replaced by an unbounded Lipschitz domain of the form $\{x \in \mathbb{R}^N : x_1 > \phi(x_2, \ldots, x_N)\}$ as in [13], then our techniques can be extended to obtain analogous results, except for those on one-dimensional symmetry in *Sect. 7.2*. Similarly, our results on (7.2) can be extended when the quarter-space $\{x \in \mathbb{R}^N : x_1 > 0, \ x_N > 0\}$ is replaced by $\{x \in \mathbb{R}^N : x_1 > \phi(x_2, \ldots, x_N), \ x_N > 0\}$, with ϕ a Lipschitz map. We note that for this case, the limiting problem as $x_1 \to \infty$ is again a half space problem over $\{x \in \mathbb{R}^N : x_N > 0\}$.

7.2 Some Basic Results

In this section, we present some basic results which will be needed in our investigation of the half- and quarter-space problems. Most of the results here are also of independent interests.

7.2.1 The Weak Sweeping Principle

Several key steps in our arguments are based on a weak sweeping principle for p-Laplacian equations, which is a variant of Serrin's sweeping principle for the Laplacian case (see [94]). Such a weak sweeping principle was used before in [38, 41] and [42]. Here we state it in a form that is convenient to use; its proof is almost identical to that of Lemma 2.7 in [38].

Proposition 7.1 *Suppose that $\mathcal{D}$ is a bounded smooth domain in $\mathbb{R}^N$, $h(x, s)$ is measurable in $x \in \mathcal{D}$, continuous in s, and for each finite interval J, there exists a continuous increasing function $L(s)$ such that $h(x, s) + L(s)$ is nondecreasing in s for $s \in J$ and $x \in \mathcal{D}$. Let u_t and v_t, $t \in [t_1, t_2]$, be functions in $W^{1,p}(\mathcal{D}) \cap C(\overline{\mathcal{D}})$ and satisfy in the weak sense,*

$$-\Delta_p u_t \geqslant h(x, u_t) + \epsilon_1(t), \quad -\Delta_p v_t \leqslant h(x, v_t) - \epsilon_2(t) \text{ in } \mathcal{D}, \ \forall t \in [t_1, t_2],$$

$$u_t \geqslant v_t + \epsilon \text{ on } \partial\mathcal{D}, \ \forall t \in [t_1, t_2],$$

where

$$\epsilon_1(t) + \epsilon_2(t) \geqslant \epsilon > 0.$$

Moreover, suppose that $u_{t_0} \geq v_{t_0}$ in $\mathcal{D}$ for some $t_0 \in [t_1, t_2]$ and $t \to u_t$, $t \to v_t$ are continuous from the finite closed interval $[t_1, t_2]$ to $C(\overline{\mathcal{D}})$. Then

$$u_t \geq v_t \quad on \ \mathcal{D}, \ \forall t \in [t_1, t_2].$$

7.2.2 Classification of One-Dimensional Solutions

In this sub-section, we classify all the solutions of the one dimensional p-Laplacian problem of the following form:

$$\Delta_p V + f(V) = 0 \text{ in } \mathbb{R}_+, \ V(0) = 0, \ V \geq, \not\equiv 0, \|V\|_\infty < +\infty. \tag{7.5}$$

We always assume that

$$\begin{cases} f : \mathbb{R}_+ \to \mathbb{R} \text{ is continuous, } f(0) \geq 0, \\ \text{and it is locally Lipschitz continuous except possibly at its zeros.} \end{cases} \tag{7.6}$$

By a solution of (7.5) we mean a function $V \in C^1(\mathbb{R}_+)$ satisfying (7.5) in the weak sense; that is

$$V(0) = 0 \text{ and } \int_0^\infty |V'|^{p-2} V' \phi' dx = \int_0^\infty f(V)\phi dx \text{ for all } \phi \in C_0^1(0, \infty).$$

Theorem 7.1 *Suppose that f satisfies (7.6), and that whenever $z \in Z_f := \{z \in \mathbb{R}_+ : f(z) = 0\}$, we have*

$$\liminf_{s \searrow z} \frac{f(s)}{(s-z)^{p-1}} > -\infty, \ \limsup_{s \nearrow z} \frac{f(s)}{(z-s)^{p-1}} < +\infty. \tag{7.7}$$

Then each solution V of (7.5) has the following properties:

(i) $V'(t) > 0 \ \forall t \in \mathbb{R}_+$ (and hence $V \in C^2(0, \infty)$),
(ii) $z_0 := V(+\infty) \in Z_f^+ := Z_f \setminus \{0\}$,
(iii) $F(z_0) = \frac{p-1}{p} V'(0)^p > 0$, $F(z) < F(z_0) \ \forall z \in [0, z_0)$,

where $F(z) = \int_0^z f(s)ds$.

Proof Let $V \in C^1(\mathbb{R}_+)$ be a solution to (7.5). Due to $f(0) \geq 0$ and (7.7), and $\|V\|_\infty < \infty$, we have $-\Delta_p V = f(V) \geq cV^{p-1}$ for some constant $c > -\infty$. Hence we can use the strong maximum principle (see Theorem 5 in [102]) to conclude that $V(t) > 0$ in $(0, +\infty)$ and $V'(0) > 0$. Thus $V'(t) > 0$ for all small positive t. It follows that either $V'(t) > 0$ for all $t > 0$, or there is a first $t_0 > 0$ such that $V'(t) > 0$ in $(0, t_0)$ and $V'(t_0) = 0$. If the latter happens, we define

$$\tilde{V}(t) := \begin{cases} V(t), & t \in [0, t_0], \\ V(2t_0 - t), & t \in (t_0, 2t_0]. \end{cases}$$

Then it is easily checked that $\tilde{V} \in C^1$ satisfies the following in the weak sense:

$$\Delta_p \tilde{V} + f(\tilde{V}) = 0 \text{ in } [0, 2t_0], \quad \tilde{V}(0) = V(0), \quad \tilde{V}'(0) = V'(0).$$

We will show in a moment that $\tilde{V} \equiv V$ on $[0, 2t_0]$, which implies $V'(2t_0) = -V'(0) < 0$ and hence V must change sign as t increases across $2t_0$, a contradiction to the assumption that V is nonnegative in $\mathbb{R}_+$. This proves (i) provided we can show $\tilde{V} \equiv V$ on $[0, 2t_0]$. We now set to show $\tilde{V} \equiv V$. We first observe that $g(V(t_0)) \neq 0$, for otherwise with $z^0 := V(t_0)$ we obtain from (7.7) that

$$f(V(t)) \leqslant c(z^0 - V(t))^{p-1} \text{ for some constant } c \text{ and all } t < t_0 \text{ and close to } t_0.$$

Hence $W := z^0 - V$ satisfies $-\Delta_p W = -f(V) \geqslant -cW^{p-1}$ and $W > 0$ for all such t. We can now apply Theorem 5 in [102] again to conclude that $W'(t_0) < 0$, i.e., $V'(t_0) > 0$, a contradiction to $V'(t_0) = 0$. We further notice that actually $f(V(t_0)) > 0$ must hold. Indeed, if $f(V(t_0)) < 0$, then from the equation we obtain $V''(t) > 0$ for all $t < t_0$ and close to t_0 (note that V is C^2 in $(0, t_0)$ since $V'(t) > 0$ in this interval). It follows that $V'(t_0) > V'(t) > 0$ for $t < t_0$ and close to t_0. This contradiction shows that $f(V(t_0)) > 0$. We show that $V'(t) \leqslant 0$ for all $t > t_0$ and close to t_0. Otherwise we can find t_n decreasing to t_0 as $n \to \infty$ such that $V'(t_n) > 0$. Fix a large n so that $f(V(t)) > 0$ for $t \in [t_0, t_n]$. By continuity we can find $\underline{t}_n \in [t_0, t_n)$ satisfying

$$V'(t) > 0 \text{ in } (\underline{t}_n, t_n], \quad V'(\underline{t}_n) = 0.$$

As before from standard elliptic regularity we know that V is C^2 in $(\underline{t}_n, t_n]$, and it satisfies in this interval

$$(p - 1)|V'(t)|^{p-2}V''(t) = -f(V(t)) < 0.$$

Thus $V''(t) < 0$ for $t \in (\underline{t}_n, t_n]$, which implies $V'(\underline{t}_n) > V'(t_n) > 0$, a contradiction. This proves what we wanted. We claim that the above conclusion can be strengthened to $V'(t) < 0$ for all $t > t_0$ and close to t_0. If $V'(t) \equiv 0$ in a small right neighborhood of t_0, then from the equation we deduce $f(V(t)) \equiv 0$ in this interval, a contradiction to $f(V(t_0)) > 0$. Therefore if the above claim is false then we can find a sequence t_n decreasing to t_0 as $n \to \infty$ such that $V'(t_n) = 0$, and another sequence s_n decreasing to t_0 such that $V'(s_n) < 0$. Fix n large so that $f(V(t)) > 0$ for $t \in [t_0, t_n]$. Then choose $s_m \in (0, t_n)$. We can now find $\bar{s}_m \in (s_m, t_n]$ such that

$$V'(t) < 0 \text{ in } [s_m, \bar{s}_m), \quad V'(\bar{s}_m) = 0.$$

Similar to above, we deduce $V''(t) < 0$ for $t \in [s_m, \bar{s}_m)$ and hence $V'(\bar{s}_m) < V'(s_m) < 0$. This contradiction proves our claim. We are now ready to show $\tilde{V} \equiv V$. Since V is C^2 in $(0, t_0)$, we have

$$\frac{p-1}{p}|V'(t)|^p + F(V(t)) \equiv constant \left(= \frac{p-1}{p}|V'(0)|^p \right) \quad \text{for } t \in [0, t_0].$$
$$(7.8)$$

Taking $t = t_0$ we deduce

$$F(z^0) = \frac{p-1}{p}|V'(0)|^p > 0.$$

Thus for $t \in (0, t_0]$ we have $F(V(t)) < F(z^0)$ which implies $F(z) < F(z^0)$ for $z \in (0, z^0)$. Moreover, it follows from (7.8) that for $t \in (0, t_0)$,

$$\frac{V'(t)}{[F(z^0) - F(V(t))]^{1/p}} = C_0 := \left(\frac{p}{p-1} \right)^{1/p}.$$

Integrating over $(0, t)$ we obtain

$$\int_0^{V(t)} [F(z^0) - F(z)]^{-1/p} dz = C_0 t, \quad t \in (0, t_0). \quad (7.9)$$

Since $F(z^0) - F(z) = f(z^0)(z^0 - z) + o(|z^0 - z|)$ near $z = z^0$, and $f(z^0) > 0$, we find that

$$M_0 := \int_0^{z^0} [F(z^0) - F(z)]^{-1/p} dz < \infty.$$

This implies that $t_0 = M_0/C_0$ and $V(t)$ for $t \in [0, t_0]$ is uniquely determined by (7.9). Let (t_0, T) be the largest interval in which $V'(t) < 0$. Then (7.8) holds in $[0, T]$ and we deduce

$$\frac{-V'(t)}{[F(z^0) - F(V(t))]^{1/p}} = C_0, \quad t \in (t_0, T).$$

Integrating over (t_0, t) we obtain

$$\int_{V(t)}^{z^0} [F(z^0) - F(z)]^{-1/p} dz = C_0(t - t_0).$$

It follows from the definition of M_0 that

$$\int_0^{V(t)} [F(z^0) - F(z)]^{-1/p} dz = M_0 - \int_{V(t)}^{z^0} [F(z^0) - F(z)]^{-1/p} dz = C_0(2t_0 - t),$$

or equivalently,

$$\int_0^{V(2t_0-t)} [F(z^0) - F(z)]^{-1/p} dz = C_0 t, \quad t \in (2t_0 - T, t_0).$$

Comparing this with (7.9) we immediately obtain $T \geqslant 2t_0$ and $V(2t_0 - t) \equiv V(t)$ in $[0, t_0]$. This completes the proof of (i). Next we prove (ii) and (iii). Since now we know that V is increasing and bounded, $z_0 := \lim_{t \to +\infty} V(t)$ is well-defined. It follows from elementary consideration and (7.5) that $f(z_0) = 0$. Letting $t \to +\infty$ in (7.8) we deduce

$$F(z_0) = \frac{p-1}{p} V'(0)^p.$$

Hence

$$\frac{p-1}{p} |V'(t)|^p = F(z_0) - F(V(t)).$$

For any $z \in (0, z_0)$, there is a unique $t_0 > 0$ such that $V(t_0) = z$, and thus

$$F(z_0) - F(z) = \frac{p-1}{p} |V'(t_0)|^p > 0.$$

$\square$

Remark 7.1 Let us note that functions satisfying (7.7) need not be locally Lipschitz at $z \in Z_f$. On the other hand, locally Lipschitz continuous functions satisfy (7.7) automatically when $p \in (1, 2]$, but when $p > 2$, locally Lipschitz continuous functions may or may not satisfy (7.7).

Theorem 7.2 *Let f be as in Theorem 7.1, and define*

$$Z_f^* := \{z_0 \in Z_f^+ : F(z) < F(z_0) \; \forall z \in [0, z_0)\}.$$

Then for every $z_0 \in Z_f^*$, (7.5) has a unique solution satisfying $V(+\infty) = z_0$.

Proof Let $z_0 \in Z_f^*$. Then from (7.7) we easily deduce that

$$F(z_0) - F(z) \leqslant C(z_0 - z)^p \text{ for some constant } C \text{ and all } z < z_0 \text{ close to } z_0.$$

This implies that

$$\int_0^{z_0} [F(z_0) - F(z)]^{-1/p} dz = +\infty.$$

It follows that the formula

$$\int_0^{V(t)} [F(z_0) - F(z)]^{-1/p} dz = C_0 t, \quad C_0 = \left(\frac{p}{p-1}\right)^{1/p},$$

uniquely defines a function $V(t)$, $t \in \mathbb{R}_+$; moreover $V(t)$ is increasing, $V(+\infty) = z_0$, and it satisfies

$$\Delta_p V + f(V) = 0 \text{ in } (0, \infty), \quad V(0) = 0, \quad V'(0) = \left(\frac{p}{p-1} F(z_0)\right)^{1/p}.$$

If $\tilde{V}$ is any solution of (7.5) satisfying $\tilde{V}(+\infty) = z_0$, then by *Theorem 7.1* we have $\tilde{V}'(t) > 0$ for all $t > 0$, and $F(z_0) = \frac{p-1}{p} \tilde{V}'(0)^p$. Thus making use of (7.8) we see that $\tilde{V}(t)$ is determined by the same integral formula used for $V(t)$ above. It follows that $\tilde{V} \equiv V$. This proves the uniqueness.   □

From the above two theorems, we immediately obtain

Corollary 7.1 *Under the assumptions on f in Theorem 7.1, the set Z_f^* and the solutions of (7.5) are in one-to-one correspondence.*

Next we focus on two types of nonlinearities f, namely (F_1) and (F_2) defined below.

(F_1): We say that f is of type (F_1), if it satisfies (7.6), is nonnegative on $\mathbb{R}_+$, and not identically zero in any open interval of $\mathbb{R}_+$.

(F_2): We say that f is of type (F_2) if it satisfies (7.6), $f(u) > 0$ in $(0, a)$, $f(u) \leqslant 0$ in $(a, +\infty)$, and is not identically zero in any open interval of $\mathbb{R}_+$.

Note that in the following two theorems, condition (7.7) is not needed.

Theorem 7.3 *Suppose that f is of type (F_1), and V is a solution of (7.5). Then $V(t) > 0$ in $(0, +\infty)$, it is nondecreasing, and has the properties in (ii) and (iii) of Theorem 7.1.*

Proof Since $f(V) \geqslant 0$, we can apply the strong maximum principle to conclude that $V(t) > 0$ in $(0, +\infty)$ and $V'(0) > 0$. Let $[0, t_0)$ be the largest interval such that $V'(t) > 0$ for $t \in [0, t_0)$. If $t_0 = \infty$, then V is C^2 in $[0, \infty)$ and the conclusions follow easily as before. So suppose now $t_0 < \infty$. Then clearly $V'(t_0) = 0$. We observe that if $V'(t_1) = 0$ for some $t_1 > t_0$ then $V'(t) \equiv 0$ on $[t_0, t_1]$. Indeed, if this is not the case, say $V'(s) \neq 0$ for some $s \in (t_0, t_1)$, then we can find a largest interval $(s_0, s_1) \subset (t_0, t_1)$ such that $V'(t) \neq 0$ in (s_0, s_1) and $V'(s_0) = V'(s_1) = 0$. It follows that $V(s_0) \neq V(s_1)$. Since V is C^2 on (s_0, s_1), we can apply (7.8) to deduce $F(V(s_0)) = F(V(s_1))$, which is a contradiction to $V(s_0) \neq V(s_1)$. Thus we have either $V'(t) \equiv 0$ for $t \geqslant t_0$ or there exists $t_1 \geqslant t_0$ such that $V'(t) = 0$ in $[t_0, t_1]$ and $V'(t) \neq 0$ for $t > t_1$. In the former case, the conclusions of the theorem follow readily. We show next that the latter case cannot happen. Otherwise,

$V(\infty) = \lim_{t \to \infty} V(t)$ exists and $V(\infty) \neq V(t_1)$. Since V is C^2 over (t_1, ∞), we can apply (7.8) to deduce $F(V(\infty)) = F(V(t_1))$, a contradiction. $\square$

Theorem 7.4 *Suppose that f is of type (F_2), and V is a solution of (7.5). Then $V(t) > 0$ in $(0, +\infty)$, it is nondecreasing, and $V(+\infty) = a$. Such V exists and is unique.*

Proof Let V be a solution of (7.5). Since $f(s) > 0$ in $(0, a)$, by the strong maximum principle we find that $V(t) > 0$ in $(0, +\infty)$ and $V'(0) > 0$. Hence $V'(t) > 0$ for all small positive t. We have two possibilities: (i) $V'(t) > 0$ for all $t > 0$, (ii) there is a first $t_0 > 0$ such that $V'(t) > 0$ in $[0, t_0)$ and $V'(t_0) = 0$. In case (i), we denote $z_0 := V(+\infty)$ and obtain, as before,

$$ f(z_0) = 0, \quad F(z_0) = \frac{p-1}{p} V'(0)^p, \quad F(V(t)) < F(z_0) \ \forall t \geqslant 0. $$

Thus $z_0 \geqslant a$. But if $z_0 > a$, then there is a unique $t_0 > 0$ such that $V(t_0) = a$ and $V(t) > a$ for $t > t_0$. Thus we obtain from (7.8) that $F(a) = F(V(t_0)) < F(z_0)$, which is impossible since $F(z_0) - F(a) = \int_a^{z_0} f(s)ds < 0$ by our assumption on f. So in case (i), we have $V(+\infty) = a$. Let us now consider case (ii). If $z_1 := V(t_0) > a$, then we can find a unique $t_1 \in (0, t_0)$ such that $V(t_1) = a$ and so, by (7.8) we obtain $F(a) = F(V(t_1)) < F(z_1)$, which is impossible as $\int_a^{z_1} f(s)ds < 0$. So we must have $V(t_0) \leqslant a$. If $V(t_0) < a$, then there is a maximal interval $I = (t_0, t_1)$ in (t_0, ∞) such that $V(t) < a$ in I. We must have $V'(t) \neq 0$ in I, since if $V'(s_0) = 0$ for some $s_0 \in I$ then we can argue as in the proof of *Theorem 7.3* above to deduce $V'(t) \equiv 0$ in (t_0, s_0), which implies $f(V(t_0)) = 0$, a contradiction to the assumption that $f(u) > 0$ in $(0, a)$. If $V'(t) > 0$ in I then for any fixed $t \in I$, $a > V(t) > V(t_0)$ which implies $F(V(t)) > F(V(t_0))$. On the other hand, since V is C^2 in I, we can apply (7.8) to deduce $F(V(t_0)) = F(V(t)) + \frac{p-1}{p}|V'(t)|^p > F(V(t))$. This contradiction shows that we must have $V'(t) < 0$ in I (note that we already know $V'(t) \neq 0$ in I). Thus $I = (t_0, \infty)$ and $V(\infty)$ exists and satisfies $V(\infty) < V(t_0) < a$. We may now apply (7.8) to obtain $F(V(\infty)) = F(V(t_0))$, which is impossible since $\int_{V(\infty)}^{V(t_0)} f(t)dt > 0$. This proves that $V(t_0) < a$ cannot occur. Thus we necessarily have $V(t_0) = a$. We show next that $V(t) \leqslant a$ for $t \geqslant t_0$. Otherwise there exists $t_1 > a$ such that $V(t_1) > a$ and $V'(t_1) > 0$. Let $[t_1, t_2)$ be the largest interval in $[t_1, \infty)$ such that $V'(t) > 0$ in $[t_1, t_2)$. Then either $t_2 = \infty$ and $V'(\infty) = \lim_{t \to \infty} V'(t) = 0$, or $t_2 < \infty$ and $V'(t_2) = 0$. In either case we have $a < V(t_1) < V(t_2)$ and $V'(t_2) = 0$. Applying (7.8) over $[t_1, t_2)$ we deduce $F(V(t_1)) = F(V(t_2))$, which is impossible since $\int_{V(t_1)}^{V(t_2)} f(t)dt < 0$. This proves that $V(t) \leqslant a$ for $t \geqslant t_0$. We are now ready to prove that $V(t) \equiv a$ for $t \geqslant t_0$. If this is not true, then there exists $t_1 > a$ such that $V(t_1) < a$ and $V'(t_1) < 0$. Similar to the argument in the last paragraph, we let $[t_1, t_2)$ be the largest interval in $[t_1, \infty)$ such that $V'(t) < 0$ in $[t_1, t_2)$. Then either $t_2 = \infty$ and $V'(\infty) = \lim_{t \to \infty} V'(t) = 0$, or $t_2 < \infty$ and $V'(t_2) = 0$. In either case we have $a > V(t_1) > V(t_2)$ and $V'(t_2) = 0$. Applying (7.8) over $[t_1, t_2)$ we deduce $F(V(t_1)) = F(V(t_2))$, which is impossible since $\int_{V(t_2)}^{V(t_1)} f(t)dt > 0$. This proves that $V(t) \equiv a$ for $t \geqslant t_0$. The existence and uniqueness of the solution of (7.5) with the above properties follows from (7.8), which shows that $V(t)$ is uniquely determined by the formula

$$\int_0^{V(t)} [F(a) - F(z)]^{-1/p} dz = C_0 t \text{ for } t > 0 \text{ such that } V(t) < a.$$

Note that this formula indicates that case (i) happens if $\int_0^a [F(a) - F(z)]^{-1/p} dz = +\infty$, and case (ii) happens when this integral is finite.   □

7.2.3   A Liouville Type Result

Let us denote $\sigma_{N,p} := (p-1)\frac{N}{N-p}$ when $N > p$, and for $N \leqslant p$, we assume that $\sigma_{N,p}$ stands for an arbitrary number in $[1, +\infty)$. As in [41], we say that f is quasi-monotone in $[0, \infty)$ if for any bounded interval $[s_1, s_2] \subset [0, \infty)$, there exists a continuous increasing function $L(s)$ such that $f(s) + L(s)$ is non-decreasing in $[s_1, s_2]$.

Theorem 7.5 *Suppose that f is of type (F_1), is quasi-monotone, and for each $z \in Z_f$,*

$$\liminf_{s \searrow z} \frac{f(s)}{(s-z)^{\sigma_{N,p}}} > 0. \tag{7.10}$$

Let u be a bounded nonnegative solution of

$$\Delta_p u + f(u) = 0 \text{ in } \mathbb{R}^N (N \geqslant 1).$$

Then u must be a constant.

Proof Let u be given as in the theorem. Denote

$$m := \inf_{\mathbb{R}^N} u \geqslant 0.$$

Since $-\Delta_p(u-m) = f(u) \geqslant 0$ and $u - m \geqslant 0$ in $\mathbb{R}^N$, we can apply the strong maximum principle to conclude that either $u \equiv m$, or $u > m$ in $\mathbb{R}^N$. We show that the second alternative cannot happen. Otherwise, we can find a sequence $x_n \in \mathbb{R}^N$ such that $u(x_n) \searrow m$ as $n \to \infty$. Without loss of generality we may assume that $u(x_n) = \min_{|x|=|x_n|} u(x)$. We claim that $f(m) = 0$. Consider the sequence of solutions $u_n(x) := u(x + x_n)$. Since $\|u_n\|_\infty = \|u\|_\infty < +\infty$, by standard elliptic regularity results [99, 100] we know that u_n is bounded in $C^{1,\mu}(K)$ for any compact subset of $\mathbb{R}^N$. Hence we may use a standard diagonal process to extract a subsequence, still denoted by u_n, such that $u_n \to v$ in $C^1_{loc}(\mathbb{R}^N)$, and v satisfies

$v \geqslant m$, $v(0) = m$ and $\Delta_p v + f(v) = 0$ in $\mathbb{R}^N$. Thus $-\Delta_p(v - m) \geqslant 0$ in $\mathbb{R}^N$ and $v(0) - m = 0$. By the strong maximum principle we conclude that $v \equiv m$ and hence $f(m) = 0$. We also claim that $|x_n| \to +\infty$ as $n \to +\infty$. Otherwise by passing to a subsequence we may assume that $x_n \to x_0$ and then the function v obtained in the last paragraph is nothing but $u(x + x_0)$, and thus we have $u \equiv m$, a contradiction to our assumption that $u > m$. By (7.10) with $z = m$, either $f(s) > 0$ for all $s > m$, or there exists $m_1 > m$ such that $f(s) > 0$ in (m, m_1) and $f(m_1) = 0$. In either case, we choose $R := |x_n|$ with large enough n such that $a := u(x_n) \in (m, m_1)$; taking $m_1 = +\infty$ when $f(s) > 0$ for all $s > m$. We next choose $\{n_k\}$ so that $R_k := |x_{n_k}|$ is an increasing sequence converging to $+\infty$, with $R_1 > R$. Now we consider the boundary value problem

$$\Delta_p w + f(w) = 0 \text{ in } R < |x| < R_k, \quad w|_{\{|x|=R\}} = a, \quad w|_{\{|x|=R_k\}} = m. \qquad (7.11)$$

For each $k \geqslant 1$, $w = u$ is an upper solution of (7.11), and $w \equiv m$ is a lower solution. Hence the standard upper and lower solution argument as in the proof of [42, Theorem 3.7] implies that (7.11) has a minimal solution w_k satisfying $m \leqslant w_k \leqslant u$ in $\{R < |x| < R_k\}$. We may now proceed as on page 1894 of [38] to conclude that

$$w_k \leqslant w_{k+1} \leqslant u \text{ when } R < |x| < R_k, \ k = 1, 2, \ldots,$$

and $w(x) := \lim_{k \to \infty} w_k(x)$ is well-defined for $|x| > R$, and it satisfies

$$m \leqslant w \leqslant u, \ \Delta_p w + f(w) = 0 \text{ for } |x| > R, \ w = a \text{ when } |x| = R.$$

Moreover, as each w_k is radially symmetric (as a minimal solution) so is w. We may then write $w = w(r)$. Denoting $\phi(r) = r^{N-1}|w'(r)|^{p-2} w'(r)$ we obtain

$$\phi'(r) = -r^{N-1} f(w) \leqslant 0 \text{ for } r \geqslant R.$$

Thus $\phi(r)$ is nonincreasing and there are two possibilities: (i) $\phi(r)$ is negative for all large r, or (ii) $\phi(r) \geqslant 0$ in $[R, +\infty)$.

If case (ii) occurs, then $w'(r) \geqslant 0$ for $r \geqslant R$, and hence for all large n so that $|x_n| > R$, we have $u(x_n) \geqslant w(|x_n|) \geqslant a > m$, a contradiction to our initial choice of x_n. If case (i) occurs, say $\phi(r) < 0$ for $r \geqslant R_0 > R$, then $w'(r) < 0$ for such r and hence $m \leqslant w(r) \leqslant m^0 < m_1$ for all $r \geqslant R^0$, where $R^0 := |x_{n_0}|$ with n_0 chosen so large that $m^0 := w(R^0) \leqslant u(x_{n_0}) < m_1$. Due to (7.10) applied to $z = m$, and the fact that $f(s) > 0$ in $(m, m^0]$, we can find $c > 0$ and $\sigma \in (p - 1, (p - 1)\frac{N}{(N-p)_+})$ such that

$$f(s) \geqslant c(s - m)^\sigma \text{ in } [m, m^0].$$

It follows that

$$-\Delta_p(w - m) \geqslant c(w - m)^\sigma, \, w - m \geqslant 0 \text{ in } \{|x| > R^0\}.$$

Denote $\tilde{w}(x) = w(|x|) - m$ and set $v(x) := \tilde{w}(c^{-p}x)$. Clearly v satisfies

$$-\Delta_p v \geqslant v^\sigma, \, v \geqslant 0 \text{ in } \{|x| \geqslant c^p R^0\}.$$

By [27] and [95, Theorem II], (see [38, Proposition 2.3]), we necessarily have $v \equiv 0$, which clearly is a contradiction. Thus we have proved that $u \equiv m$. □

Remark 7.2 Condition (7.10) is sharp. For each $\xi > \sigma_{N,2} = N/(N - 2)$, there are examples with $f(s) = cs^\xi$ for small positive s, such that the problem $\Delta u + f(u) = 0$ has a non-constant positive solution in $\mathbb{R}^N$ which decays to 0 at infinity; see [38, p. 1892] for more details.

Remark 7.3 For the case $p = 2$ *Theorem 7.5* was previously proved in *Chap. 6*, see *Theorem 6.5*. There, however, a more restrictive condition on f was used instead of (7.10), namely

$$\liminf_{s \searrow z} \frac{f(s)}{s - z} > 0.$$

7.3   Half- and Quarter-Space Problems

7.3.1   *Asymptotic Convergence in Half-Spaces*

We start by introducing some notations. For any closed set $D \subset \mathbb{R}^N$, the space $C_b^1(D)$ consists of functions $u : D \to \mathbb{R}$ such that

$$\|u\|_{C_b^1(D)} := \sup_{x \in D} \|u\|_{C^1(\overline{B_1(x)} \cap D)} < +\infty.$$

Theorem 7.6 *Suppose that f is of type (F_1), is quasi-monotone, its zeros are isolated and for each $z \in Z_f^+$, (7.10) holds. Moreover, assume that*

$$\liminf_{s \searrow 0} \frac{f(s)}{s^{p-1}} > 0. \tag{7.12}$$

Let u be any nontrivial nonnegative bounded solution of (7.1). Then there exists $z \in Z_f^+$ such that

$$\lim_{h \to \infty} u(x_1 + h, x_2, \ldots, x_N) = z \text{ in } C_b^1([A, +\infty) \times \mathbb{R}^{N-1})$$

for every $A \in \mathbb{R}$.

Proof By the assumptions on f, $f(s)$ is positive on $(0, \delta_0]$ for some $\delta_0 > 0$. Therefore we can use the strong maximum principle, much as before, to conclude that $u > 0$ in $(0, +\infty) \times \mathbb{R}^{N-1}$. Moreover, using (7.12) we can find some $\eta > 0$ such that

$$f(s) \geqslant \eta s^{p-1} \ \forall s \in [0, \delta_0]. \tag{7.13}$$

We divide the proof below into several steps.

Step 1 $m := \lim_{A \to +\infty} \inf_{[A, +\infty) \times \mathbb{R}^{N-1}} u \geqslant \delta_0$. It is well-known that for any bounded domain $\mathcal{D}$ of $\mathbb{R}^N$,

$$\lambda_1(\mathcal{D}) := \inf \left\{ \int_{\mathcal{D}} |Du|^p dx : u \in W_0^{1,p}(\mathcal{D}), \ \int_{\mathcal{D}} |u|^p dx > 0 \right\}$$

is achieved by some positive function ϕ which satisfies

$$-\Delta_p \phi = \lambda_1(\mathcal{D}) \phi^{p-1} \text{ in } \mathcal{D}, \ \phi|_{\partial \mathcal{D}} = 0.$$

Moreover, $\lambda_1(\mathcal{D}) > 0$, and such ϕ is unique if we require $\|\phi\|_\infty = 1$. Take $\mathcal{D} = B_1(0)$, then it is well-known from the rearrangement theory (see [K]) that ϕ is radially symmetric and $\phi(0) = \|\phi\|_\infty = 1$. We now fix $\lambda > 0$ large enough such that

$$\lambda^{-p} \lambda_1(B_1(0)) < \eta/2.$$

For arbitrary $x_0 \in [\lambda + 1, +\infty) \times \mathbb{R}^{N-1}$, since $u > 0$ in the closed ball $\overline{B_\lambda(x_0)}$ which is contained in $[1, +\infty) \times \mathbb{R}^{N-1}$, there exists $\delta \in (0, \delta_0)$ such that

$$u(x) \geqslant \delta, \ \forall x \in \overline{B_\lambda(x_0)}.$$

We now let

$$t_1 = \delta, \ t_2 = \delta_0,$$

and for $t \in [t_1, t_2]$, we define

$$v_t(x) = t\phi(\lambda^{-1}(x - x_0)), \ x \in B_\lambda(x_0).$$

Clearly,

$$0 \leqslant v_t(x) \leqslant \delta_0, \ \forall x \in B_\lambda(x_0), \ \forall t \in [t_1, t_2].$$

Let $\delta_1 \in (0, 1)$ be so small that $t_2 \phi(x) < \delta/2$ whenever $1 \geqslant |x| \geqslant 1 - \delta_1$. Then

$$v_t(x) \leqslant \delta/2, \ \forall x \in \partial B_{\lambda(1-\delta_1)}(x_0), \ \forall t \in [t_1, t_2].$$

Moreover, by the definition of v_t and (7.13), we obtain, for $t \in [t_1, t_2]$,

$$-\Delta_p v_t = \lambda^{-p} \lambda_1(B_1(0)) v_t^{p-1} \leqslant (1/2) \eta v_t^{p-1}$$

$$\leqslant \eta v_t^{p-1} - \zeta \leqslant f(v_t) - \zeta, \ \forall x \in B_{\lambda(1-\delta_1)}(x_0),$$

where

$$\zeta = \min_{x \in B_{\lambda(1-\delta_1)}(x_0)} (1/2) \eta v_{t_1}^{p-1} > 0.$$

Therefore, in view of $v_{t_1} \leqslant \delta \leqslant u$ in $B_{\lambda(1-\delta_1)}(x_0)$, we can apply Proposition 7.1 with $\mathcal{D} = B_{\lambda(1-\delta_1)}(x_0)$, $\epsilon = \min\{\delta/2, \zeta\}$ to conclude that

$$u \geqslant v_t, \ \forall x \in B_{\lambda(1-\delta_1)}(x_0), \ \forall t \in [t_1, t_2].$$

In particular, $u(x_0) \geqslant v_{t_2}(x_0) = \delta_0$. Since $x_0 \in [\lambda + 1, +\infty) \times \mathbb{R}^{N-1}$ is arbitrary, this implies that

$$\lim_{A \to +\infty} \inf_{[A,+\infty) \times \mathbb{R}^{N-1}} u \geqslant \delta_0.$$

Step 2 For every $A \in \mathbb{R}$,

$$u(x_1 + h, x_2, \ldots, x_N) \to m \text{ as } h \to +\infty$$

uniformly with respect to $(x_1, x_2, \ldots, x_N) \in [A, +\infty) \times \mathbb{R}^{N-1}$. By Step 1 above,

$$M := \lim_{A \to +\infty} \sup_{[A,+\infty) \times \mathbb{R}^{N-1}} u \geqslant m > 0.$$

We show next that $f(M) = 0$. This follows from a standard consideration as described below. Choose a sequence $x^n = (x_1^n, x_2^n, \ldots, x_N^n)$ such that $x_1^n \to +\infty$ and $u(x^n) \to M$ as $n \to +\infty$. Then by passing to a subsequence, $u_n(x) = u(x + x_n)$ converges in $C_{loc}^1(\mathbb{R}^N)$ to a function U which satisfies

$$\Delta_p U + f(U) = 0, \ \|u\|_\infty \geqslant U \geqslant 0 \text{ in } \mathbb{R}^N, \ U(0) = M.$$

We can now apply *Theorem 7.5* to conclude that $U \equiv M$, which infers $f(M) = 0$. Next we set to show that

$$u(x_1 + h, x_2, \ldots, x_N) \to M \text{ as } h \to +\infty$$

uniformly with respect to $(x_1, x_2, \ldots, x_N) \in [A, +\infty) \times \mathbb{R}^{N-1}$ for every $A \in \mathbb{R}$. Clearly this implies $M = m$. We will prove the above conclusion by applying the weak sweeping principle. The construction of the required lower solution is rather involved; for clarity we divide the remaining arguments into two sub-steps.

Sub-step 2.1 Construction of a lower solution. To construct a suitable lower solution which can be used to bound u from below via the weak sweeping principle, we first modify f suitably and then use the modified f to form a Dirichlet problem over a ball. The required lower solution will be a positive solution of this auxiliary Dirichlet problem. The modification of f is needed in order to create a gap from f, so that the condition $\epsilon_1(t) + \epsilon_2(t) \geq \epsilon$ in *Proposition 7.1* is met; see (7.14) below. Since the zeros of f are isolated, we can find $0 < M_0 < M$ such that $f(s) > 0$ in $[M_0, M)$. Define

$$F_1(s) = \int_s^M f(t)dt.$$

Clearly $F_1(s) > 0$ in $[0, M)$. For any small $\epsilon > 0$, we consider

$$g(s) = g_\epsilon(s) := f(s) - \epsilon s^\sigma \text{ in } [0, M],$$

where $\sigma = \max\{1, \sigma_{N,p}\}$ with $\sigma_{N,p}$ as given in (7.10) in the case $f(0) = 0$, and $\sigma = 1$ when $f(0) > 0$. There exists $M_\epsilon \in (M_0, M)$ such that $g(M_\epsilon) = 0$ and $g(s) > 0$ in $[M_0, M_\epsilon)$. Set

$$G(s) = G_\epsilon(s) := \int_s^{M_\epsilon} g(t)dt.$$

Clearly $G(s) > 0$ in $[M_0, M_\epsilon)$, and $M_\epsilon \to M$ as $\epsilon \to 0$. Since $G_\epsilon(s) \to F_1(s)$ uniformly in $[0, M]$ as $\epsilon \to 0$, and $F_1(s) \geq F_1(M_0) > 0$ in $[0, M_0]$, we thus find that there exists $\epsilon_0 > 0$ sufficiently small such that for each $\epsilon \in (0, \epsilon_0]$,

$$\begin{cases} M_\epsilon - \epsilon > M_0, \\ G_\epsilon(s) > 0 \text{ in } [0, M_\epsilon), \\ G_\epsilon(s) \geq G_\epsilon(M_\epsilon - \epsilon) \text{ for } s \in [0, M_\epsilon - \epsilon), \\ G_\epsilon(s) \text{ is decreasing in } [M_0, M_\epsilon). \end{cases}$$

Let us also notice that due to (7.10), we always have $f(s) > g_\epsilon(s) > 0$ for small positive s, say $s \in (0, s_0)$, and s_0 can be chosen independent of $\epsilon \in (0, \epsilon_0]$. Set

$$\tilde{g}(s) = \begin{cases} g(0) & \text{for } s < 0, \\ g(s) & \text{for } s \in [0, M_\epsilon], \\ 0 & \text{for } s > M_\epsilon, \end{cases}$$

and

$$\tilde{G}(s) = \int_s^{M_\epsilon} \tilde{g}(t)dt.$$

Clearly $\tilde{G}(s) \geq 0$ for all $s \in \mathbb{R}$. Motivated by *Lemma 6.2* in *Chap. 6*, we consider the functional

$$I_r(v) = \frac{1}{p}\int_{B_r(0)} |\nabla v|^p + \int_{B_r(0)} \tilde{G}(v)$$

for all $v \in H_0^p(B_r(0))$. It is well-known that a critical point of I_r corresponds to a weak solution of

$$\Delta_p v + \tilde{g}(v) = 0 \text{ in } B_r(0), \quad v|_{\partial B_r(0)} = 0.$$

Since $\tilde{g} \geq 0$ in $(-\infty, 0]$ and $\tilde{g} = 0$ for $s \geq M_\epsilon$, by the weak maximum principle, any such solution satisfies $0 \leq v \leq M_\epsilon$. Consequently for any such solution we have $\tilde{g}(v) = g(v)$. Moreover, by elliptic regularity for p-Laplacian equations [99, 100] we know that such a solution also belongs to $C^{1,\alpha}(\overline{B}_r(0))$. Let us observe that the functional I_r is well-defined and is coercive. Thus by standard argument we know that it has a minimizer v_r, which is a critical point of I_r and thus, as discussed above, is a nonnegative solution to

$$\Delta_p v_r + g(v_r) = 0 \text{ in } B_r(0), \quad v_r|_{\partial B_r(0)} = 0.$$

Since v_r is a minimizer, by well-known rearrangement theory (see, e.g. [70]) it must be radially symmetric and decreasing away from the center of the domain. Thus $0 \leq v_r(|x|) \leq v_r(0) \leq M_\epsilon$ in $B_r(0)$. We claim that there exists $r > 0$ such that $v_r(0) \geq M_\epsilon - \epsilon$. Otherwise $v_r \leq M_\epsilon - \epsilon$ for all $r > 0$. Hence, recalling $G(s) \geq G(M_\epsilon - \epsilon)$ for $s \in [0, M_\epsilon - \epsilon]$, we obtain

$$I_r(v_r) \geq \int_{B_r(0)} G(v_r) \geq \int_{B_r(0)} G(M_\epsilon - \epsilon) = \alpha_N r^N G(M_\epsilon - \epsilon), \ \forall r > 0,$$

where α_N stands for the volume of $B_1(0)$. On the other hand, for $r > 1$ define

$$w_r(x) = \begin{cases} M_\epsilon & \text{for } |x| < r - 1, \\ M_\epsilon(r - |x|) & \text{for } r - 1 \leq |x| \leq r. \end{cases}$$

This function belongs to $W_0^{1,p}(B_r(0))$, and $|\nabla w_r|^p$ and $G(w_r)$ are supported on the annulus $\{r - 1 \leq |x| \leq r\}$. Thus there exists a constant C independent of r such that

$$I_r(w_r) \leq C[r^N - (r-1)^N] \ \forall r > 1.$$

Since v_r is the minimizer of I_r, we have $I_r(v_r) \leqslant I_r(w_r)$. Thus

$$\alpha_N G(M_\epsilon - \epsilon) r^N \leqslant C[r^N - (r-1)^N] \quad \forall r > 1.$$

Since $G(M_\epsilon - \epsilon) > 0$, the above inequality does not hold for large r. This contradiction shows that $v_r(0) \geqslant M_\epsilon - \epsilon$ for all large r. We will use v_r as the lower solution in the weak sweeping principle.

Sub-step 2.2 Completion of the proof by the weak sweeping principle. Fix $\epsilon \in (0, \epsilon_0]$ and then choose $r > 0$ such that $v_r(0) \geqslant M_\epsilon - \epsilon$. Let x^* be an arbitrary point in $[r+1, +\infty) \times \mathbb{R}^{N-1}$. Let x^n be the sequence used earlier with the properties $x_1^n \to +\infty$ and $u(x^n) \to M$. We recall that $u_n(x) := u(x - x^n) \to M$ in $C^1_{loc}(\mathbb{R}^N)$. Thus we can find a point $x^0 = (x_1^0, x_2^0, \ldots, x_N^0)$ with $x_1^0 > r+1$ such that $u(x) \geqslant M_\epsilon$ in $B_r(x^0)$. We now define $x^t = tx^0 + (1-t)x^*$ and $u^t(x) = u(x + x^t)$. Clearly $x_1^t \geqslant r+1$ and thus $B_r(x^t) \subset [1, +\infty) \times \mathbb{R}^{N-1}$ for all $t \in [0,1]$. Since $u > 0$ on the compact set $\cup_{t \in [0,1]} \overline{B_r(x^t)}$, we can find $\delta > 0$ such that $u \geqslant \delta$ on this set. Let $r_1 \in (0, r)$ be chosen so that $v_r \in (0, \delta/2)$ on $\{|x| = r_1\}$. Denote $\mathcal{D} := B_{r_1}(0)$. Then we have, for $t \in [0, 1]$,

$$v_r + \delta/2 \leqslant u^t \quad \text{on } \partial \mathcal{D}$$

and

$$-\Delta_p v_r = g(v_r) \leqslant f(v_r) - \zeta, \quad -\Delta_p u^t = f(u^t) \text{ in } \mathcal{D},$$

where

$$\zeta := \inf_{x \in B_{r_1}(0)} [f(v_r(x)) - g(v_r(x))] > 0. \tag{7.14}$$

Moreover, $u^0 \geqslant M_\epsilon \geqslant v_r$ on $\mathcal{D}$. Thus we can apply Proposition 7.1 to conclude that $u^t \geqslant v_r$ in $\mathcal{D}$ for all $t \in [0, 1]$. In particular, $u(x^*) \geqslant v_r(0) \geqslant M_\epsilon - \epsilon$. Since $\epsilon \in (0, \epsilon_0]$ and $x^* \in [r+1, +\infty) \times \mathbb{R}^{N-1}$ are arbitrary, and $M_\epsilon - \epsilon \to M$ as $\epsilon \to 0$, this implies that

$$\lim_{A \to +\infty} \inf_{[A, +\infty) \times \mathbb{R}^{N-1}} u \geqslant M.$$

In view of the definition of M, the above inequality implies the required conclusion in Step 2. Finally we note that the uniform convergence proved in Step 2 implies convergence in $C_b^1([A, +\infty) \times \mathbb{R}^{N-1})$ for any $A \in \mathbb{R}$ by standard elliptic estimates.
$\square$

Theorem 7.7 *Suppose that f is a type (F_2) function which is quasi-monotone, (7.12) holds and for each $z \in Z_f^+\backslash\{a\}$,*

$$\limsup_{s \nearrow z} \frac{f(s)}{(z-s)^{\sigma_{N,p}}} < 0. \qquad (7.15)$$

In addition, we assume that the zeros of f are isolated. Let u be any nontrivial nonnegative bounded solution of (7.1). Then there exists $z \in Z_f^+$ such that

$$\lim_{h \to \infty} u(x_1 + h, x_2, \ldots, x_N) = z \text{ in } C_b^1([A, +\infty) \times \mathbb{R}^{N-1})$$

for every $A \in \mathbb{R}$.

Proof Since $f(s) > 0$ in $(0, a)$, we can apply the strong maximum principle in the set $\{0 \leqslant u < a\}$ to conclude that u is either identically 0 or it is positive in $(0, +\infty) \times \mathbb{R}^{N-1}$. Since u is nontrivial, we must have $u > 0$. We divide the proof below into three steps.

Step 1 $m := \lim_{A \to +\infty} \inf_{[A, +\infty) \times \mathbb{R}^{N-1}} u \geqslant a$. For any $\epsilon \in (0, a)$, by our assumptions on f, there exists $\eta > 0$ such that

$$f(s) \geqslant \eta s^{p-1} \ \forall s \in [0, a - \epsilon].$$

Hence we can repeat Step 1 in the proof of *Theorem 7.6*, with $\delta_0 = a - \epsilon$, to conclude that

$$m = \lim_{A \to +\infty} \inf_{[A, +\infty) \times \mathbb{R}^{N-1}} u \geqslant a - \epsilon.$$

Letting $\epsilon \to 0$ we obtain $m \geqslant a$.

Step 2 $m \in Z_f^+$ and $u(x + x^n) \to m$ in $C_{loc}^1(\mathbb{R}^N)$ along some sequence x^n with $x_N^n \to +\infty$. Let $x^n = (x_1^n, \ldots, x_N^n)$ be a sequence in $\mathbb{R}^N$ such that $x_1^n \to +\infty$ and $u(x^n) \to m$ as $n \to +\infty$. Up to extraction of a subsequence, the functions $u_n(x) = u(x + x^n)$ converge in $C_{loc}^1(\mathbb{R}^N)$ to a bounded solution U of $\Delta_p u + f(u) = 0$ in $\mathbb{R}^N$ with the properties that $m \leqslant U$ and $U(0) = m$. Clearly we also have $U \leqslant \|u\|_\infty < M := \|u\|_\infty + 1$. It follows that $v := M - U$ is a bounded nonnegative solution of

$$\Delta_p v + \tilde{f}(v) = 0 \text{ in } \mathbb{R}^N,$$

where

$$\tilde{f}(s) = \begin{cases} -f(M-s) & \text{for } 0 \leqslant s \leqslant M - m, \\ (s - M + m)^{p-1} - f(m) & \text{for } s > M - m. \end{cases}$$

By our assumptions on f, we see that $\tilde{f}$ satisfies all the conditions of *Theorem 7.5*. Hence we must have $v \equiv$ constant. It follows that $U \equiv m$ and hence $f(m) = 0$.

Step 3 For every $A \in \mathbb{R}$,

$$\lim_{h \to +\infty} u(x_1 + h, x_2, \ldots, x_n) = m \text{ uniformly in } [A, +\infty) \times \mathbb{R}^{N-1}.$$

Let $\tilde{f}$ be defined as above and denote $\tilde{M} := M - m$. As in Step 2 of the proof of *Theorem 7.6*, for any $\epsilon > 0$ we can find $r > 0$ and a radially symmetric function $\tilde{v}_r$ satisfying

$$0 \leqslant \tilde{v}_r \leqslant \tilde{v}_r(0) \leqslant \tilde{M}_\epsilon, \ \tilde{v}_r(0) \geqslant \tilde{M}_\epsilon - \epsilon,$$

and

$$-\Delta_p \tilde{v}_r = g(\tilde{v}_r) \text{ in } B_r(0), \ \tilde{v}_r = 0 \text{ on } \partial B_r(0),$$

with $g(s) = \tilde{f}(s) - \epsilon s^\sigma < \tilde{f}(s)$ in $(0, \tilde{M}_\epsilon]$, where $\tilde{M}_\epsilon < \tilde{M}$ and converges to $\tilde{M}$ as $\epsilon \to 0$. Now clearly $v_r := M - \tilde{v}_r$ satisfies

$$-\Delta_p v_r = -g(M - v_r) \text{ in } B_r(0), \ v_r = M \text{ on } \partial B_r(0).$$

Since $-g(M - s) > f(s)$ for $s \in (0, \tilde{M}_\epsilon]$, and $u \leqslant M - 1$ in $\mathbb{R}_+ \times \mathbb{R}^{N-1}$, and $u < M - \tilde{M}_\epsilon$ in some $B_r(x^0)$ (by the conclusion in Step 2 and the fact that $M - \tilde{M}_\epsilon > m$), we can use the weak sweeping principle to show that $u \leqslant v_r(0) \leqslant M - \tilde{M}_\epsilon$ in $[r + 1, +\infty) \times \mathbb{R}^{N-1}$. It follows that

$$\lim_{A \to +\infty} \sup_{[A, +\infty) \times \mathbb{R}^{N-1}} u \leqslant M - \tilde{M}_\epsilon.$$

Letting $\epsilon \to 0$, we deduce

$$\lim_{A \to +\infty} \sup_{[A, +\infty) \times \mathbb{R}^{N-1}} u \leqslant M - \tilde{M} = m.$$

In view of the definition of m, we obtain

$$\lim_{h \to +\infty} u(x_1 + h, x_2, \ldots, x_N) = m$$

uniformly in $[A, +\infty) \times \mathbb{R}^{N-1}$ for any $A \in \mathbb{R}$. As in *Theorem 7.6*, the convergence in the C_b^1 norm is a consequence of the above uniform convergence and standard elliptic estimates. $\qquad \square$

7.3.2   One-Dimensional Symmetry in Half-Spaces

For treating the quarter-space problems, we will need some one-dimensional symmetry results for the half-space problem (7.4). We want to stress that due to the lack of a strong comparison principle for general p-Laplacian equations, such partial symmetry results are not readily available as in the Laplacian case treated in *Chap. 6*. In Theorem 1.8 of the recent paper [58], a one-dimensional symmetry result has been established for the case $N = 2, 3$, with $1 + \frac{N}{N+2} < p \leqslant 2$, and with $f(u)$ locally Lipschitz continuous, and satisfying $f(u) > 0$ for $u < a$, $f(u) < 0$ for $u > a$.

Theorem 7.8 *Suppose that f is of type (F_1) which is quasi-monotone, and moreover (7.12) holds, the zeros of f are isolated, and at each of its zero f satisfies (7.7). Let u be a nonnegative bounded solution of (7.4) satisfying $\partial_{x_N} u \geqslant 0$. Then either $u \equiv 0$ or $u(x) \equiv V(x_N)$, where V is the positive solution of the one dimensional problem (7.5) satisfying $V(\infty) = \|u\|_\infty$.*

Proof Due to (7.12), we can apply the strong maximum principle to conclude that either $u \equiv 0$ or $u > 0$ for $x_N > 0$. So we only need to consider the case $u > 0$. By *Theorem 7.6*,

$$\lim_{x_N \to \infty} u(x', x_N) = m \qquad (7.16)$$

uniformly in $x' \in \mathbb{R}^{N-1}$ for some positive constant $m \in Z_f^+$. Since $\partial_{x_N} u \geqslant 0$, we necessarily have $u(x) \leqslant m$ for all $x \in \Omega := \mathbb{R}^{N-1} \times \mathbb{R}_+$. Thus, since $f(m) = 0$, $\bar{u} \equiv m$ is an upper solution of (7.4) satisfying $u \leqslant \bar{u}$. By standard upper and lower solution argument we know that (7.4) has a solution in the order interval $[u, \bar{u}]$. We show next that there is a maximal solution u^* in this order interval in the sense that any solution v in $[u, \bar{u}]$ satisfies $v \leqslant u^*$. For this purpose, we choose a sequence of increasing numbers $R_n \to \infty$ and define

$$B_n^+ = \{x \in \mathbb{R}^N : |x| < R_n, \ x_N > 0\}.$$

Set $\phi_n(x) = \max\{0, m(|x| - R_n + 1)\}$. We now consider the auxiliary problem

$$-\Delta_p u = f(u) \text{ in } B_n^+, \quad u = \phi_n \text{ on } \partial B_n^+. \qquad (7.17)$$

Clearly $u|_{B_n^+}$ is a lower solution to (7.17) and m is an upper solution to (7.17). Hence by standard upper and lower solution argument (7.17) has a maximal solution u_n in the order interval $[u, m]$. One then easily sees that $u_n \geqslant u_{n+1}$ on B_n^+ and $u^* := \lim_{n \to \infty} u_n$ is a solution of (7.4). Clearly $u^* \in [u, \bar{u}]$. If v is any solution of (7.4) in $[u, \bar{u}]$, then it is evident that $v|_{B_n^+}$ is a lower solution of (7.17) and it follows that $u_n \geqslant v$ (since the standard upper and lower solution argument implies the existence of a solution in $[v, m]$ which is less than or equal to u_n). It follows that $v \leqslant u^*$. Thus u^* is the maximal solution in $[u, \bar{u}]$. The maximality implies that u^* is a function of x_N only. Indeed, if this is not true, then there exist two points $((x')^1, x_N^0)$ and $((x')^2, x_N^0)$ in $\mathbb{R}^{N-1} \times \mathbb{R}_+$ such that

$$u^*((x')^1, x_N^0) < u^*((x')^2, x_N^0).$$

Define $\tilde{u}^*(x) = u^*(x' + (x')^2 - (x')^1, x_N)$. Then $\tilde{u}^*$ is a solution of (7.4) and

$$\underline{v}(x) := \max\{u^*(x), \tilde{u}^*(x)\} \geqslant, \neq u^*(x).$$

Moreover, $\underline{v}$ is a lower solution to (7.4) satisfying $\underline{v} \leqslant \bar{u} = m$. Hence (7.4) has a solution v^* in the order interval $[\underline{v}, \bar{u}]$. It follows that v^* is a solution in $[u, \bar{u}]$ which satisfies $v^* \geqslant, \neq u^*$, a contradiction to the maximality of u^*. Thus u^* is a positive solution of (7.5). Since $u \leqslant u^* \leqslant m$, we necessarily have $u^*(\infty) = m$. Next we construct a suitable lower solution which is below u. Since the zeros of f are isolated, there exists a small $\delta > 0$ such that $f(u) > 0$ in $[m - \delta, m)$. We now choose $m_\epsilon \in (m - \delta, m)$ and then modify $f(u)$ over $[0, m]$ to obtain a new function $f_\epsilon(u) \geqslant 0$ so that

$$f_\epsilon(u) = f(u) \text{ in } [0, m - \delta], 0 < f_\epsilon(u) \leqslant f(u) \text{ in } (m - \delta, m_\epsilon), \ f_\epsilon(m_\epsilon) = 0.$$

Since u^* is a positive solution of (7.5), by *Theorem 7.1* we see that $F(z) < F(m)$ for $z \in [0, m)$. In view of the above construction of f_ϵ, it follows that $F_\epsilon(z) < F_\epsilon(m_\epsilon)$ for $z \in [0, m_\epsilon)$. Thus we can apply *Theorem 7.2* to see that the following 1-d problem

$$-\Delta_p v = f_\epsilon(v) \text{ in } \mathbb{R}_+, \ v(0) = 0, v(\infty) = m_\epsilon$$

has a unique positive solution v_ϵ. For any given $R > 0$, define

$$V_{\epsilon,R}(t) = \begin{cases} 0, & t \in [0, R], \\ v_\epsilon(t - R), & t \in [R, \infty). \end{cases}$$

It is easily checked that $V_{\epsilon,R}$ is a lower solution of (7.4). Using (7.16) and $m_\epsilon < m$, we can find a large $R = R_\epsilon$ so that $u(x) > m_\epsilon$ for $x_N \geqslant R_\epsilon$. Hence in view of $v_\epsilon(x_N) \leqslant m_\epsilon$ we find that $V_{\epsilon,R_\epsilon}(x_N) \leqslant u(x)$ for all $x_N \geqslant 0$. Thus we may define $\underline{u}_\epsilon = V_{\epsilon,R_\epsilon}$, and $\underline{u}_\epsilon$ is a lower solution of (7.4) that satisfies $\underline{u}_\epsilon \leqslant u$. By an analogous argument to the one used above for u^*, we can show that in the order interval $[\underline{u}_\epsilon, u]$ the problem (7.4) has a minimal solution u_*, and moreover the minimality of u_* implies that u_* is a function of x_N only. Thus u_* is a positive solution of (7.5), and $u_*(\infty)$ is a zero of $f(u)$ contained in $[m_\epsilon, m]$. But m is the only zero of f in this range. Hence $u_*(\infty) = m$. By *Theorem 7.2*, (7.5) has a unique solution with limit m at infinity. Thus we necessarily have $u_* = u^*$. Since $u_* \leqslant u \leqslant u^*$, this implies that u is a function of x_N only. This completes the proof. $\qquad \square$

Remark 7.4 From the proof we easily see that the condition $\partial_{x_N} u \geqslant 0$ in *Theorem 7.5* can be replaced by

$$\|u\|_\infty \leqslant M := \overline{\lim}_{x_N \to \infty} \sup_{x' \in \mathbb{R}^{N-1}} u(x', x_N),$$

which is a consequence of $\partial_{x_N} u \geqslant 0$. We conjecture that these extra conditions are unnecessary.

The following result is an extension of Theorem 1.1 in [42], where instead of (F_2), it was assumed that $f(s) > 0$ in $(0, a)$ and $f(s) < 0$ in $(a, +\infty)$. See also Theorem 1.8 of [58].

Theorem 7.9 *Suppose that f is of type (F_2), is quasi-monotone and (7.12) holds. Let u be a nonnegative bounded solution of (7.4). Then either $u \equiv 0$ or $u(x) \equiv V(x_N)$, where V is the unique one dimensional solution determined by Theorem 7.4.*

Proof Since $f(s) > 0$ in $(0, a)$, we can apply the strong maximum principle in the set $\{0 \leqslant u < a\}$ to conclude that either $u \equiv 0$ or $u > 0$ in $\mathbb{R}^{N-1} \times (0, +\infty)$. So it suffices to consider the case that $u > 0$. Let $M := \|u\|_\infty \in (0, +\infty)$. Since (F_2) holds, either there exists $b \geqslant M$ such that $f(b) = 0$, or $f(s) < 0$ for all $s \geqslant M$. In the latter case, we can find $\hat{a} \geqslant a$ such that $f(\hat{a}) = 0$ and $f(s) < 0$ for $s > \hat{a}$. Then one can use the weak sweeping principle as in the proof of Proposition 2.2 of [41] to deduce that $u \leqslant \hat{a}$ and hence $M \leqslant \hat{a}$. Thus we can always find a constant $b \geqslant M$ such that $f(b) = 0$. Clearly we must have $b \geqslant a$. Now we can repeat the arguments used to prove Lemma 3.2 of [42] to conclude that (7.4) has a maximal positive solution $\overline{u}$ among all bounded positive solutions. We can also repeat the argument in the proof of Lemma 3.1 of [42] to show that (7.4) has a minimal positive solution $\underline{u}$ among all bounded positive solutions. Thus we have in particular that $\underline{u} \leqslant u \leqslant \overline{u}$. As in [42], being the maximal and minimal solution of (7.4), $\overline{u}$ and $\underline{u}$ must be functions of x_N only, and thus are positive solutions of (7.5). By Theorem 7.4, $\overline{u} = \underline{u} = V$. Thus $u \equiv V$.   □

Remark 7.5 It is interesting to compare *Theorem 7.9 above with [13, Theorem 1.1]. Here we do not require f to be non-increasing in a small left neighborhood of a, but we require that f is not identically zero in any open interval.*

7.3.3 Asymptotic Convergence in Quarter-Spaces

Theorem 7.10 *Suppose that f is of type (F_1) satisfying all the conditions in Theorem 7.8. Let u be any nontrivial nonnegative bounded solution of (7.2) with $\partial_{x_N} u \geqslant 0$. Then*

$$\lim_{h\to\infty} u(x_1 + h, x_2, \ldots, x_N) = V(x_N) \text{ in } C_b^1([A, +\infty) \times \mathbb{R}^{N-2} \times \mathbb{R}_+) \quad (7.18)$$

for every $A \in \mathbb{R}$, where V is a solution of the one dimensional problem (7.5).

Proof Since u is nontrivial, we can apply the strong maximum principle to conclude that $u > 0$. Let

$$m = \lim_{A\to+\infty} \inf_{[A,+\infty)\times\mathbb{R}^{N-2}\times[A,+\infty)} u.$$

Then a simple modification of the arguments in Steps 1 and 2 of the proof of *Theorem 7.6* shows that

$$\lim_{h\to+\infty} u(x_1 + h, x_2, \ldots, x_{N-1}, x_N + h) = m \in Z_f^+ \quad (7.19)$$

uniformly in $[A, +\infty) \times \mathbb{R}^{N-2} \times [A, +\infty)$ for every $A \in \mathbb{R}$. Now let x_1^n be any sequence of positive numbers converging to $+\infty$ as $n \to +\infty$, and y^n be any sequence in $\mathbb{R}^{N-2}$. Define $u_n(x_1, y, x_N) = u(x_1 + x_1^n, y + y^n, x_N)$. By standard elliptic estimates we can find a subsequence of u_n that converges to some v in $C_{loc}^1(\mathbb{R}^{N-2} \times \mathbb{R}_+)$, and v is a bounded nonnegative solution of

$$\Delta_p v + f(v) = 0 \text{ in } \mathbb{R}^{N-1} \times \mathbb{R}_+, \quad v = 0 \text{ on } R^{N-1} \times \{0\}.$$

In view of (7.19), we have $\lim_{x_N\to\infty} v(x', x_N) = m$ uniformly for $x' \in \mathbb{R}^{N-1}$. Moreover, due to $\partial_{x_N} u \geq 0$ we have $\partial_{x_N} v \geq 0$. Hence $\|v\|_\infty = m$. We may now apply *Theorem 7.8* to conclude that $v \equiv V$ where V is the unique solution of (7.5) satisfying $V(\infty) = m$. The uniqueness of V implies that the entire sequence u_n converges to V. Moreover, in view of (7.19) and $V(\infty) = m$, we find that

$$\lim_{x_1\to+\infty} u(x_1, x_2, \ldots, x_N) = V(x_N)$$

uniformly for $(x_2, \ldots, x_N) \in \mathbb{R}^{N-2} \times \mathbb{R}_+$. From this and standard elliptic estimates we see that (7.18) holds. $\square$

Remark 7.6 The condition $\partial_{x_N} u \geq 0$ in *Theorem 7.10* is required only in order to be able to apply *Theorem 7.8*. Thus if we can drop the corresponding condition in *Theorem 7.8*, then we can remove this condition here.

Theorem 7.11 *Suppose that f is of type (F_2), is quasi-monotone, and (7.12) holds. Let u be any nontrivial nonnegative bounded solution of (7.2). Then*

$$\lim_{R\to\infty} \inf_{[R,\infty)\times\mathbb{R}^{N-2}\times[R,\infty)} u \geq a$$

and

$$\lim_{h\to\infty} u(x_1+h, x_2,\ldots, x_N) = V(x_N) \text{ in } C_b^1([A, +\infty)\times \mathbb{R}^{N-2}\times [0, B]) \qquad (7.20)$$

for every $A \in \mathbb{R}$ and $B > 0$, where V is the unique one dimensional solution determined by Theorem 7.4. If we assume further that the zeros of f are isolated and (7.15) holds for each $z \in Z_f^+\backslash\{a\}$, then (7.20) can be strengthened to

$$\lim_{h\to\infty} u(x_1 + h, x_2,\ldots, x_N) = V(x_N) \text{ in } C_b^1([A, +\infty)\times \mathbb{R}^{N-2}\times \mathbb{R}_+) \qquad (7.21)$$

for every $A \in \mathbb{R}$, which implies that

$$\lim_{R\to\infty}\inf_{[R,\infty)\times\mathbb{R}^{N-2}\times[R,\infty)} u = a.$$

Proof Since $f(s) > 0$ in $(0, a)$ and (7.12) holds, we can apply the strong maximum principle in the set $\{0 \leqslant u < a\}$ to conclude that u is either identically 0 or it is positive in $(0, +\infty)\times \mathbb{R}^{N-2}\times (0, +\infty)$. Thus $u > 0$ in $(0, +\infty)\times \mathbb{R}^{N-2}\times (0, +\infty)$. By (7.12), for any $\epsilon \in (0, a)$ we can find $\eta > 0$ such that (7.13) holds with $\delta_0 = a - \epsilon$. Then by using the weak sweeping principle in the same way as in Step 1 of the proof of *Theorem 7.6*, we can show that

$$m = \lim_{A\to+\infty}\inf_{[A,+\infty)\times\mathbb{R}^{N-2}\times[A,+\infty)} u \geqslant a - \epsilon.$$

Letting $\epsilon \to 0$ we deduce

$$m = \lim_{A\to+\infty}\inf_{[A,+\infty)\times\mathbb{R}^{N-2}\times[A,+\infty)} u \geqslant a. \qquad (7.22)$$

Now let x_1^n be any sequence of positive numbers converging to $+\infty$ as $n \to +\infty$, and y^n be any sequence in $\mathbb{R}^{N-2}$. Define $u_n(x_1, y, x_N) = u(x_1 + x_1^n, y + y^n, x_N)$. By standard elliptic estimates we can find a subsequence of u_n that converges to some v in $C_{loc}^1(\mathbb{R}^{N-2}\times \mathbb{R}_+)$, and v is a bounded nonnegative solution of

$$\Delta_p v + f(v) = 0 \text{ in } \mathbb{R}^{N-1}\times \mathbb{R}_+, \quad v = 0 \text{ on } R^{N-1}\times \{0\}.$$

By (7.22) we know that $v \not\equiv 0$. Hence we can apply *Theorem 7.9* to conclude that $v(x) \equiv V(x_N)$, where V is the unique solution determined by *Theorem 7.4*. The uniqueness of v implies that the limit does not depend on the sequences $\{x_1^n\}$ and $\{y^n\}$. This proves (7.20). Under the extra assumptions, the arguments used in Steps 2 and 3 of the proof of *Theorem 7.7* can be easily modified to show that (7.19) holds. Thus we necessarily have $m = a$ and

$$\lim_{x_1 \to +\infty} u(x_1, x_2, \ldots, x_N) = V(x_N)$$

uniformly for $(x_2, \ldots, x_N) \in \mathbb{R}^{N-2} \times \mathbb{R}_+$. From this and standard elliptic estimates we see that (7.21) holds. □

7.4 Comparison with Standard Laplacian ($p = 2$)

Consider the quasilinear elliptic problem

$$\Delta_p u + f(u) = 0 \text{ in } Q, \ u = 0 \text{ on } \partial Q, \tag{7.23}$$

where $Q = (0, \infty) \times (0, \infty) \times \mathbb{R}^{N-2}$ is a quarter-space in $\mathbb{R}^N$ ($N \geqslant 2$), $\Delta_p u = \operatorname{div}(|\nabla u|^{p-2} \nabla u)$ is the usual p-Laplacian operator with $p > 1$, and $f \geqslant 0$ is a continuous function over $\mathbb{R}_+ := [0, \infty)$. If further f is locally Lipschitz continuous in $\mathbb{R}_+$, and

$$\lim_{s \searrow z} \frac{f(s)}{s - z} > 0 \text{ whenever } f(z) = 0, \tag{7.24}$$

then it follows from *Theorem 6.2* that, in the special case $p = 2$, for every bounded positive solution of (7.23), one has

$$\lim_{x_1 \to \infty} u(x_1, x_2, \ldots, x_N) = V_z(x_2), \tag{7.25}$$

and

$$\lim_{x_2 \to \infty} u(x_1, x_2, \ldots, x_N) = V_{\tilde{z}}(x_1), \tag{7.26}$$

where $z, \tilde{z} \in (0, \infty)$ are zeros of f, and whenever z_0 is a positive zero of f, V_{z_0} denotes the unique positive solution of the one-dimensional problem

$$\Delta_p V + f(V) = 0 \text{ in } \mathbb{R}_+, \ V(0) = 0, \ V(t) > 0 \text{ for } t > 0, \ V(\infty) = z_0. \tag{7.27}$$

This result was extended to the general case $p > 1$ in *Theorem 7.8* (see previous section) under suitable conditions on f and the extra assumption that $\partial_{x_i} u \geqslant 0$ in Q for $i = 1, 2$. (In *Chap. 6*, it was actually assumed that $u = 0$ for $x_2 = 0$ and $u \geqslant, \not\equiv 0$ for $x_1 = 0$. However, this latter condition is only used to guarantee that the solution is positive in Q. The conclusions there remain valid for positive solutions satisfying $u = 0$ on ∂Q.) We assume that

$$\{z > 0 : f(z) = 0\} = \{z_1, \ldots, z_k\}, \tag{7.28}$$

and prove that for each z_i, $i = 1, \ldots, k$, (7.23) has a bounded positive solution satisfying

$$\lim_{x_1 \to \infty} u(x_1, x_2, \ldots, x_N) = V_{z_i}(x_2), \quad \lim_{x_2 \to \infty} u(x_1, x_2, \ldots, x_N) = V_{z_i}(x_1).$$
(7.29)

Moreover, if $p = 2$, we prove that (7.23) has no other bounded positive solutions. Therefore, in (7.25) and (7.26), we necessarily have $z = \tilde{z}$. If further f is non-increasing in $(z_i - \epsilon, z_i)$, then (7.23) has a unique bounded positive solution satisfying (7.29). When $p = 2$, the existence of a positive solution of (7.23) satisfying (7.29) was established in *Sect. 2.3*, where the special case $k = N = 2$ was considered.

7.5   Existence Result

In this section, we construct a bounded positive solution of (7.23) satisfying (7.29). We need to be more precise about the conditions imposed on f. We assume that (7.28) holds,

$$\begin{cases} f : \mathbb{R}_+ \to \mathbb{R} \text{ is continuous, nonnegative and} \\ \text{locally Lipschitz continuous except possibly at } \{z_1, \ldots, z_k\}, \end{cases}$$
(7.30)

and for $i = 1, \ldots, k$,

$$\liminf_{s \searrow z_i} \frac{f(s)}{(s - z_i)^{\sigma_{N,p}}} > 0, \quad \limsup_{s \nearrow z_i} \frac{f(s)}{(z_i - s)^{p-1}} < +\infty,$$
(7.31)

where

$$\sigma_{N,p} = (p - 1)\frac{N}{N - p} \text{ if } N > p,$$

and $\sigma_{N,p}$ stands for an arbitrary number in $[1, \infty)$ if $N \leqslant p$. Moreover, we assume

$$\liminf_{s \searrow 0} \frac{f(s)}{s^{p-1}} > 0 \text{ if } f(0) = 0.$$
(7.32)

Let us note that since f is nonnegative, we automatically have

$$\liminf_{s \searrow z_i} \frac{f(s)}{(s - z_i)^{p-1}} \geqslant 0.$$

Moreover, in the special case $p = 2$, the second condition in (7.31) is automatically satisfied if f is locally Lipschitz continuous, and the first condition in (7.31) is less restrictive than (7.24). By *Theorems 7.1* and *7.2* from *Sect. 7.2*, we have the following result.

Proposition 7.2 *Let f satisfy* (7.28), (7.30), (7.31) *and* (7.32). *Then for every z_i, $i = 1, \ldots, k$,* (7.27) *has a unique solution with z_0 replaced by z_i. Moreover, the unique solution is a strictly increasing function.*

To construct a positive solution of (7.23) with the desired properties, a key step is the following result.

Lemma 7.1 *With f as in Proposition 7.2, for each z_i and any given small $\delta > 0$, there exists $R = R_\delta > 0$ and a function $v \in W_0^{1,p}(B)$, with $B = B_R := \{x \in \mathbb{R}^N : |x| < R\}$, satisfying*

(i) $\Delta_p v + f(v) \geq 0$ in B, $v = 0$ on ∂B,
(ii) $0 < v < z_i$ in B,
(iii) $v(x_1, x_2, \ldots, x_N) \leq \min\{V_{z_i}(x_1 + R + 1), V_{z_i}(x_2 + R + 1)\}$ in B,
(iv) $\sup_B v \geq z_i - \delta$.

Proof We define $g(s) = f(s) - \epsilon s^\sigma$ for small $\epsilon > 0$ as in sub-step 2.1 of the proof of *Theorem 7.6*. The argument there (with $M = z_i$) shows that for $R = R_\epsilon$ large enough, there exists a positive solution to

$$\Delta_p v + g(v) = 0 \text{ in } B_R, \quad v = 0 \text{ on } \partial B_R,$$

which satisfies $v(0) = \sup_{B_R} v \in (z_i - \delta, z_i)$ provided that $\epsilon > 0$ is small enough. Thus v has properties (i), (ii) and (iv). It remains to prove (iii). We may use the week sweeping principle as in sub-step 2.2 of the proof of *Theorem 7.6*, with $u = V_{z_i}(x_1)$ or $V_{z_i}(x_2)$. It results that $v(x_1, x_2, \ldots, x_N) \leq V_{z_i}(x_1 + R + 1)$ and $v(x_1, x_2, \ldots, x_N) \leq V_{z_i}(x_2 + R + 1)$.   □

Theorem 7.12 *Let f satisfy* (7.28), (7.30), (7.31) *and* (7.32). *Then for each z_i,* (7.23) *has a bounded positive solution u satisfying* (7.29).

Proof Let $\delta > 0$ be small enough such that $f(s) > 0$ in $[z_i - \delta, z_i)$. Then let $R = R_\delta$ and v be given by Lemma (7.1). Fix $x_0 \in \mathbb{R}^N$ such that the ball $B_{R+1}(x_0) := \{x \in \mathbb{R}^N : |x - x_0| < R + 1\}$ is contained in Q. Then define

$$v_{x_0}(x) = \begin{cases} v(x - x_0) & \text{if } x \in B_R(x_0), \\ 0 & \text{otherwise} \end{cases}$$

Since $f(0) \geq 0$, it is clear that v_{x_0} is a subsolution of (7.23). Define

$$\underline{u} = \sup_{B_{R+1}(x_0) \subset Q} v_{x_0}.$$

Then $\underline{u}$ is again a subsolution of (7.23), and it satisfies

$$\underline{u}(x) \geq z_i - \delta \text{ when } x_1 \geq R+1, \ x_2 \geq R+1. \tag{7.33}$$

Define

$$\bar{u} = \min\{V_{z_i}(x_1), \ V_{z_i}(x_2)\}.$$

Then $\bar{u}$ is a supersolution to (7.23), and by Lemma 7.1, we have $\bar{u} \geq \underline{u}$ in Q. Therefore we can apply the standard sub- and supersolution argument to conclude that (7.23) has a positive solution u satisfying

$$\underline{u} \leq u \leq \bar{u} \text{ in } Q.$$

As in the proof of *Theorem 7.10*, we find that

$$\lim_{h \to \infty} u(x_1 + h, x_2 + h, \dots, x_N) = m$$

and m is a positive zero of f. By (7.33) and the choice of δ, we necessarily have $m = z_i$. On the other hand, we have $u \leq \bar{u} < z_i$ in Q. Therefore we are able to use *Remark 7.6* in the proof of *Theorem 7.10* to conclude that

$$\lim_{x_1 \to \infty} u(x_1, x_2, \dots, x_N) = V_{z_i}(x_2)$$

uniformly for $(x_2, \dots, x_N) \in \mathbb{R}_+ \times \mathbb{R}^{N-2}$. We similarly have

$$\lim_{x_2 \to \infty} u(x_1, x_2, \dots, x_N) = V_{z_i}(x_1)$$

uniformly for $(x_1, x_3, \dots, x_N) \in \mathbb{R}_+ \times \mathbb{R}^{N-2}$. Thus (7.29) holds. $\quad\square$

Bibliography

1. A. V. Babin and M. I. Vishik. *Attractors of evolution equations. Transl. from the Russian by A. V. Babin.* English. Amsterdam etc.: North-Holland, 1992, pp. x + 532.
2. R. A. Adams. *Sobolev spaces.* English. Pure and Applied Mathematics, 65. A Series of Monographs and Textbooks. New York-San Francisco-London: Academic Press, Inc., a subsidiary of Harcourt Brace Jovanovich, Publishers. XVIII, 268 p. $ 24.50 (1975). 1975.
3. S. Agmon, A. Douglas, and L. Nirenberg. "Estimates near the boundary for solutions of elliptic partial differential equations satisfying general boundary conditions. I." English. In: *Commun. Pure Appl. Math.* 12 (1959), pp. 623–727.
4. M. Agranovich. "Elliptic singular integro-differential operators." Russian. In: *Usp. Mat. Nauk* 20.5(125) (1965), pp. 3–120.
5. L. Ambrosio and X. Cabré. "Entire solutions of semilinear elliptic equations in R3 and a conjecture of De Giorgi." English. In: *J. Am. Math. Soc.* 13.4 (2000), pp. 725–739.
6. S. Angenent. "Uniqueness of the solution of a semilinear boundary value problem." English. In: *Math. Ann.* 272 (1985), pp. 129–138.
7. N. Aronszajn and E. Gagliardo. "Interpolation spaces and interpolation methods." English. In: *Ann. Mat. Pura Appl. (4)* 68 (1965), pp. 51–117.
8. A. V. Babin. "Attractor of the generalized semigroup generated by an elliptic equation in a cylindrical domain". In: *Russian Acad. Sci. Izv. Math.* 44 (1995), pp. 207–223.
9. A. V. Babin. "Inertial manifolds for travelling-wave solutions of reaction-diffusion systems". In: *Communications on Pure and Applied Mathematics* 48.2 (), pp. 167–198. eprint: https://onlinelibrary.wiley.com/doi/pdf/10.1002/cpa.3160480205.
10. A. V. Babin. "Symmetry of instabilities for scalar equations in symmetric domains." English. In: *J. Differ. Equations* 123.1 (1995), pp. 122–152.
11. A. V. Babin and G. R. Sell. "Attractors of non-autonomous parabolic equations and their symmetry properties." English. In: *J. Differ. Equations* 160.1 (2000), 1–50, art. no. jdeq.1999.3654.
12. J. Ball. "Continuity properties and global attractors of generalized semi-flows and the Navier-Stokes equations." English. In: *J. Nonlinear Sci.* 7.5 (1997), pp. 475–502.

© Springer Nature Switzerland AG 2018
M. Efendiev, *Symmetrization and Stabilization of Solutions of Nonlinear Elliptic Equations*, Fields Institute Monographs 36, https://doi.org/10.1007/978-3-319-98407-0

13. H. Berestycki, L. Caffarelli, and L. Nirenberg. "Monotonicity for elliptic equations in unbounded Lipschitz domains." English. In: *Commun. Pure Appl. Math.* 50.11 (1997), pp. 1089–1111.

14. H. Berestycki, L. Caffarelli, and L. Nirenberg. "Symmetry for elliptic equations in a half space." English. In: *Boundary value problems for partial differential equations and applications. Dedicated to Enrico Magenes on the occasion of his 70th birthday.* Paris: Masson, 1993, pp. 27–42.

15. H. Berestycki, M. A. Efendiev, and S. Zelik. "Dynamical approach for positive solutions of semilinear elliptic problems in unbounded domains." English. In: *Nonlinear Bound. Value Probl.* 14 (2004), pp. 3–15.

16. H. Berestycki, F. Hamel, and R. Monneau. "One-dimensional symmetry of bounded entire solutions of some elliptic equations." English. In: *Duke Math. J.* 103.3 (2000), pp. 375–396.

17. H. Berestycki and L. Nirenberg. "Monotonicity, symmetry and antisymmetry of solutions of semilinear elliptic equations." English. In: *J. Geom. Phys.* 5.2 (1988), pp. 237–275.

18. H. Berestycki and L. Nirenberg. "On the method of moving planes and the sliding method." English. In: *Bol. Soc. Bras. Mat., Nova Sér.* 22.1 (1991), pp. 1–37.

19. H. Berestycki and L. Nirenberg. *Some qualitative properties of solutions of semilinear elliptic equations in cylindrical domains.* English. Analysis, et cetera, Res. Pap. in Honor of J. Moser's 60th Birthd., 115–164 (1990). 1990.

20. H. Berestycki, L. Nirenberg, and S. Varadhan. "The principal eigenvalue and maximum principle for second-order elliptic operators in general domains." English. In: *Commun. Pure Appl. Math.* 47.1 (1994), pp. 47–92.

21. H. Berestycki, L. Caffarelli, and L. Nirenberg. "Further qualitative properties for elliptic equations in unbounded domains." English. In: *Ann. Sc. Norm. Super. Pisa, Cl. Sci., IV. Ser.* 25.1-2 (1997), pp. 69–94.

22. H. Berestycki, F. Hamel, and L. Rossi. "Liouville-type results for semi-linear elliptic equations in unbounded domains." English. In: *Ann. Mat. Pura Appl. (4)* 186.3 (2007), pp. 469–507.

23. H. Berestycki and P.-L. Lions. "Nonlinear scalar field equations. I: Existence of a ground state." English. In: *Arch. Ration. Mech. Anal.* 82 (1983), pp. 313–345.

24. H. Berestycki, H. Matano, and F. Cois Hamel. "Bistable traveling waves around an obstacle." English. In: *Commun. Pure Appl. Math.* 62.6 (2009), pp. 729–788.

25. H. Berestycki and L. Nirenberg. "Travelling fronts in cylinders." English. In: *Ann. Inst. Henri Poincaé, Anal. Non Linéaire* 9.5 (1992), pp. 497–572.

26. O. Besov, V. Il'in, and S. Nikol'skij. *Integral'nye predstavleniya funktsij i teoremy vlozheniya.* Russian. 2nd ed., rev. and compl. Moskva: Nauka. Fizmatlit, 1996, p. 480.

27. M.-F. Bidaut-Véron and S. Pohozaev. "Nonexistence results and estimates for some nonlinear elliptic problems." English. In: *J. Anal. Math.* 84 (2001), pp. 1–49.

28. H. Brézis. "Semilinear equations in $\mathbb{R}^N$ without condition at infinity." English. In: *Appl. Math. Optim.* 12 (1984), pp. 271–282.

29. P. Brunovský, X. Mora, P. Poláčik, and J. Solà-Morales. "Asymptotic behavior of solutions of semilinear elliptic equations on an unbounded strip." English. In: *Acta Math. Univ. Comen., New Ser.* 60.2 (1991), pp. 163–183.

30. J. Busca and P. Felmer. "Qualitative properties of some bounded positive solutions to scalar field equations." English. In: *Calc. Var. Partial Differ. Equ.* 13.2 (2001), pp. 191–211.

31. A. Calsina, X. Mora, and J. Solà-Morales. "The dynamical approach to elliptic problems in cylindrical domains, and a study of their parabolic singular limit." English. In: *J. Differ. Equations* 102.2 (1993), pp. 244–304.

32. V. V. Chepyzhov and M. I. Vishik. "Evolution equations and their trajectory attractors." English. In: *J. Math. Pures Appl. (9)* 76.10 (1997), pp. 913–964.

33. V. V. Chepyzhov and M. I. Vishik. "Non-autonomous evolutionary equations with translation-compact symbols and their attractors." English. In: *C. R. Acad. Sci., Paris, Sér. I* 321.2 (1995), pp. 153–158.

34. V. V. Chepyzhov and M. I. Vishik. "Trajectory attractors for evolution equations." English. In: *C. R. Acad. Sci., Paris, Sér. I* 321.10 (1995), pp. 1309–1314.

35. V. Chepyzhov and M. Vishik. *Attractors for equations of mathematical physics.* English. Providence, RI: American Mathematical Society (AMS), 2002, pp. xi +363.

36. V. Chepyzhov and M. Vishik. "Attractors of non-autonomous dynamical systems and their dimension." English. In: *J. Math. Pures Appl. (9)* 73.3 (1994), pp. 279–333.

37. P. Clément and G. Sweers. "Existence and multiplicity results for a semilinear elliptic eigenvalue problem." English. In: *Ann. Sc. Norm. Super. Pisa, Cl. Sci., IV. Ser.* 14.1 (1987), pp. 97–121.

38. E. N. Dancer and Y. Du. "Some remarks on Liouville type results for quasilinear elliptic equations." English. In: *Proc. Am. Math. Soc.* 131.6 (2003), pp. 1891–1899.

39. E. Dancer. "Some notes on the method of moving planes." English. In: *Bull. Aust. Math. Soc.* 46.3 (1992), pp. 423–432.

40. M. Dauge. *Elliptic boundary value problems on corner domains. Smoothness and asymptotics of solutions.* English. Berlin etc.: Springer-Verlag, 1988, pp. viii + 259.

41. Y. Du and Z. Guo. "Liouville type results and eventual flatness of positive solutions for *p*-Laplacian equations." English. In: *Adv. Differ. Equ.* 7.12 (2002), pp. 1479–1512.

42. Y. Du and Z. Guo. "Symmetry for elliptic equations in a half-space without strong maximum principle." English. In: *Proc. R. Soc. Edinb., Sect. A, Math.* 134.2 (2004), pp. 259–269.

43. L. Dupaigne and A. Farina. "Liouville theorems for stable solutions of semilinear elliptic equations with convex nonlinearities." English. In: Nonlinear Anal., *Theory Methods Appl., Ser. A, Theory Methods* 70.8 (2009), pp. 2882–2888.

44. L. Dupaigne. *Stable solutions of elliptic partial differential equations.* English. Boca Raton, FL: CRC Press, 2011, pp. xiv + 321.

45. L. Dupaigne and A. Farina. "Stable solutions of $-\Delta u = f(u)$ in $\mathbb{R}^N$." English. In: *J. Eur. Math. Soc. (JEMS)* 12.4 (2010), pp. 855–882.

46. M. A. Efendiev. *Attractors for degenerate parabolic type equations.* English. Providence, RI: American Mathematical Society (AMS); Madrid: Real Sociedad Matemática Española (RSME), 2013, pp. x + 221.

47. M. A. Efendiev. *Evolution equations arising in the modelling of life sciences.* English. New York, NY: Birkhäuser/Springer, 2013, pp. xii + 217.

48. M. A. Efendiev. *Finite and infinite dimensional attractors for evolution equations of mathematical physics.* English. Tokyo: Gakkōtosho, 2010, pp. ii + 239.

49. M. A. Efendiev. *Fredholm structures, topological invariant and applications.* English. Springfield, MO: American Institute of Mathematical Sciences, 2009, pp. x + 205.

50. M. A. Efendiev. "On an elliptic attractor in an asymptotically symmetric unbounded domain in $\mathbb{R}^4_+$". In: *Bull. Lond. Math. Soc.* 39.6 (2007), pp. 911–920.

51. M. A. Efendiev and F. Hamel. "Asymptotic behavior of solutions of semilinear elliptic equations in unbounded domains: two approaches." English. In: *Adv. Math.* 228.2 (2011), pp. 1237–1261.

52. M. A. Efendiev and S. V. Zelik. "The attractor for a nonlinear reaction-diffusion system in an unbounded domain." English. In: *Commun. Pure Appl. Math.* 54.6 (2001), pp. 625–688.

53. M. Esteban and P.-L. Lions. "Existence and non-existence results for semilinear elliptic problems in unbounded domains." English. In: *Proc. R. Soc. Edinb., Sect. A, Math.* 93 (1982), pp. 1–14.

54. A. Farina. "Finite energy solutions, quantization effects and Liouville-type results for a variant of the Ginzburg-Landau systems in $\mathbb{R}^K$." English. In: *Differ. Integral Equ.* 11.6 (1998), pp. 875–893.

55. A. Farina. "Liouville-type results for solutions of $-\Delta u = |u|^{p-1}u$ on unbounded domains of $\mathbb{R}^N$." English. In: *C. R., Math., Acad. Sci. Paris* 341.7 (2005), pp. 415–418.

56. A. Farina. "Liouville-type theorems for elliptic problems." English. In: *Handbook of differential equations: Stationary partial differential equations. Vol. IV.* Amsterdam: Elsevier/North Holland, 2007, pp. 61–116.

57. A. Farina. "On the classification of solutions of the Lane-Emden equation on unbounded domains of $\mathbb{R}^N$." English. In: *J. Math. Pures Appl. (9)* 87.5 (2007), pp. 537–561.

58. A. Farina, L. Montoro, and B. Sciunzi. "Monotonicity and one-dimensional symmetry for solutions of $-\Delta_p u = f(u)$ in half-spaces." English. In: *Calc. Var. Partial Differ. Equ.* 43.1–2 (2012), pp. 123–145.

59. A. Farina, B. Sciunzi, and E. Valdinoci. "Bernstein and De Giorgi type problems: new results via a geometric approach." English. In: *Ann. Sc. Norm. Super. Pisa, Cl. Sci. (5)* 7.4 (2008), pp. 741–791.

60. A. Farina and E. Valdinoci. "Flattening results for elliptic PDEs in unbounded domains with applications to overdetermined problems." English. In: *Arch. Ration. Mech. Anal.* 195.3 (2010), pp. 1025–1058.

61. B. Fiedler, A. Scheel, and M. I. Vishik. "Large patterns of elliptic systems in infinite cylinders." English. In: *J. Math. Pures Appl. (9)* 77.9 (1998), pp. 879–907.

62. G. Fischer. "Zentrumsmannigfaltigkeiten bei elliptischen Differentialgleichungen". In: *Math. Nachr.* 115 (1984), pp. 137–157.

63. P. M. Fitzpatrick and J. Pejsachowicz. *An extension of the Leray-Schauder degree for fully nonlinear elliptic problems*. English. Nonlinear functional analysis and its applications, Proc. Summer Res. Inst., Berkeley/Calif. 1983, Proc. Symp. Pure Math. 45/1, 425–438 (1986). 1986.

64. L. Fraenkel. *An introduction to maximum principles and symmetry in elliptic problems. Reprint of the 2000 hardback edition. English. Reprint of the 2000 hardback edition.* Cambridge: Cambridge University Press, 2011, pp. x + 340.

65. B. Gidas, W.-M. Ni, and L. Nirenberg. "Symmetry and related properties via the maximum principle." English. In: *Commun. Math. Phys.* 68 (1979), pp. 209–243.

66. J. Hadamard. *Lectures on Cauchy's problem in linear partial differential equations*. Dover Publications, New York, 1953, pp. iv + 316.

67. J. K. Hale. *Asymptotic behavior of dissipative systems*. English. Providence, RI: American Mathematical Society, 1988, pp. ix + 198.

68. P. Hess and P. Poláčik. "Symmetry and convergence properties for non- negative solutions of nonautonomous reaction-diffusion problems." English. In: *Proc. R. Soc. Edinb., Sect. A, Math.* 124.3 (1994), pp. 573–587.

69. V. Hutson, J. S. Pym, and M. J. Cloud. *Applications of functional analysis and operator theory. 2nd ed.* English. 2nd ed. Amsterdam: Elsevier, 2005, pp. xiv + 426.

70. B. Kawohl. *Rearrangements and convexity of level sets in PDE*. English. Lecture Notes in Mathematics. 1150. Berlin etc.: Springer-Verlag. V, 136 p. DM 21.50 (1985). 1985.

71. N. Kenmochi and N. Yamazaki. "Global attractors for multivalued flows associated with subdifferentials." English. In: *Elliptic and parabolic problems. Proceedings of the 4th European conference, Rolduc, Netherlands, June 18–22, 2001 and Gaeta, Italy, September 24–28, 2001*. Singapore: World Scientific, 2002, pp. 135–144.

72. K. Kirchgaessner. "Wave-solutions of reversible systems and applications." English. In: *J. Differ. Equations* 45 (1982), pp. 113–127.

73. A. Kolmogorov and V. Tikhomirov. "ε-entropy and ε-capacity of sets in function spaces." English. In: *Transl., Ser. 2, Am. Math. Soc.* 17 (1959), pp. 227–364.

74. M. Krasnosel'skij. *Topological methods in the theory of nonlinear integral equations. Translated by A.H. Armstrong.* English. International Series of Monographs on Pure and Applied Mathematics. 45. Oxford etc.: Pergamon Press. XI, 395 p. (1964). 1964.

75. M. K. Kwong. "Uniqueness of positive solutions of $\Delta u - u + u^p = 0$ in R^n." English. In: *Arch. Ration. Mech. Anal.* 105.3 (1989), pp. 243–266.

76. O. Ladyzhenskaya and N. Ural'tseva. *Linear and quasilinear elliptic equations*. Mathematics in Science and Engineering. Elsevier Science, 1968.

77. O. Ladyzhenskaya. *Attractors for semigroups and evolution equations*. Lezioni Lincee. [Lincei Lectures]. Cambridge University Press, Cambridge, 1991, pp. xii+73.

78. J. Lions and E. Magenes. *Problèmes aux limites non homogenes et applications. Vol. 1, 2.* French. Paris: Dunod 1: XIX, 372 p.; 2: XV, 251 p. (1968). 1968.

79. Y. B. Lopatinskiĭ. "On a method of reducing boundary problems for a system of differential equations of elliptic type to regular integral equations". In: *Ukrain. Mat. Ž.* 5 (1953), pp. 123–151.

80. A. McNabb. "Strong comparison theorems for elliptic equations of second order." English. In: *J. Math. Mech.* 10 (1961), pp. 431–440.

81. A. Mielke and S. Zelik. "Infinite-dimensional trajectory attractors of elliptic boundary value problems in cylindrical domains." English. In: *Russ. Math. Surv.* 57.4 (2002), pp. 753–784.

82. A. Mielke. "Essential manifolds for an elliptic problem in an infinite strip". In: *J. Differential Equations* 110.2 (1994), pp. 322–355.

83. S. A. Nazarov and B. A. Plamenevsky. *Elliptic problems in domains with piecewise smooth boundaries*. English. Berlin: de Gruyter, 1994, pp. vii + 525.

84. S. Nikol'skij. *Approximation of functions of several variables and imbed- ding theorems. (Priblizhenie funktsij mnogikh peremennykh i teoremy vlozheniya). 2nd ed., rev. and suppl.* Russian. Moskva: "Nauka". 455 p. R. 2.00 (1977). 1977.

85. W. Rudin. *Functional analysis. 2nd ed.* English. 2nd ed. New York, NY: McGraw-Hill, 1991, pp. xviii + 424.

86. H. H. Schaefer. *Topological vector spaces. 4th corr. printing.* English. Graduate Texts in Mathematics, 3. New York-Heidelberg-Berlin: Springer- Verlag. XI, 294 p. DM 48.00; $ 28.40 (1980). 1980.

87. A. Scheel. "Existence of fast traveling waves for some parabolic equations: a dynamical systems approach". In: *J. Dynam. Differential Equations* 8.4 (1996), pp. 469–547.

88. B.-W. Schulze. "Boundary value problems and edge pseudo-differential operators." English. In: *Microlocal analysis and spectral theory. Proceedings of the NATO Advanced Study Institute, Il Ciocco, Castelvecchio Pascoli (Lucca), Italy, 23 September-3 October 1996.* Dordrecht: Kluwer Academic Publishers, 1997, pp. 165–226.

89. B.-W. Schulze. *Boundary value problems and singular pseudo-differential operators.* English. Chichester: John Wiley & Sons, 1998, pp. xvi + 637.

90. B.-W. Schulze. *Pseudo-differential operators on manifolds with edges.* English. Partial differential equations, Proc. Symp., Holzhau/GDR 1988, Teubner-Texte Math. 112, 259–288 (1989). 1989.

91. B. Schulze, M. Vishik, I. Witt, and S. Zelik. "The trajectory attractor for a nonlinear elliptic system in a cylindrical domain with piecewise smooth boundary". In: *Rend. Accad. Naz. Sci. XL Mem. Mat. Appl.* 5.23 (1999), pp. 125–166.

92. G. R. Sell. "Global attractors for the three-dimensional Navier-Stokes equations." English. In: *J. Dyn. Differ. Equations* 8.1 (1996), pp. 1–33.

93. J. Serrin. "A symmetry problem in potential theory." English. In: *Arch. Ration. Mech. Anal.* 43 (1971), pp. 304–318.

94. J. Serrin. "Nonlinear elliptic equations of second order". Lectures at AMS Symposium on Partial Differential Equations, Berkeley. 1971.

95. J. Serrin and H. Zou. "Cauchy-Liouville and universal boundedness theorems for quasilinear elliptic equations and inequalities." English. In: *Acta Math.* 189.1 (2002), pp. 79–142.

96. I. V. Skrypnik. *Nonlinear elliptic boundary value problems. Transl. from the Russian by Jiri Jarnik.* English. Teubner-Texte zur Mathematik, Bd. 91. Leipzig: BSB B. G. Teubner Verlagsgesellschaft. 232 p. M 24.00 (1986). 1986.

97. G. Sweers. *Maximum principles, a start.* 2000.

98. R. Temam. *Infinite-dimensional dynamical systems in mechanics and physics. 2nd ed.* English. 2nd ed. New York, NY: Springer, 1997, pp. xxi + 648.

99. P. Tolksdorf. "On the Dirichlet problem for quasilinear equations in domains with conical boundary points." English. In: *Commun. Partial Differ. Equations* 8 (1983), pp. 773–817.

100. P. Tolksdorf. "Regularity for a more general class of quasilinear equations." English. In: *J. Differ. Equations* 51 (1984), pp. 126–150.

101. H. Triebel. *Interpolation theory, function spaces, differential operators. 2nd rev. a. enl. ed.* English. 2nd rev. a. enl. ed. Leipzig: Barth, 1995, p. 532.

102. J. Vázquez. "A strong maximum principle for some quasilinear elliptic equations." English. In: *Appl. Math. Optim.* 12 (1984), pp. 191–202.

103. M. Vishik and S. Zelik. "The trajectory attractor of a nonlinear elliptic system in a cylindrical domain." English. In: *Sb. Math.* 187.12 (1996), 1755–1789 (1996); translation from mat. sb. 187, no. 12, 21–56.

104. A. Vol'pert and S. Khudyaev. *Analysis in classes of discontinuous functions and equations of mathematical physics.* English. Mechanics: Analysis, 8. Dordrecht - Boston - Lancaster: Martinus Nijhoff Publishers, a member of the Kluwer Academic Publishers Group. XVIII, 678 p. Dfl. 340.00; $ 117.50; £86.50 (1985). 1985.

105. K. Yosida. *Functional analysis. Repr. of the 6th ed.* English. Repr. of the 6th ed. Berlin: Springer-Verlag, 1994, pp. xiv + 506.

106. E. Zeidler. *Nonlinear functional analysis and its applications. Volume I: Fixed-point theorems. Translated from the German by Peter R. Wadsack. 2. corr. printing.* English. 2. corr. printing. New York: Springer-Verlag, 1993, pp. xxiii + 909.

107. S. Zelik. "The dynamics of fast non-autonomous travelling waves and homogenization." English. In: *Actes des journées "Jeunes numériciens" en l'honneur du 60ème anniversaire du Professeur Roger Temam, Poitiers, France, Mars 9–10, 2000.* Poitou-Charentes: Atlantique, 2001, pp. 131–142.

108. S. Zelik. "A trajectory attractor of a nonlinear elliptic system in an unbounded domain." English. In: *Math. Notes* 63.1 (1998), 120–123 (1998); translation from mat. zametki 63, no. 1, 135–138.

Printed in the United States
By Bookmasters